Archives of Virology

Supplementum 3

B. Liess, V. Moennig, J. Pohlenz, G. Trautwein (eds.)

Ruminant Pestivirus Infections

Virology, Pathogenesis, and Perspectives of Prophylaxis

Springer-Verlag Wien New York

Dr. B. Liess
Dr. V. Moennig
Dr. J. Pohlenz
Dr. G. Trautwein
Tierärztliche Hochschule Hannover
Hannover, Federal Republic of Germany

Printed on acid-free paper

With 78 Figures

ISSN 0939-1983
ISBN-13:978-3-211-82279-1   e-ISBN-13:978-3-7091-9153-8
DOI: 10.1007/978-3-7091-9153-8

Foreword

During the past four decades pestivirology has expanded with progress. In addition to hog cholera the discovery of various clinical manifestations in ruminants has demonstrated the economic importance of this group of infections. The impact of bovine viral diarrhea (BVD) and border disease infections was aggravated in association with changes in farm management and introduction of modern techniques in animal reproduction. For many years mucosal disease, one clinical sequel of BVD virus infection remained enigmatic, since Koch's postulates could not be fulfilled and researchers failed to reproduce the disease when inoculating susceptible cattle. In addition, not all consequences of intrauterine infection were understood.

Virological studies yielded information on the morphological, physico-chemical, and antigenic properties of pestiviruses. Early data justified the classification of pestiviruses as a genus of non-arthropod-borne togaviruses. Since pestiviruses are difficult to work with, progress in the understanding of virus and disease gradually came to a standstill because conventional techniques failed to yield further insights.

About ten years ago interest in pestivirology was revived by strong impulses of modern biotechnology and a breakthrough in pathogenesis research, i.e. in vitro translation of BVD viral proteins and the experimental reproduction of mucosal disease in cattle. In this context the importance of cytopathic and noncytopathic biotypes of BVD virus was reassessed. During the following years our knowledge about the pathogenesis, replication, antigenic structure and genomic organization expanded rapidly. In order to summarize and discuss these exciting developments, an international community of pestivirus researchers came together in June 1990 in Hannover (Germany) for the symposium "Ruminant Pestivirus Infections: Virology, Pathogenesis and Perspectives of Prophylaxis".

The meeting was intended to assemble clinicians epidemiologists, pathologists, virologists, and molecular biologists. The current knowledge in these fields was presented and extensively discussed. Apart from reviews on different aspects of pestivirology novel results were contributed. This volume of "Archives of Virology" contains most papers presented at the meeting, thereby covering the full heterogeneous scope of activities in pestivirology. The goal of the symposium was a better understanding of the pathogenesis and epidemiology of pestivirus infections in cattle, sheep and goats and the discussion of perspectives for effective control strategies.

The organizers are thankful to Jutta Jedzini, Jutta Hombach, Ingeborg Engert and Helmut Schulz for their efficient help. In addition we are indebted

to the following companies and institutions for generous financial support of
the symposium:

Bayer AG	Intervet/AKZO
Biochrom	Menno-Chemie GmbH
Biospa GmbH	Norden Laboratories GmbH
Boehringer Vetmedica GmbH	Rhone Merieux
Dr. Bommeli AG	Sebak GmbH
Coopers Tierarzneimittel	Sera Tech Gmbh
Deutsche Bank AG	TAD Pharmazeutische Werke
Hoechst Veterinär GmbH	Virbac GmbH

B. Liess   V. Moennig   J. Pohlenz   G. Trautwein

Contents

Poster presentations

Arch Virol (1991) [Suppl 3]: 1–5

Pestiviruses—taxonomic perspectives

M. C. Horzinek

Dept. Infect. Dis. & Immunol., Veterinary Faculty, Institute of Virology, Utrecht,
The Netherlands

Accepted March 14, 1991

Summary. The history of pestivirus taxonomy is surprisingly consistent: almost 30 years ago it was recognized that pestiviruses are structurally akin to flaviviruses, and recent nucleotide sequence data have confirmed this resemblance at the level of genome organization. For other enveloped positive stranded RNA viruses with ikosahedral nucleocapsids e.g. equine arteritis and lactate dehydrogenase virus of mice a taxonomic dilemma is encountered; while virions resemble ("non-arthropod-borne") togaviruses, the replication via a nested set of subgenomic RNAs is corona and torovirus-like. Pestiviruses, flaviviruses and the hepatitis C virus group have been assigned generic status in the Flaviviridae family.

Key words: Pestivirus, taxonomy.

For a long time, viral taxonomy was based exclusively on structural features of the infectious particle. Three alternative criteria were used: the type of nucleic acid (DNA or RNA), the lipoprotein envelope (present or absent) and the nucleocapsid symmetry type (icosahedral or helical). While the first two features of a virus are easily established—using nucleic acid inhibitors and organic solvents or detergents, respectively—identification of the symmetry type in enveloped virions is less straightforward. It involves electron microscopic analysis after selective removal of the virion membrane, an approach that has led to the identification of icosahedral capsids in the arthropod-borne alphaviruses [10] and in other, then unclassified small enveloped viruses without notorious arthropod transmission [11]. These included rubella virus (which has remained in the Togaviridae family), the pestiviruses discussed in this volume, and the equine arteritis/lactic dehydrogenase viruses.

In recent years, there is a tendency to abandon strictly structural criteria of classification and to make use of other properties, e.g. the replication strategy of viruses. This is a rational approach if taxonomy is to reflect evolutionary relationships. After all, more constraints, more selective pressure may be expected to operate at the level of the extracellular virus particle, its proteins and antigenic determinants than e.g. at the level of mRNA transcription; the mode of replication is probably more conserved during evolution.

In 1973, I have coined the term "pestiviruses" to congregate two antigenically related enveloped RNA viruses: hog cholera virus (HCV) and BVDV [8]. A third animal pathogen, the border disease virus of sheep, was found later to be a close relative of BVDV. In its Fourth Report published in 1982 for the Virology Division of the International Union of Microbiological Societies (IUMS), the International Committee on Taxonomy of Viruses (ICTV) has adopted this nomenclature and has assigned generic status to the pestiviruses, with BVDV as the prototype [12]. Pestiviruses (from lat. *pestis suum* = hog cholera or 'classical' swine fever, an economically important disease) are amongst the smallest enveloped animal RNA viruses (measuring about 40 nm in diameter) and possess a nucleocapsid of non-helical, probably icosahedral symmetry [9]. They share these traits with the numerous flaviviruses, of which the mosquito-transmitted yellow fever virus is the prototype (the pestiviruses are non-arthropod-borne). Previously, also the flaviviruses held generic status in this family, but when details of their molecular structure, replication strategy and gene sequence became known in the early 1980s, the Togavirus Study Group acknowledged the differences as fundamental and established the new family Flaviviridae—with flavivirus then its only genus [20].

In view of the recent progress in the description of the molecular features of pestiviruses, the discussion of virus classification has been re-opened. The first molecular data which suggested that pestiviruses are distinct from members of the Togaviridae family relate to characteristics of the virus-specific RNA. In infected cells, only a single high molecular weight species was found which lacked a 3' poly(A) tract [14, 15; R. Moormann, personal communication]. No subgenomic RNA was detected at any time after infection. These properties distinguish pestiviruses from togaviruses—of which both the alpha- and rubiviruses possess one subgenomic RNA—and suggest a similarity to flaviviruses. Much of the nucleotide sequence and genetic organization of BVDV is known, and further comparisons with flaviviruses have been made [2]. With the exception of several short but significant stretches of identical amino acids within two of the putative nonstructural proteins, no extended regions of homology exist between BVDV and representatives of the three antigenic subgroups of mosquito-borne flaviviruses. Nevertheless, the molecular layout of the BVDV and flavivirus genomes is strikingly similar. Comparison of the arrangement of the protein coding domains along both genomes and the hydropathic features of their amino acid sequences revealed pronounced similarities. Based on these comparisons, it was suggested that the Pestiviruses no longer be grouped in the Togaviridae family, but rather be considered a genus within the Flaviviridae [2]. A joint proposal by the Togavirus and Flavivirus Study Groups read: "To remove the genus Pestivirus form the Family TOGAVIRIDAE and place it in the family FLAVIVIRIDAE." This proposal has been accepted at the last assembly of the ICTV during the VIIIth International Congress of Virology in Berlin. In the forthcoming Report of the ICTV (Francki et al., 1991, in press) a third virus will be found allotted to the FLAVIVIRIDAE: the recently cloned and sequenced hepatitis C virus. Although neither visualized by electron microscopy nor analyzed by conventional means, the 7310 base genome shows an organization akin to flaviviruses [1]. It can be expected that additional members of animal 'hepatitis C virus group' members will be discovered.

The implications of this proposition are immediately provocative. Considering the parallel organization of their genomes, analogous polypeptides may be predicted to possess similar biologic functions. Examples corroborating this hypothesis have been given in our recent review on molecular advances in pestivirus research [3]. Drawing analogies to the flaviviruses may help in experimental designs to resolve the issue of structural vs. nonstructural proteins for pestiviruses. Certainly insights beyond those in molecular biology may be gained from further comparisons between the three groups of viruses (flavi-, pesti- and hepatitis C-viruses). Thus the aspect of human infections with pestiviruses has recently obtained special attention. In 41% to 87% of human sera from Europe, Africa, Central and South America antibodies against BVDV have been found in persons with frequent contacts to bovines [19]; the evidence has been confirmed by antigen detection and by polymerase chain reaction amplification of pestiviral sequences in faeces from children with gastrointestinal disease [21].

Non-arthropod-borne togaviruses were last reviewed in 1981 (with flaviviruses at that time still belonging to the Togaviridae family), when virtually no molecular data were available [9]. Meanwhile, the expanding knowledge has made yet another morphological classification untenable: equine arteritis virus (EAV) which has been listed as a possible member of the Togaviridae family [12] replicates via multiple subgenomic mRNAs [18]; we have recently found that they form a 3'-coterminal nested set [4], like in coronaviruses and toroviruses [16]. However, by having the ≤ 70 nm size and possessing an enveloped icosahedral nucleocapsid, EAV does fulfill conventional structural criteria of a togavirus. Equine arteritis virus is morphologically and structurally similar to lactic dehydrogenase virus (LDV [9]) and also its genome organization was recently found to be quite similar [6]. In all three groups, the coronaviruses, toroviruses and arteriviruses the polymerase gene contains two overlapping open reading frames which are expressed as a fusion protein by a process of ribosomal frame-shifting; the predicted tertiary RNA structure (a pseudoknot) in the frameshift-directing region is similar. The data obtained in Utrecht on the organization and translation of the polymerase gene of Berne virus (the torovirus prototype) provide additional evidence that toro-, corona- and arteriviruses are evolutionarily more related to each other than to any other family of positive-stranded RNA viruses. They can be considered to form a third 'superfamily' [7, 17], perhaps an order, as has recently been decided by the ICTV for three families of negative-stranded RNA viruses: the paramyxo-, rhabdo- and filoviridae (Order Mononegavirales [13]).

The presence of "incompatible" taxonomic elements in one virus (e.g. EAV showing the togavirus characteristics of an icosahedral capsid together with the nested set of subgenomic RNAs of the third 'superfamily') indicates that our traditional concept of a structural classification of viruses is certainly too narrow. In retrospect, it must be considered fortuitous that modern analytic methods have confirmed what morphology and scarce physico-chemical data have suggested already in the 1960s [5]: that pestiviruses are flavivirus-like.

Acknowledgements

The author should like to thank Marc Collett and Volker Moennig for many discussions and suggestions, most of which have been included in a recent review article on pestiviruses [3].

References

1. Choo QL, Kuo G, Weiner AJ, Overby LR, Bradley DW, Houghton M (1989) Isolation of a cDNA clone derived from a blood-borne non-A, non-B viral hepatitis genome. Science 244: 359–362
2. Collett MS, Anderson DK, Retzel E (1988) Comparisons of the pestivirus bovine viral diarrhoea virus with members of Flaviviridae. J Gen Virol 69: 2637–2643
3. Collett MS, Moennig V, Horzinek MC (1989) Recent advances in pestivirus research. J Gen Virol 70: 253–266
4. De Vries AAF, Chirnside ED, Bredenbeek PJ, Gravestein LA, Horzinek MC, Spaan WJM (1990) All subgenomic mRNAs of equine arteritis virus contain a common leader sequence. Nucl Acids Res 18: 3241–3247
5. Dinter Z (1963) Relationship between bovine virus diarrhoea virus and hog cholera virus. Zentralbl Bakteriol Parasitenk Abt 1 Orig 188: 475–486
6. Godeny EK, Brinton MA (1990) Sequence analysis and genome organization of lactate dehydrogenase-elevating virus (LDV) Abstracts VIII Intern Cong Virol Berlin P52-016: 376
7. Goldbach R, Wellink J (1988) Evolution of plus-stranded RNA viruses. Intervirology 29: 260–267
8. Horzinek MC (1973) The structure of togaviruses. Prog Med Virol 16: 109–156
9. Horzinek MC (1981) Non-arthropod-borne togaviruses. Academic Press, London
10. Horzinek MC, Mussgay M (1969) Studies on the nucleocapsid structure of a group A arbovirus. J Virol 4: 514–520

11. Horzinek MC, Maess J, Laufs R (1967) Studies on the substructure of togaviruses. II. Analysis of equine arteritis, rubella, bovine viral diarrhoea and hog cholera viruses. Arch Ges Virusforsch 33: 306–318
12. Matthews REF (1982) Classification and nomenclature of viruses. 4th Report of the International Committee on Taxonomy of Viruses. Karger, Basel
13. Pringle CR (1991) The order Mononegavirales. Arch Virol 117: 137–140
14. Purchio AF, Larson R, Collett MS (1983) Characterization of virus-specific RNA synthesized in bovine cells infected with bovine viral diarrhea virus. J Virol 48: 320–324
15. Renard A, Guiot C, Schmetz D, Dagenais L, Pastoretp P, Dina D, Martial JA (1985) Molecular cloning of bovine diarrhea viral diarrhea virus sequences. DNA 4: 429–438
16. Snijder EJ, Ederveen J, Spaan WJM, Weiss M, Horzinek MC (1988) Characterization of Berne virus genomic and messenger RNAs. J Gen Virol 69: 2135–2144
17. Strauss JH, Strauss EG (1988) Evolution of RNA viruses. Ann Rev Microbiol 42: 657–683
18. Van Berlo MF, Spaan WJM, Horzinek MC (1987) Replication of equine arteritis virus (EAV): a comparative review. In: The molecular biology of the positive strand RNA Viruses. Academic Press, London, p 105–115
19. Verhulst A, Van Opdenbosch E, Wellemans G, Giangaspero M (1988) Susceptibility of man to bovine viral diarrhea virus. Lancet ii: 110
20. Westaway EG, Brinton MA, Gaidamovich SYA, Horzinek MC, Igarashi A, Kääriäinen L, Lvov DK, Porterfield JS, Russell PK, Trent DW (1985) Flaviviridae. Intervirology 24: 183–192
21. Yolken R, Leister F, Eiden J (1990) RNA analyses of pestivirus isolated from humans. Abstracts VIII Intern Cong Virol Berlin, W21-002: 50

Author's address: Prof. Dr. M. C. Horzinek, Institute of Virology, Dept. Infect. Dis. & Immunol., Veterinary Faculty, Yalelaan 1 (de Uithof), 3508 TD Utrecht, The Netherlands.

Arch Virol (1991) [Suppl 3]: 7–18

Molecular characterization of hog cholera virus

T. Rümenapf, G. Meyers, R. Stark, and **H.-J. Thiel**

Federal Research Centre for Virus Diseases of Animals

Accepted March 14, 1991

Summary. An efficient tissue culture system was established which allowed to obtain substantial quantities of hog cholera virus (HCV) from the cell free tissue culture supernatant. After preparation of viral RNA and cDNA synthesis, the complete HCV genome was cloned and sequenced. Comparison with published BVDV sequences revealed a surprisingly high homology between HCV and BVDV at both the nucleotide and the amino acid level. In addition host cellular sequences were identified in BVDV genomes.

The genomic localization of HCV glycoproteins was determined by the use of sequence specific antisera directed against bacterial fusion proteins. The order on the HCV genome was determined as follows: N-gp44/48-gp33-gp55-C. HCV gp33 and HCV gp55 were shown to be intracellularly linked by disulfide bridges.

A cDNA fragment covering the genomic region that encodes the structural proteins of HCV was inserted into a vaccinia recombination vector. Expression studies with vaccinia/HCV recombinants led to identification of HCV specific glycoproteins which migrated on sodium dodecyl sulfate-gels identically to glycoproteins precipitated from HCV-infected cells. The vaccinia virus/HCV recombinant that expressed all four structural proteins induced virus-neutralizing antibodies in mice and swine. After immunization of pigs with this recombinant virus, full protection against a lethal challenge with HCV was achieved. A construct that lacked most of the HCV gp55 gene failed to induce neutralizing antibodies but induced protective immunity.

Key words: Hog cholera virus, pestivirus, comparison of pestiviruses, glycoproteins of hog cholera virus, hog cholera virus recombinant vaccine, neutralization of hog cholera virus.

Introduction

Hog cholera (HC) or classical swine fever (CSF) is assumed to be the economically most important viral epidemic of swine worldwide [3]. The causative agent of HC, the hog cholera virus (HCV) is an enveloped virus with a diameter of 40–50 nm, which contains a single stranded infectious RNA [25]. HCV was found to be serologically related to bovine viral diarrhea virus (BVDV) [11, 13]. Together with the border disease virus (BDV) of sheep these three viruses were classified as genus Pestivirus within the Togaviridae [41].

Typically, the highly contagious disease in its acute form is characterized by high body temperature and high mortality. Primary virus replication is assumed to take place in the tonsills after oral or nasal infection. From here virus spread into the lymphoid system occurs followed by a viremia around day 5 after infection. Thereafter virus can be isolated from almost all tissues. Diffuse and petechial haemorrhages on the serosa as well as haemorrhagic and necrotic lymphnodes are typical lesions observed post mortem. While "classical" HC which is easy to diagnose vanishes other forms of the disease appear which are apparently due to low virulent virus strains. These are often responsible for breeding problems which may have several alternative etiologies and such infections with low virulent HCV strains are therefore difficult to diagnose. Recognition of "atypical" and subclinical HCV infections is crucial for maintenance of eradication and control programs which are administered in most countries where HC appears. Some countries got rid of HC by vaccination programs and/or stamping out regimens of infected animals and herds (e.g. USA, Great Britain). Eradication appears to be very difficult and cost extensive in some european countries (e.g. Belgium, Germany, The Netherlands). High economic losses require new approaches for disease control which should benefit from efforts directed towards characterization of HCV at the molecular level. However, until recently such information was not available. This paper summarizes the current knowledge about molecular biology of hog cholera virus.

Viral RNA

The first steps towards molecular analysis of HCV were directed at identification and characterization of the viral nucleic acid. Early investigations described the HCV genome as an infectious single stranded RNA genome with a sedimentation coefficient of $S = 40$–45 [15] and a calculated size of 4×10^6 D [15, 16]. Recent experiments employing agarose gel electrophoresis after metabolic labeling led to identification of the genomic HCV RNA which represents a molecule of about 12.5 kb (Fig. 1) [26, 34]. This actually was the same size as reported for BVDV [9, 30]. Unlike other togaviruses subgenomic mRNAs could not be detected in cells infected with either BVDV or HCV [26, 27, 33, 34] and the pestiviral RNA genomes lack polyA tracts at their 3′ ends [9, 22, 29].

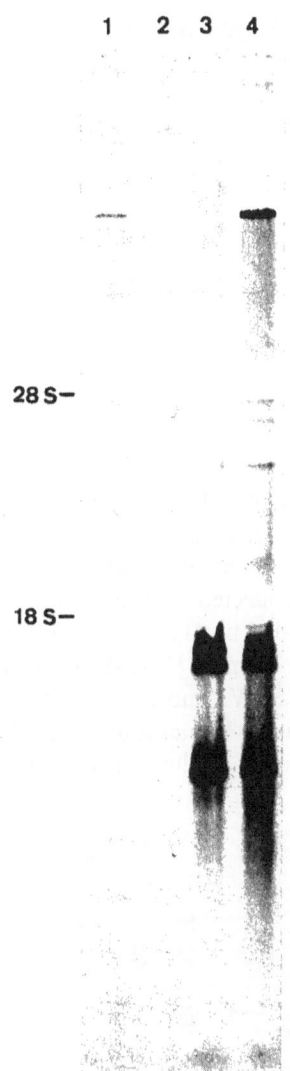

Fig. 1. RNA from pelleted virions and cells infected with HCV strain Alfort. Cells were treated with actinomycin D and metabolically labeled with [^{3}H] uridine. RNA was prepared [6], denatured by glyoxal and separated on an 0.8% agarose gel containing 5.5% formaldehyde [5]. Bands were visualized after fluorography. Lane 1: RNA from pelleted HCV; lane 2: RNA from pelleted material after mock infection; lane 3: RNA from non-infected 38A$_1$D cells; lane 4: RNA from HCV-infected 38A$_1$D cells

Tissue culture system

A serious problem in pestivirus research is the difficulty to obtain reasonable amounts of virus. HCV replicates in tissue culture cells of various species, preferentially porcine cells, without a cytopathogenic effect [25]. In addition to low titres of infectious virus in cell extracts and especially tissue culture supernatants virions are hard to purify by physical means like equilibrium-centrifugation due to their low buoyant density (Fig. 2.) [18, 25]. The establishment of a more efficient tissue culture system for HCV propagation considerably facilitated further examination. A cell line termed 38A$_1$D which is of porcine lymphoma origin [37] turned out to provide HCV titres (strain Alfort [1]) in the supernatant of 5×10^7 TCID$_{50}$/ml which is almost 10x the amount obtained with the standard cell line PK15 (Fig. 3) [33, 34].

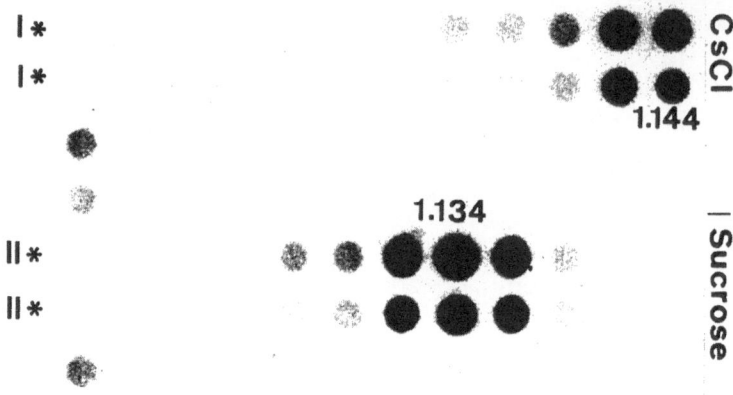

Fig. 2. Buoyant density of HCV in CsCl (I) and sucrose gradients (II) determined by dot blot. Pelleted virus obtained from 100 ml supernatant of infected $38A_1D$ cells was layered onto linear 5–30% CsCl or 10–40% sucrose gradients (two gradients of each) and centrifuged for 18 h at 150.000 g in a SW41 rotor. 18 fractions were collected from the bottom of each tube and spotted on nitrocellulose (bottom fraction indicated by *) after denaturation in $7 \times$ SSC, 7% formaldehyde and 0.25% NP40 at 60 °C for 15 min. Hybridization was performed using a nick translated 4.5 kb cDNA fragment of HCV. Buoyant density was calculated by determination of the refractive index of the solutions. Values indicated below dots (CsCl) or above dots (sucrose) refer to mean buoyant density of the respective HCV containing fractions

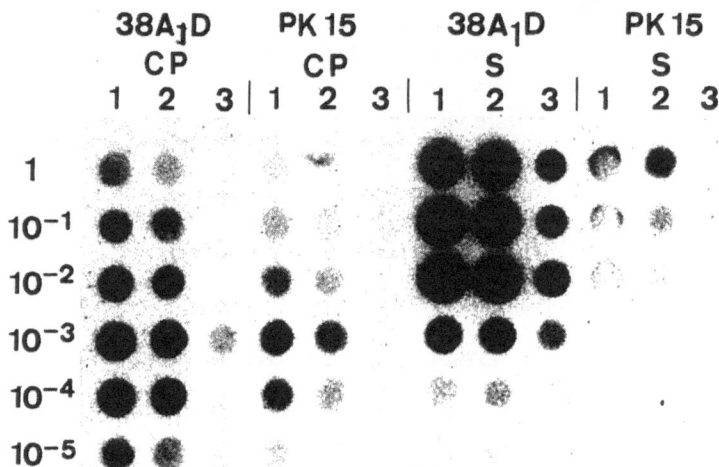

Fig. 3. Comparison of virus yields from $38A_1D$ and PK15 cells by dot blot analysis. Equal amounts of cells were infected with HCV (strain Alfort) using different multiplicities of infection (indicated on the left) and incubated for 48 h. Crude RNA from cells (CP) and virions collected from the supernatant (S) was dotted on nitrocellulose after denaturation in $7 \times$ SSC, 7% formaldehyde, 0.25% NP40 for 15 min, 60 °C. Material on the dots corresponds to number of cells (CP); lane 1: 4×10^5; lane 2: 2×10^5; lane 3: 4×10^4; volume of supernatant (S); lane 1: 0.2 ml; lane 2: 0.1 ml; lane 3: 0.02 ml). A 4.5 kb HCV cDNA fragment was employed as probe after radioactive labeling by nick translation

Other successful approaches to improve HCV propagation were described recently using porcine cells growing on microcarrier beads [4].

Cloning and sequencing of the genome

Characterization of a viral genome at the nucleotide level is a prerequisite for profound investigation of the virus encoded proteins. First, viral RNA was prepared from virions collected from the tissue culture supernatant of infected $38A_1D$ cells (up to 20 ng RNA/ml supernatant). This material served as template for reverse transcription using random primers and gave rise to different HCV cDNA libraries [22, 33, 34]. The crucial step in molecular cloning, however, is the identification of specific cDNA clones. For BVDV this problem was overcome using single stranded radiolabeled probes synthesized by reverse transcription of gradient- or gel-purified BVDV RNA [9, 29].

Screening of the HCV cDNA libraries was performed using two different approaches. One approach used a degenerated oligonucleotide as a probe which was synthesized according to sequences encompassing the highly conserved Gly-Asp-Asp motif present in all RNA dependant RNA polymerases of positive stranded RNA viruses including BVDV and flaviviruses [17, 22]. For the second approach an autologous polyspecific anti-HCV serum was raised in a goat [33, 34]. These antibodies were employed to screen a cDNA library established in the expression vector lambda gt11 for detection of HCV specific fusion proteins. While the oligonucleotide approach resulted in isolation of cDNA clones from the 3' region of the genome a clone from the 5' end was detected by antibodies. Subsequently the entire genome was cloned and sequenced with the exception of the utmost 3' and 5' ends. The HCV genome was found to consist of 12.284 nucleotides and to contain a single large open reading frame encompassing 3898 codons [22]. Compared to both published BVDV sequences the HCV genome was shorter by about 200 nucleotides. This was obviously due to host cellular sequences inserted in the BVDV genomes [23]. Sequence comparison of HCV and both published BVDV sequences revealed a surprisingly high homology at both the nucleotide and the amino acid level. Overall the homology was 66% with regard to nucleotides and 85% with regard to amino acids (Fig. 4) [22]. This high degree of similarity provided evidence for the close relationship between both viruses.

Taxonomy

There is general agreement that pestiviruses do not belong to the family Togaviridae. It has been proposed repeatedly to group pestiviruses into the Flaviviridae [7, 10]. With regard to strategy of translation and genome

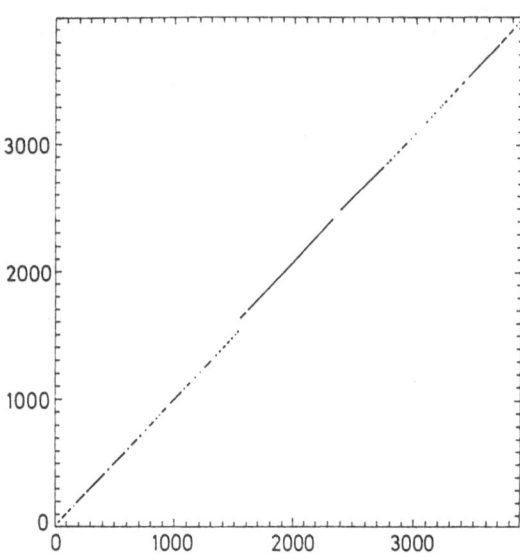

Fig. 4. Dot matrix comparison of the deduced amino acid sequences of HCV (abscissa) and the NADL strain of BVDV (ordinate). Parameters: window = 12; stringency = 14

organization there is similarity between pestiviruses and flaviviruses [7, 9, 32]. However, there are major differences concerning in particular different aspects of the putative virus structural proteins (40 and manuscript submitted for publication). In this context it is interesting to mention that pestiviruses share amino acid sequence similarity not only with flaviviruses but also with hepatitis C virus as well as certain plant viruses [24]. The final classification of the genus pestivirus should be postponed until more data not only about pestiviruses but also about hepatitis C virus are available.

Characterization of HCV glycoproteins

One major aim of our HCV research was the development of a suitable vaccine. In this regard analysis of the HCV induced glycoproteins was of major interest. Using hyperimmune sera against HCV, three virus specific glycosylated proteins (gp33, gp44/48 and gp55) were precipitated from HCV infected cells (Fig. 5) [33, 34, 36]. It is generally assumed that these glycoproteins are structural components of the virion. However, the actual molecular composition of virions from the genus pestivirus remains to be determined (manuscript submitted for publication).

The genome of pestiviruses apparently serves as mRNA [28] and is translated into a polyprotein which is processed by the action of unknown viral and host cellular proteases. Thus there are no apparent structures at the RNA level which would indicate the borders of the proteins encoded by the respective genes. Analysis of the topographical correlation between gene product and gene required serological reagents of defined specificity. For mapping of the glycoprotein genes HCV cDNA fragments were expressed in E. coli; sera obtained from rabbits immunized with these fusion proteins were employed in immunoprecipitation and Western blot analyses [36].

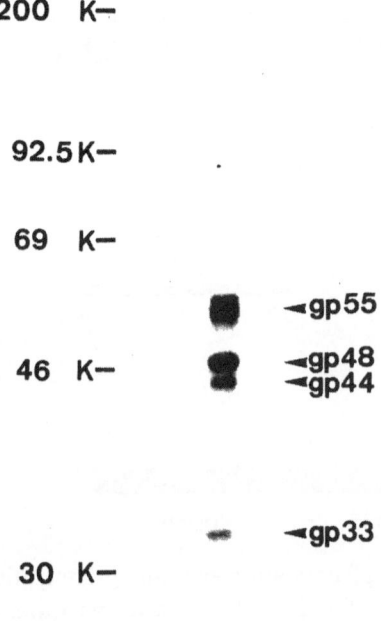

Fig. 5. Glycoproteins precipitated from HCV-infected cells. Lanes 1 and 2: noninfected 38A$_1$D cells; lanes 3 and 4: HCV-infected 38A$_1$D cells. Cells were metabolically labeled with [^{3}H] glucosamine and extracts incubated with either goat anti-HCV serum (lanes 1 and 3) or preimmune goat serum (lanes 2 and 4). Immunoprecipitates were analyzed by 10% SDS-PAGE and visualized by fluorography

From these experiments it was evident that all three glycoproteins were encoded within nucleotides 1100 to 3600 in the order NH2– gp44/48/ gp33/ gp55 –COOH (Fig. 6). The genomic region encompassing nucleotides 364 to 1100 probably encodes the HCV nucleocapsid (core) protein which has not been demonstrated yet. For BVDV a basic 20 kD molecule was detected which corresponded to this genomic region [8]. The genome organization within the region encoding the putative structural proteins of HCV is similar to the one reported for BVDV [36]. For processing between the glyco-proteins recognition sites of host cell proteases such as signalases or golgi-apparatus associated proteases were suggested [36]. Cleavage may occur between gp44/48 and gp33 at RR/S (amino acids 488–490), between gp33 and gp55 at A/E (708/709) and also at A/E (1067/1068) between gp55 and the following non identified protein. However, without determination of the amino-terminal sequences of the mature proteins the cleavage sites remain speculative.

One particular topic in pestivirus research deals with analysis of HCV gp55 and the analogous BVDV gp53 since this glycoprotein was shown to represent an antigen which induces virus neutralizing antibodies. This evidence was obtained from experiments with monoclonal antibodies (MABs); all MABs with neutralizing activity against the respective pestivirus recognized HCV gp55 or the analogous BVDV glycoprotein [2, 14, 38, 40].

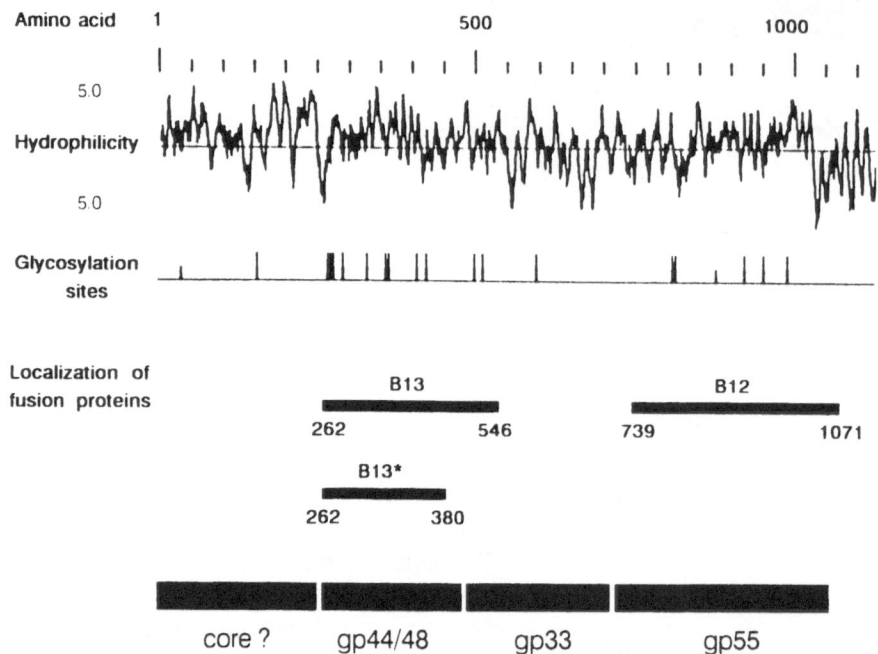

Fig. 6. Hydrophilicity plot, distribution of possible N- glycosylation sites and organization of the HCV polyprotein encompassing amino acids 1–1100 [22]. The tentative positions of the structural proteins are indicated (bottom line) together with the location of clones expressed as fusion proteins in bacteria [36]. Hydrophilicity is displayed by positive values (max = 5). Computer analysis was performed using GCG software on a VAX computer [12]

Interestingly, immunoprecipitation analyses with extracts of HCV infected cells revealed that gp55 and gp33 always coprecipitated [36, 39, 40]. In Western blot experiments neutralizing MABs recognized only gp55 [40]. Further insight was obtained using SDS-PAGE under nonreducing conditions which led to identification of molecules with apparent molecular weights of 55 kD, 68 kD and 90–100 kD [39, 40]. Reduction of disulfide bonds revealed that the 68 kD molecule represents a heterodimer between gp55 and gp33 [40]. In BVDV infected cells disulfide linked heterodimers between the analogous BVDV glycoproteins, gp53 and gp25, were demonstrated [40]. Preliminary evidence indicates that such heterodimers are also present in HCV virions (manuscript submitted for publication).

Eukaryotic expression of HCV glycoproteins

Understanding the biosynthesis, function and immunogenicity of the putative structural glycoproteins of HCV is essential for development of a recombinant vaccine. Such investigations can be greatly facilitated by heterologous expression of cloned cDNA fragments. For expression of glycoproteins a suitable eukaryotic expression vector system is required. In addition, for examination of the immunogenicity of each of the expressed proteins immunization trials should be easy to perform, preferentially in the natural host. According to these requirements the vaccinia vector system was employed for expression of HCV structural proteins. Due to the unique features of vaccinia virus i.e. replication in the cytoplasm, large cloning capacity and extremely broad host range it has evolved to an important tool in virus research [19, 20, 21].

For investigation of properly performed biosynthesis a HCV cDNA fragment spanning the entire region encoding the putative structural proteins of HCV (nucleotides 364–4000) was cloned into a vaccinia recombination vector [21, 31]. The respective recombinant vaccinia virus was generated within the cell by homologous recombination between the plasmid bearing the HCV insert and vaccinia genomic DNA [19, 20].

Proteins expressed from this HCV vaccinia recombinant (termed VAC3.8) were found to react specifically with anti-HCV antibodies in immunofluorescence and radioimmunoprecipitation [33]. After fluorography the respective molecules appeared to be authentic to HCV with regard to size and glycosylation (Fig. 7).

VAC3.8 was further employed for immunization trials in mice and pigs. In mice induction of HCV and vaccinia neutralizing antibodies was demonstrated after infection with VAC3.8 while after inoculation of vaccinia wildtype (WR) only vaccinia neutralizing antibodies could be detected. Immunization of pigs revealed similar results with respect to the induction of

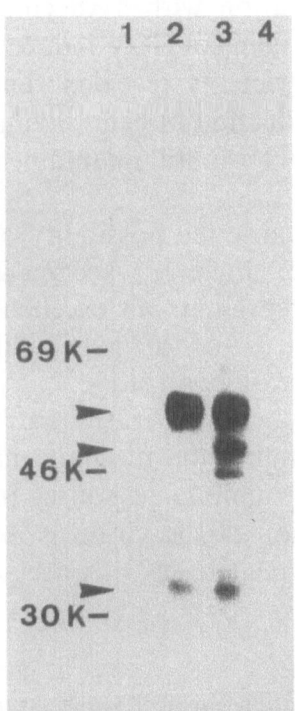

Fig. 7. HCV specific proteins expressed by recombinant vaccinia VAC3.8. CVI cells were metabolically labeled with a mixture of [^{35}S] cysteine/methionine after infection with VAC3.8. Extract was immunoprecipitated using MAB A18 anti-gp55 (lane 2), polyspecific goat anti-HCV serum (lane 3), pre-immune-goat serum (lane 4) and control MAB anti-FMDV (lane 1). Arrows indicate the positions of gp55, gp44/48 and gp33. Precipitates were visualized by fluorography after separation by 10% SDS-PAGE

Immunization of pigs with Vaccinia - HCV recombinants

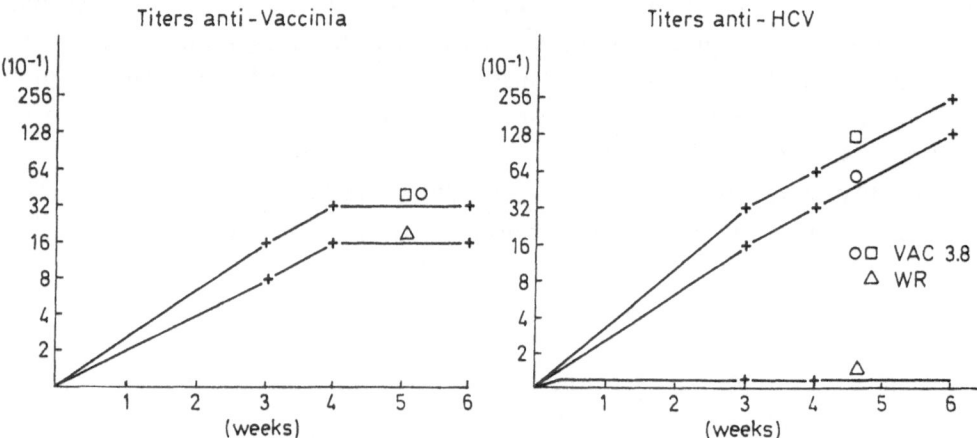

Fig. 8. Seroconversion of pigs immunized with VAC3.8 (n = 2) and vaccinia WR strain (n = 1). Vaccinia virus was administered intradermally, intraperitoneally and intravenously (5×10^7 pfu, each inoculation). Titres of neutralizing antibodies (refer to full neutralization) against vaccinia (left) and HCV (right) were determined 21 and 28 days after infection

neutralizing antibodies. After challenge infection of the pigs with a lethal dose of HCV Alfort animals immunized with VAC3.8 remained absolutely unaffected while the WR immunized pigs died around day 12 after challenge infection (Fig. 8).

These experiments showed that single immunization with a vaccinia HCV recombinant expressing the structural glycoproteins of HCV induced a protective immunity in swine. Preliminary experiments revealed that deletion of the gp55 encoding sequence abolished induction of neutralizing antibodies while at least partial protection against HC was still maintained [35].

The results of the animal experiments point towards the possibility of vaccination against HC with a recombinant vaccine. However, vaccinia is probably not suitable for practical purposes, but represents an excellent model system. Other vectors like swine pox virus or in particular pseudorabies virus may be more appropriate for disease control.

Molecular characterization of HCV and also of other pestiviruses is still at an early stage. We are far away from understanding the mechanisms involved in pathogenesis, virulence and protective immune response. In particular, investigation of the molecular composition of the virion is an urgent task with respect to all pestiviruses and also related RNA viruses.

References

1. Aynaud JM (1976) Characteristics of a new live virus vaccine against swine fever prepared in tissue culture at low temperature: the Thiverval strain. Comm Eur Communities EUR Rep 5486: 93–96

2. Bolin S, Moennig V, Kelso Gourley NE, Ridpath J (1988) Monoclonal antibodies with neutralizing activity segregate isolates of bovine viral diarrhea virus into groups. Arch Virol 99: 117–123

3. Brown F (1989) How do vaccinia-vectored vaccines fit into animal immunization programmes? In: Murphy FA (ed) Vaccinia vectored vaccines-risks and benefits. Res Virol 140: 463–491

4. Caij A, De Smet A, Dubois N, Koenen F (1989) High titre hog cholera virus production on cytodex 3 microcarrier cultures. Arch Virol 105: 113–118

5. Carmichael GC, Mc Master GK (1980) The analysis of nucleic acids in gels using glyoxal and acridin orange. In: Grossman L, Moldave K (eds) Methods of enzymology. Academic Press, New York, 165: 380–391

6. Chirgwin JM, Przybyla AE, MacDonald RJ, Rutter WJ (1979) Isolation of biologically active ribonucleic acid from sources enriched in ribonuclease. Biochemistry 18: 5294–5299

7. Collett MS, Anderson DK, Retzel E (1988) Comparison of the pestivirus bovine viral diarrhea virus with members of the flaviviridae. J Gen Virol 69: 2637–2643

8. Collett MS, Larson R, Belzer SK, Retzel E (1988) Proteins encoded by bovine viral diarrhea virus; the genomic organization of a pestivirus. Virology 165: 200–208

9. Collett MS, Larson R, Gold C, Strick D, Anderson DK, Purchio AF (1988) Molecular cloning and nucleotide sequence of the pestivirus bovine viral diarrhea virus. Virology 165: 191–199

10. Collett MS, Moennig V, Horzinek MC (1989) Recent advances in pestivirus research. J Gen Virol 70: 253–266

11. Darbyshire JH (1960) A serological relationship between swine fever and mucosal disease of cattle. Vet Rec 72: 331

12. Devereux J, Haeberli P, Smithies O (1984) A comprehensive set of sequence analysis programms for the VAX. Nucl Acids Res 12: 387–395

13. Dinter Z (1963) Relationship between bovine virus diarrhea virus and hog cholera virus. Zentralbl Mikrobiol Hyg A 188: 475–486

14. Donis RO, Corapi W, Dubovi EJ (1988) Neutralizing monoclonal antibodies to bovine diarrhoea virus bind to the 56k to 58k glycoprotein. J Gen Virol 69: 1549–1554

15. Enzmann PJ (1988) Molecular biology of the virus. In: Liess B (ed) Classical swine fever and related infections. Martinus Nijhoff Publishing, Boston Dordrecht Lancaster, pp 81–99

16. Enzmann PJ, Rehberg H (1977) The structural components of hog cholera virus. Z Naturforsch 32c: 456–458

17. Kamer G, Argos P (1984) Primary structural comparison of RNA-dependent polymerases from plant, animal and bacterial viruses. Nucl Acids Res 12: 7269–7282

18. Laude H (1979) Nonarbo-Togaviridae: comparative hydrodynamic properties of the pestivirus genus. Arch Virol 62: 347–352

19. Mackett M, Smith GL, Moss B (1982) Vaccinia virus: a selectable eucaryotic cloning and expression vector. Proc Natl Acad Sci USA 79: 7415–7419

20. Mackett M, Smith GL, Moss B (1984) General method for production and selection of infectious vaccinia virus recombinants expressing foreign genes. J Virol 49: 857–864

21. Mackett M, Yilma TY, Rose JA, Moss B (1984) Vaccinia virus recombinants: Expression of VSV genes and protective immunization of mice and cattle. Science 227: 433–435

22. Meyers G, Rümenapf T, Thiel H-J (1989) Molecular cloning and nucleotide sequence of the genome of hog cholera virus. Virology 171: 555–567

23. Meyers G, Rümenapf T, Thiel H-J (1989) Ubiquitin in a togavirus. Nature 341: 491

24. Miller RH, Purcell RH (1990) Hepatitis C virus shares amino acid sequence similarity

with pestiviruses and flaviviruses as well as members of two plant virus supergroups. Proc Natl Acad Sci USA 87: 2057–2061

25. Moennig V (1988) Characteristics of the virus. In: Liess B (ed) Classical swine fever and related infections. Martinus Nijhoff Publishing, Boston Dordrecht Lancaster, pp 55–80

26. Moormann RJM, Hulst MM (1988) Hog cholera virus: identification and characterization of the viral RNA and the virus-specific RNA synthesized in infected swine kidney cells. Virus Res 11: 281–291

27. Purchio AF, Larson R, Collett MS (1984) Characterization of virus specific RNA synthezised in bovine cells infected with bovine viral diarrhea virus. J Virol 48: 320–324

28. Purchio AF, Larson R, Torborg LL, Collett MS (1984) Cell free translation of bovine viral diarrhea virus. J Virol 52: 973–975

29. Renard A, Dino D, Martial J (1987) J European Patent Application number 86870095.6. Publication number 02.08672

30. Renard A, Guiot C, Schmetz D, Dagenais L, Pastoret P-P, Dina D, Martial JA (1985) Molecular cloning of bovine viral diarrhea viral sequences. DNA 4: 429–438

31. Rice CM, Franke CA, Strauss JH, Hruby DE (1985) Expression of Sindbis structural proteins via recombinant vaccinia virus: synthesis, processing and incorporation into mature Sindbis virions. J Virol 56: 227–239

32. Rice CM, Strauss EG, Strauss JH (1986) Structure of the flavivirus genome. In: Schlesinger S, Schlesinger MJ (eds) The togaviridae and flaviviridae. Plenum Publishing Corp, New York

33. Rümenapf T (1990) Klonierung, Sequenzierung und Expression des Genoms des Virus der Klassischen Schweinepest. Thesis, University of Giessen, Giessen, Federal Republic of Germany

34. Rümenapf T, Meyers G, Stark R, Thiel H-J (1989) Hog cholera virus—characterization of specific antiserum and identification of cDNA clones. Virology 171: 18–27

35. Rümenapf T, Stark R, Meyers G, Thiel H-J (1991) Structural proteins of hog cholera virus expressed by vaccinia virus: Further characterization and induction of protective immunity. J Virol 65: 589–597

36. Stark R, Rümenapf T, Meyers G, Thiel H-J (1990) Genomic localization of hog cholera virus glycoproteins. Virology 174: 286–289

37. Strandström H, Veijalainen P, Moennig V, Hunsmann G, Schwarz H, Schäfer W (1974) C-type particles produced by a permanent cell line from a leucemic pig. I. Origin and properties of the host cells and some evidence for occurence of C-type-like particles. Virology 57: 175–178

38. Weensvoort G (1989) Topographical and functional mapping of epitopes on hog cholera virus with monoclonal antibodies. J Gen Virol 70: 2865–2876

39. Weensvoort G, Boonstra J, Bodzinga BG (1990) Immunoaffinity purification and characterization of the envelope protein E1 of hog cholera virus. J Gen Virol 71: 531–540

40. Weiland E, Stark R, Haas B, Rümenapf T, Meyers G, Thiel H-J (1990) Pestivirus glycoprotein which induces neutralizing antibodies forms part of a disulfide-linked heterodimer. J Virol 65: 589–597

41. Westaway EG, Brinton MA, Gaidamowich SY, Horzinek MC, Igarashi A, Kääriäinen L, Lvov DK, Porterfield JS, Russel PK, Trent DW (1985) Togaviridae. Intervirology 24: 125–139

Author's address: H.-J. Thiel, Federal Research Centre for Virus Diseases of Animals, P. O. Box 1149, D-W-7400 Tübingen, Federal Republic of Germany.

Arch Virol (1991) [Suppl 3]: 19–27
© Springer-Verlag 1991

Bovine viral diarrhea virus genomic organization

Marc S. Collett[1,2], **MaryAnn Wiskerchen**[1,a], **Ellan Welniak**[2,b], and **Susan K. Belzer**[2,c]

[1] MedImmune, Inc., Gaithersburg, Maryland, and
[2] College of Veterinary Medicine, The University of Minnesota, Department of Large Animal Clinical Sciences, St. Paul, Minnesota USA

Accepted March 14, 1991

Summary. In previous work, we developed a preliminary description of the genetic organization of the prototypic pestivirus bovine viral diarrhea virus (BVDV). In order to refine this genetic map and to further elucidate the gene products and expression strategy of this virus, we have generated a broad panel of sequence-specific antibody reagents. Use of these reagents not only allowed the identification of several previously undescribed viral polypeptides, but when used in in vivo pulse-chase experiments, they identified precursor polyproteins and processing intermediates. Data generated from these studies provide a more accurate and complete view of viral gene organization, as well as insight into several aspects of protein processing and the gene expression strategy employed by this pestivirus. These experiments also revealed varying stability and turnover rates for the mature BVDV proteins. These latter results have implications for the functional roles of certain gene products.

Key words: Pestivirus, viral proteins, polyprotein precursors, protein processing, gene expression.

Introduction

Bovine viral diarrhea virus (BVDV) represents the prototypic member of the genus Pestivirus. Hog cholera virus (HCV) and border disease virus (BDV)

---

[a] Present address: Virology Division, USAMRIID, Fort Detrick, Frederick, Maryland 21701, USA.

[b] Present address: Solvay Animal Health, Inc., Mendota Heights, Minnesota 55120, USA.

[c] Present address: The University of Minnesota, College of Veterinary Medicine, Department of Veterinary Pathobiology, St. Paul, Minnesota 55108, USA.

of sheep are the other known members of this group. The pestivirus genome consists of a single molecule of single-stranded, positive polarity RNA of about 12–13 kilobases (see review, [4]). The genomes of two isolates of BVDV (NADL and Osloss) and two isolates of HCV (Alfort and Brescia) have been molecularly cloned and sequenced (3, 7–9). Computer-assisted sequence comparisons among these genomes have revealed a high degree of relatedness, as well as certain interesting regions of divergence.

Numerous reports have appeared describing pestivirus-specific poly-peptides found either in virus-infected cells or in enriched virus preparations (4). However, these studies were unable to distinguish between 'virus-encoded', 'virus-induced', and 'virus-associated' polypeptides. Using the molecularly cloned genome of BVDV (NADL), we set out to define the complete complement of polypeptides encoded by the pestiviral RNA.

The BVDV genome possesses a single large open reading frame (ORF) extending the length of the RNA capable of encoding about 449 kilodaltons (kDa) of protein. Using sequence-specific antiserum reagents generated against recombinant bacterial fusion proteins possessing select regions of the BVDV ORF, in combination with radioimmunoprecipitation analyses, we were able to identify specific viral proteins and position their coding regions along the ORF (2). The results from these studies allowed us to produce a preliminary map of the genetic organization of BVDV (2). However, this first map provided only general boundaries for the various BVDV poly-peptides. Furthermore, there were regions of the ORF for which gene products had not yet been assigned or identified.

Here, we summarize data from the further use of these antisera, as well as new antibody reagents generated against synthetic peptides representing specific sequences within the BVDV ORF. The results serve to refine our previous genetic map, and moreover, have identified previously undescribed viral polypeptides. Additionally, pulse-chase analyses of viral protein synthesis in infected cells have provided further clarification of the protein coding organization of BVDV, as well as insight into the possible mechanisms of gene expression and protein biogenesis by this pestivirus.

Materials and methods

The cytopathic NADL isolate of BVDV was propagated in MDBK cells as previously described (2, 3). Procedures for radiolabeling of virus-infected cell cultures with ^{35}S-methionine and for radioimmunoprecipitation have been detailed previously (2). Briefly, MDBK cells were infected with BVDV (NADL) at a multiplicity of 0.5, and at approximately 18 hours post-infection, cell cultures were incubated in methionine-free medium for one hour. The medium was then replaced with fresh methionine-free medium supplemented with ^{35}S-methionine at 100–200μCi per ml. For standard radioimmunoprecipitation experiments, the cultures were incubated for 2 hours. For pulse-chase experiments, parallel cultures were incubated with ^{35}S-methionine for 15 minutes (pulse time), at which point cultures were either harvested and quickly frozen (pulse-labeled cells) or they were washed several times with warm, nonradioactive medium containing 100 times the normal level of methionine and then incubated in this medium further supplemented with 30 µg/ml cyclo-heximide. After 1, 2, and 6 hours (chase times), cultures were harvested and frozen. Cell lysates were prepared in SDS lysis buffer (6), cleared, and aliquoted for immunoprecipitation. Between 5 to 12 µl of various rabbit antisera were used in each precipitation. After collection of immune complexes onto Pansorbin (Calbiochem-Behring, La Jolla, Calif. USA) and washing, the precipitated proteins were released from the bacteria in SDS sample buffer. Samples were fractionated on SDS-containing 10–18% gradient polyacrylamide gels, and the gels were fluorographed and exposed to X-ray film. Antisera to *E. coli* β-galactosidase fusion proteins have been previously described (2). Antisera to synthetic peptides representing various sequences within the BVDV ORF were prepared by procedures previously reported (10).

Results

To assist in the identification of BVDV-encoded proteins and in the description of the gene expression strategy of pestiviruses, we have concentrated on the generation of sequence-specific antiserum reagents. Previously, we described the use of *E. coli*-produced fusion proteins for production of such reagents (2). Here, we have added to our arsenal of antibody reagents antisera to synthetic peptides representing specific sequences within the BVDV ORF. Antipeptide antisera represent reagents of higher specificity than the antisera to the bacterial fusion proteins in that antibodies are directed to only a 15 to 20 amino acid segment of the ORF. Furthermore, the chemically synthesized peptides offer an alternative to the biological production of immunogens; an important advantage for particular regions of the ORF that may be poorly expressed by recombinant means. All of these reagents are schematically summarized in Fig. 1B.

Each of these antisera was used to immunoprecipitate radiolabeled BVDV-infected MDBK cell lysates. In initial experiments, virus-infected cell cultures were radiolabeled with ^{35}S-methionine for 2 hours. Cells were then harvested and lysates were prepared for immunoprecipitation. The results from gel analyses showed all sera, with the exceptions of antisera to peptides 241 and 2621 and *E. coli* protein C26, immunoprecipitated proteins specifically from BVDV-infected cells, not present in mock-infected cells (data not shown). Reasons for the failure of the three antisera to recognize BVDV-specific proteins are not clear at this time.

In general, the data from the use of the antipeptide antisera consistent with previous protein alignments. For example, anti-697 antiserum recognized gp53, confirming the general position of the N-terminal boundary of this glycoprotein (Fig. 1). However, in the case of anti-1130 antiserum, its ability to immunoprecipitate p125 and p54 served to significantly reposition (move to the left relative to the position in our previous map) the amino terminal boundary of the coding region for these polypeptides. This shift was confirmed by the lack of reactivity of anti-2396 antiserum to p125

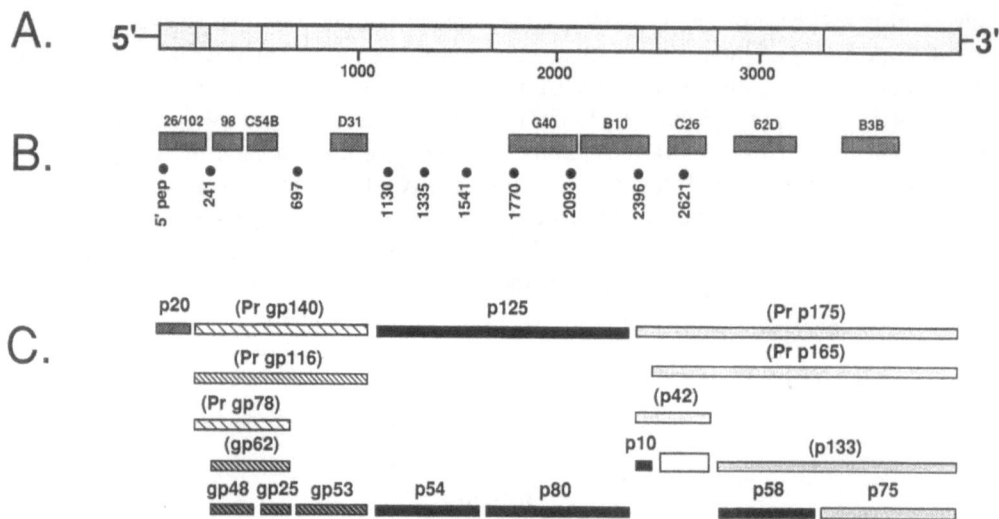

Fig. 1. Organization of the BVDV genome. A. Schematic representation of the BVDV genome, showing the 5′ noncoding region, the large ORF (shaded box), and the 3′ noncoding region. The vertical lines within the large ORF indicate approximate positions of proteolytic cleavage sites. B. Regions of the ORF represented in various *E. coli* fusion proteins (shaded boxes) or as synthetic peptides (dots), to which rabbit antisera were generated. C. Approximate positioning within the ORF of various polypeptides identified by the sequence-specific antibody reagents in immunoprecipitation analyses

or p80, establishing the carboxy terminal limit of the coding position for these proteins. This change eliminates a region for which we previously speculated encoded an unidentified product (2). Antiserum 2396 further served to identify two new BVDV-encoded proteins: p42 and p10, and in doing so, filled in the last protein coding gap of our earlier map (see below).

Throughout these analyses, as well as in our previous studies, we often observed faint, minor high molecular weight radiolabeled polypeptide bands immunoprecipitated by various antisera. Indeed, as previously discussed (2), the genomic structure of BVDV suggests viral protein biogenesis involves the proteolytic processing of a large polyprotein precursor(s). Such processing is likely to involve both cotranslational and post-translational cleavage events. Thus, from the cotranslational cleavage of the primary translation product of the viral genome, there may be various short-lived intermediate polyproteins which are then further processed to give rise to mature viral proteins.

To investigate this proposed scheme of events and to establish the precursor-product relationships suggested among the various BVDV proteins, we conducted a pulse-chase experiment on virus-infected cells and analyzed the polypeptides immunoprecipitated with select antisera. Parallel virus-infected cultures were radiolabeled with ^{35}S-methionine for 15 minutes (pulse time), at which point cultures were either harvested and quickly frozen (pulse-labeled cells) or they were washed several times in nonradioactive medium containing 100 times the normal level of methionine and then incubated in this medium further supplemented with cycloheximide. After 1, 2, and 6 hours (chase times), cultures were harvested and frozen. Cell lysates were prepared and immunoprecipitated with the various BVDV sequence-specific antisera. A gel analysis of the immunoprecipitated polypeptides from this pulse-chase experiment is presented in Fig. 2. Table 1 summarizes the BVDV-specific polypeptides recognized by each of the antisera shown in Fig. 1B.

Antibodies to the first 15 amino acids of the ORF (anti-5′ pep antiserum) recognized only the p20 protein; with this antiserum, there was no evidence of a higher molecular weight precursor protein in the pulse-labeled sample. In contrast, 26/102 antiserum, previously shown to immunoprecipitate p20 (2), was found with the pulse-labeled sample to immunoprecipitate several additional larger polypeptides. These same higher molecular weight poly-peptides were also recognized by antisera 98, C54B, and 697. Endo-glycosidase treatment established these polypeptides to be glycoproteins (data not shown). These latter antisera revealed the apparent rapid transition of a glycoprotein precursor Prgp140 to Prgp116. Prgp116 gave rise to Prgp78 and the mature, stable viral glycoprotein gp53. Prgp78 appeared then to be processed to yield gp62, likely involving the proteolytic removal of an amino-terminal portion of Prgp78 since reactivity with antiserum 26/102 was lost. Finally, as previously suggested (2), gp62 gave rise to the stable viral glycoproteins gp48 and gp25.

Proteins encoded from the central portion of the BVDV ORF (p125, p54, and p80) showed a very different pattern of expression. All three poly-peptides appeared during the 15 minute pulse and remained stable there-after. The protein half-lives ($T_{1/2}$) for all three proteins were greater than 6 hours. There was no significant decay of the presumed precursor p125 to its putative processed products p54 and p80.

The carboxy-terminal 39% of the ORF appears served by the precursor polyprotein Prp175, immunoprecipitated by antisera 2396, 62D, and B3B. Prp175 rapidly yielded p10 and Prp165, and p42 and p133. p42 was short-lived, and is likely a processing precursor to p10 and a yet-to-be identified gene product (32 kDa). Similarly, Prp165 represents an alternative intermediate to this unidentified product and to p133. p133, with a half-life of less than 60 minutes, was processed to yield two mature products with distinct characteristics: the stable p58 protein ($T_{1/2}$ greater than 6 hours) and the rapidly turned-over p75 protein ($T_{1/2}$ less than 60 minutes).

Discussion

Put together, the above data allow construction of a more complete protein organizational map for the BVDV genome, in which several precursor polypeptides may be included (Fig. 1C). We are now able to account for the

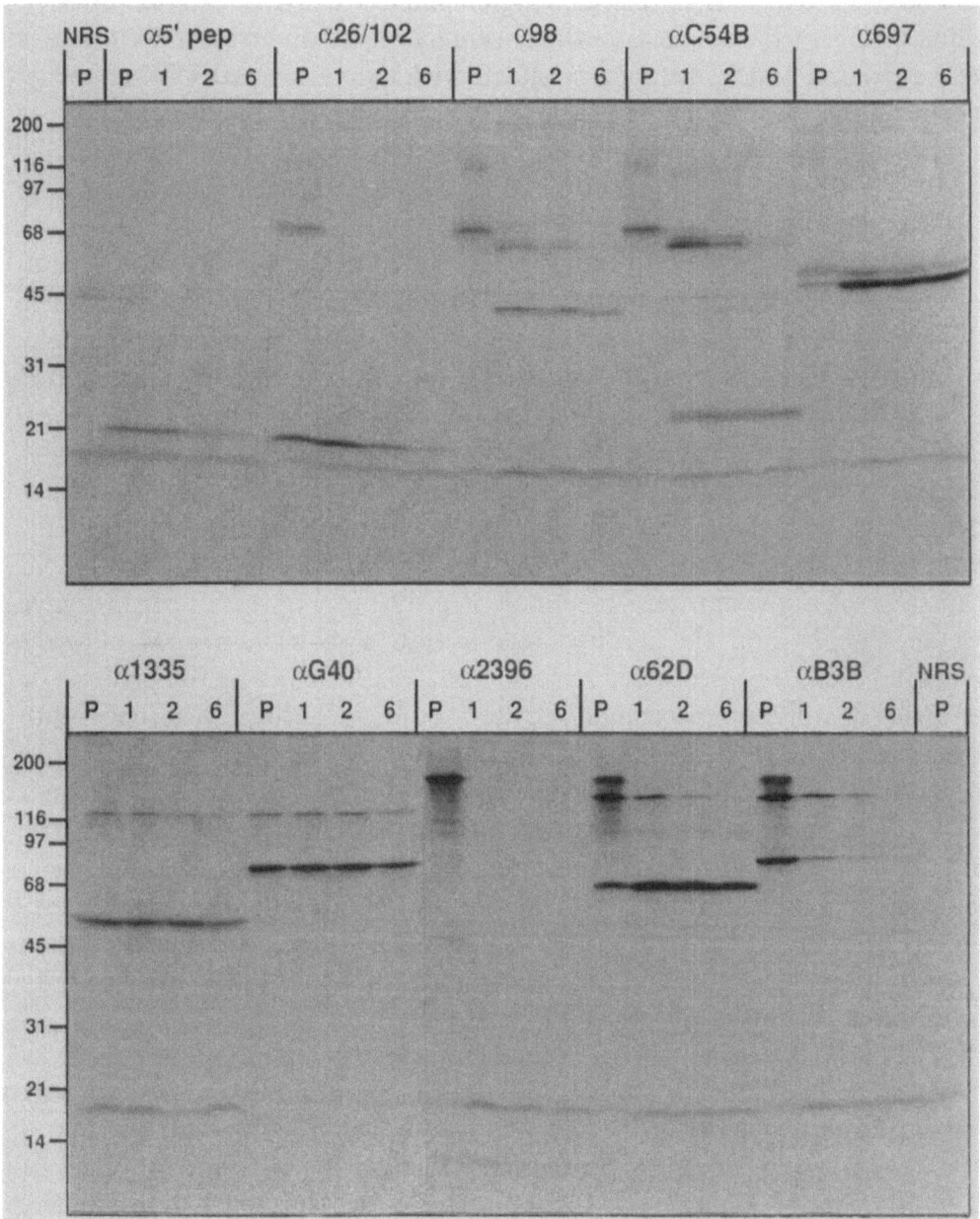

Fig. 2. Pulse-chase radioimmunoprecipitation of BVDV-infected cell lysates. The cytopathic NADL isolate of BVDV was used to infect MDBK cells. Parallel cultures were pulse radiolabeled (P) and chased for 1, 2, and 6 hours, as described in the text. Cleared cell lysates were prepared and immunoprecipitated with the indicated antisera; the precipitated polypeptides were analyzed on SDS-containing 10–18% gradient polyacrylamide gels, all as previously described (2, 6). The numbers at the left indicate approximate molecular masses (in kDa). NRS, normal rabbit serum. The protein band migrating at approximately 17 kDa in all tracks is a nonspecific polypeptide unrelated to BVDV proteins

Table 1. BVDV sequence-specific antiserum reagents and the proteins
they recognize

Immunogen/Antiserum	Polypeptides Immunoprecipitated
5′ pep (2–16)[1]	p20
26/102	p20, Prgp140, Prgp116, Prgp78
241 (241–255)	none detected
98	Prgp140, Prgp116, Prgp78, gp62, gp48
C54B	Prgp140, Prgp116, Prgp78, gp62, gp25
697 (697–711)	Prgp140, Prgp116, gp53
D31	Prgp140, Prgp116, gp53
1130 (1130–1150)	p125, p54
1335 (1335–1351)	p125, p54
1541 (1541–1559)	p125, p54
1770 (1770–1787)	p125, p80
G40	p125, p80
2093 (2093–2112)	p125, p80
B10	p125, p80, Prp175, p42, p10
2396 (2396–2414)	Prp175, p42, p10
C26	none detected
2621 (2621–2635)	none detected
62D	Prp175, Prp165, p133, p58
B3B	Prp175, Prp165, p133, p75

[1] Numbers in parentheses indicate the amino acid coordinates of the
BVDV ORF represented in the peptide.

entire coding capacity of the BVDV ORF with the exception of identification of a putative 32 kDa polypeptide processing product of p42. From the pulse-chase experiment results, the following protein expression scheme can be envisioned. A primary translation product initiated from the first AUG codon of the ORF immediately yields the p20 protein via a cotranslational proteolytic cleavage event. Primary translation continues on to similarly produce three immediate precursor polyproteins that encompass the remainder of the genome: Prgp140, p125, and Prp175. Prgp140 and Prp175 appear then to be proteolytically processed in a stepwise fashion to yield mature viral proteins. In contrast, the p125 region seems to follow a different pathway, in that all products of this region of the genome appear rapidly (within the pulse-labeling period) and are stable thereafter. The nature of the mechanisms and/or kinetics involved here are unclear at present. This apparently distinct means for the biogenesis of p125, p54, and p80 deserves further investigation.

From the data presented here, there are at least 10 proteolytic processing sites involved in the biogenesis of the complete complement of BVDV proteins (Fig. 1A). There are likely to be additional sites not depicted in

Fig. 1A. In fact, our data suggest the production of gp53 (which is actually two polypeptides collectively referred to as gp53; see Fig. 2, α697) involves two alternative cleavage sites at its carboxy terminus. As protein analyses become more detailed (e.g., amino- and carboxy-terminal protein sequence determinations), additional proteolytic trimming of BVDV polypeptides may be revealed.

The proteinases affecting these cleavages are likely to be of both viral and cellular origin. The immediate appearance of p20 suggests its release from a primary translation product may be autocatalytic. Indeed, as will be presented elsewhere, we have established p20 as a viral proteinase responsible for its own release from the primary translation product [11]. Elsewhere in the pestivirus genome, using computer-assisted sequence comparisons, homologies have been found between the BVDV p80 protein and a class of serine proteinases, leading to the prediction that p80 might also be a viral protease (1, 5). This prediction has been experimentally confirmed. We have demonstrated the BVDV p80 protein indeed possesses proteolytic activity [12]. There are certainly cellular proteinases, and may be additional yet-to-be identified virus-encoded proteases, involved in the production of BVDV proteins. Their identification will further clarify the processes involved in pestivirus protein expression.

In addition to elucidating various aspects of the process of BVDV protein expression, the pulse-chase experiment described here also revealed information on the polypeptide half-lifes of the mature viral proteins within virus-infected cells. The glycoproteins (gp48, gp25, gp53), p125, p54, p80, and p58 were all very stable ($T_{1/2}$ greater than 6 hours); p10 was relatively stable ($T_{1/2}$ about 6 hours); p20, less stable ($T_{1/2}$ about 110 minutes); and p75, unstable ($T_{1/2}$ less than 60 minutes). These protein stability characteristics are likely to have functional consequences for the process and regulation of virus replication. One obvious implication is that the mature viral protein with the shortest half-life, or highest turnover rate, could represent the rate-limiting or regulating component for virus replication. It is interesting that p75 has this characteristic, and that it represents the putative RNA-dependent RNA polymerase of the virus (2).

Future studies of the functional roles of pestivirus proteins in the gene expression and replication may well assist in our understanding aspects of pathogenesis by members of this virus group.

Acknowledgements

This work was supported in part by USDA grants 87-CRCR-1-2586 and 88-37266-4195.

References

1. Bazan JF, Fletterick RJ (1989) Detection of a trypsin-like serine protease domain in flaviviruses and pestiviruses. Virology 171: 637–639

2. Collett MS, Larson R, Belzer SK, Retzel E (1988) Proteins encoded by bovine viral diarrhea virus: the genomic organization of a Pestivirus. Virology 165: 200–208
3. Collett MS, Larson R, Gold C, Strick D, Anderson DK, Purchio AF (1988) Molecular cloning and nucleotide sequence of the Pestivirus bovine viral diarrhea virus. Virology 165: 191–199
4. Collett MS, Moennig V, Horzinek M (1989) Recent advances in pestivirus research. J Gen Virol 70: 253–266
5. Gorbalenya AE, Donchenko AP, Koonin V, Blinov VM (1989) N-terminal domains of putative helicases of flavi- and pestiviruses may be serine proteases. Nucl Acids Res 17: 3889–3897
6. Keegan K, Collett MS (1986) Use of bacterial expression cloning to define the amino acid sequences of antigenic determinants on the G2 glycoprotein of Rift Valley fever virus. J Virol 58: 263–270
7. Meyers G, Rumenapf T, Thiel H-J (1989) Molecular cloning and nucleotide sequence of the genome of hog cholera virus. Virology 171: 555–567
8. Moormann JMR, Warmerdam PAM, van der Meer B, Schaaper WMM, Wensvoort G, Hulst MM (1990) Molecular cloning and nucleotide sequence of hog cholera virus strain Brescia and mapping of the genomic region encoding envelope protein E1. Virology 177: 184–198
9. Renard A, Dino D, Martial J (1987) Vaccines and diagnostics derived from bovine diarrhea virus. European Patent Application number 86870095.6. Publication number 0208672
10. Suzich JA, Collett MS (1988) Rift Valley fever virus M segment: cell-free transcription and translation of virus-complementary RNA. Virology 164: 478–486
11. Wiskerchen M, Belzer SK, Collett MS (1991) Pestivirus gene expression: The first protein product of the bovine viral diarrhea virus large open reading frame, p20, possesses proteolytic activity. J Virol 65: 000–000
12. Wiskerchen M, Collett MS (1991) Pestivirus gene expression: Protein p80 of bovine viral diarrhea virus is a proteinase involved in polyprotein processing. Virology 184: 000–000

Authors' address: Marc S. Collett, MedImmune, Inc., Gaithersburg, Maryland 20878, U.S.A.

Arch Virol (1991) [Suppl 3]: 29–40

Bovine viral diarrhea virus proteins and their antigenic analyses

R. O. Donis[1], **W. V. Corapi**[2a], and **E. J. Dubovi**[2]

[1] Department of Veterinary Science, University of Nebraska, Lincoln, Nebraska, and
[2] Diagnostic Laboratory, NYS College of Veterinary Medicine, Ithaca, New York, U.S.A.

Accepted March 14, 1991

Summary. Bovine Viral Diarrhea Virus (BVDV) polypeptides present in infected cells are the result of the processing of the polyprotein translated from the large single open reading frame of the BVDV genomic RNA. The presence of these proteins in infected cells was studied by radiolabeling under hypertonic conditions and with the aid of radioimmunoprecipitation. The genomic mapping of these polypeptides suggests a complex pattern of processing which involves cellular and viral proteases. The consistent absence of 80k in noncytopathic isolates of BVDV suggests that the processing of the viral polyprotein is different in cytopathic and noncytopathic biotypes of BVDV. The antigenic structure of BVDV was studied with a panel of monoclonal antibodies (MABs) prepared against the Singer isolate of BVDV. Neutralizing MABs were found to bind the 56–58k polypeptide, providing evidence that this glycoprotein is present on the surface of the virion and carries neutralization epitopes. Antigenic analyses with the panel of MABs reveals extensive antigenic heterogeneity among BVDV field isolates. MABs were used to determine the frequency of neutralization escape mutants in stocks of BVDV. Plaque-purified BVDV stocks contain neutralization escape mutants with a frequency of $10^{-2.47}$.

Key words: Pestivirus proteins, monoclonal antibodies, neutralization, mutation rate.

Introduction

The bovine pestivirus, Bovine Viral Diarrhea Virus, or BVDV, is the single most important viral pathogen of cattle in developed countries. In developing nations, other viral pathogens such as foot and mouth disease virus,

---

[a] Present address: Department of Microbiology, Parasitology and Immunology, NYS College of Veterinary Medicine, Ithaca, NY 14853, U.S.A.

unfortunately continue to inflict even more devastating economic damage. Every year, losses due to BVDV infection, in its many forms, amount to the monetary equivalent of 0.5% to 1% of all cattle [14, 27]. The cattle population of the world is well in excess of 700 million head. Thus, losses to BVDV could mean the loss of as many as 5 million head of cattle annually [10, 35]. The bulk of damage inflicted by BVDV to the industry stems from fetal infection. Fetal infection is preventable by vaccination, therefore, the economic damage caused by BVDV is unacceptably high. Both veterinarians and producers have been considerably reluctant to use BVDV vaccines. The reasons for this phenomenon include: 1) Poor understanding of the epidemiology and pathogenesis of BVDV infections and the ensuing diseases; 2) Difficulties associated with laboratory and clinical diagnosis of BVDV infection which lead to an underestimation of the problems it causes; 3) Adverse effects due to improper use of the vaccines; 4) Vaccines of questionable quality [2, 27, 28, 33, 37, 44].

Hopefully, the continuing pestivirus research carried out in many laboratories around the world will help reverse the current situation and lead to the implementation of rational control practices to reduce the economic impact of BVDV infection.

Biochemistry of viral proteins: biotype differences

The first isolation of BVDV in 1957 was carried out in bovine primary cell cultures [32]. The virus replicated in these cells without causing any cytopathic changes. During that same year, Underdahl et al. reported the isolation of an agent from cattle which induced cell vacuolation and lysis upon replication [43]. Except for the differences in the mode of interaction with the host cells, the two types of BVDV isolates appeared to be identical [24]. Furthermore, infection of cells with noncytopathic BVDV induced cellular resistance to the lytic activity of cytopathic BVDV [25]. Consequently, the two types of isolates were considered to be two different biotypes of the same virus [24, 30, 31]. More recently it was found that these biotype differences play a pivotal role in the pathogenesis of mucosal disease [4, 6].

A brief discussion of the replication strategy of BVDV will aid the interpretation of viral protein profiles in infected cells. Attachment of a BVDV virion to the susceptible target cell is followed by virus entry and uncoating of the viral genome. The single RNA molecule that makes up the genome is subsequently translated by the ribosomes into a hypothetical protein of almost 4,000 amino acid residues. However, such a giant protein has not been observed in infected cells. Instead, at least 9 viral proteins are generated from a single open reading frame present in the genomic RNA molecule [8, 9]. Proteolytic processing is necessarily one of the mechanisms whereby the translation of a single open reading frame results in several proteins. Two major classes of proteases are involved in the cleavage of the viral polyprotein resulting from translation of the entire genomic RNA: viral and cellular peptidases [42, 46]. These two types of proteases reside in different compartments and are likely to meet their substrate(s) at different times after their synthesis. In addition, they probably have different rates of catalysis and cleavage site specificity. These cleavages of the polyprotein will result in a complex pattern of intermediate precursor polypeptides before the final processed proteins are obtained [42, 46]. This information will become important in the interpretation of the patterns of BVDV polypeptides present in infected cells.

We have used two different approaches to identify viral polypeptides in BVDV-infected cells: one direct and the other immunological. In both cases we relied in the use of radioactive amino acid precursors to label polypeptide backbones or in radioactive sugars to label carbohydrate moieties present in glycoproteins; the radiolabeled polypeptides are then separated by electrophoresis in denaturing polyacrylamide gels.

The direct approach to identify viral polypeptides has been very successful with viruses (i.e. polio) that effectively inhibit the translation of cellular mRNA: the radioactive precursor is incorporated almost exclusively into viral polypeptides. BVDV infection does not result in a significant cellular mRNA translational shutoff [19]. To inhibit the background of cellular polypeptide synthesis we took advantage of a still incompletely understood phenomenon: the differential susceptibility of viral and cellular mRNA translation to a hypertonic environment [34]. In the presence of 150 mM excess NaCl in the culture medium, initiation of translation of cellular messages is almost completely suppressed yet BVDV genomic RNA translation proceeds [19]. Thus this hypertonic initiation block (HIB) of cellular mRNA allowed us to identify in a direct fashion the viral polypeptides present in infected cells. The polypeptides we were able to identify in cells infected with the Singer (cytopathic) isolate of BVDV using HIB conditions to label cells with radioactive amino acids were: 165k, 135k, 118k, 80k, 65k, 56–58k, 48k, 37k, 25k and 19k (k = molecular mass of polypeptide in kilodaltons, as determined by migration in SDS/PAGE). Similar experiments using radioactive sugars led to the identification of 56–58k and 48k as glycosylated polypeptides [17, 19].

The second approach to identify viral polypeptides in infected cells makes use of antibodies present in hyperimmune serum to select radiolabeled polypeptides from a complex mixture. The technique, known as radioimmunoprecipitation (RIP), does not require the use of selective inhibitors: cells are labeled with a radioactive precursor and lysed with detergent. Viral polypeptides are then separated from cell polypeptides present in the lysate by binding to the specific antibodies attached to a solid phase [26, 29]. Polypeptides that fail to elicit antibody production in the

animal used to produce the hyperimmune serum will not be identified: this
constitutes the major limitation of the RIP technique.

RIP of cell lysates prepared from cells infected with the cytopathic Singer
isolate of BVDV and labeled with radioactive amino acids led to the
identification of the following viral polypeptides: 165k, 118k, 80k, 75k,
56–58k, 48k, 37k and 25k. Of these, 118k, 75k, 56–58k, 48k and 25k were
also shown to contain attached oligosaccharides by labeling with radioactive mannose [17, 19].

When identical procedures are carried out with cells infected with a non-
cytopathic isolate of BVDV such as NY-1, the pattern of polypeptides suffers
a single drastic change: the 80k polypeptide is absent. This difference proved
to be a general phenomenon: all noncytopathic BVDV isolates tested to date
lack the 80k polypeptide [18].

Mapping viral proteins in the BVDV genome

Since the publication of the entire nucleotide sequence of the open reading
frame of the BVDV genome by Collett et al. [9] and the accompanying
studies by the same group on the mapping of the BVDV polypeptides with
anti-fusion protein antisera we have been able to better understand the
genomic organization of this virus [7, 8]. One of the first obstacles that will
confuse the newcomers to the field of pestivirology is the differences in the
polypeptide nomenclature used by different authors. To aid the readers in
the endeavor of comparing the literature on BVDV proteins, a tabulation of
the polypeptides identified by us and by Collett et al. indicating the detection
method can be found in Table 1 [8, 17–19]. Hopefully, in the near future
a more rational and functional nomenclature, reflecting the close relationship of pestiviruses with flaviviruses, will be developed and agreed upon by
researchers in the field.

To aid in reconciling the results we obtained in our studies [17–19]
with those of Collett et al. [8] the following comments are in order (refer to
Table 1 for nomenclature):

1. The polypeptide we identified as 118k is in fact a doublet: it represents
the co-migration of p125 (nonglycosylated precursor of p80 and p54) and
p118 (glycosylated precursor of the glycoproteins).
2. p118 (glycosylated) is the precursor of gp62 and gp53. gp62 is then
cleaved to yield gp48 and gp25. All these polypeptides have been identified
in our studies.
3. The cleavage of p118 to yield gp62 and gp53 is carried out by cellular
signal peptidase in the lumen of the endoplasmic reticulum. Cleavage of
gp62 is likely to be catalyzed by the same enzyme.
4. p20, which we called p19, is not found by RIP, presumably due to the
absence of antibody against this polypeptide in hyperimmune bovine serum.
It can be readily identified by HIB labeling.

Table 1. Nomenclature of BVDV Proteins in infected cells

| Collett et al. [8] | Donis et al. [17–19] | Detection | |
		HIB	RIP
Processed proteins			
p20	19k	+	−
gp48	48k	+	+
gp25	25k	+	+/−
gp53	56–58k	+	+
p125(NC)	118k	+	+
p80(CP)	80k	+	+
(38k)	37k	+	+
Precursor proteins		Detection	
gp118	118k	−	+
gp62	65/75k	−	+
p125(CP)	118k	+	+
p133	135k	+	+
(p165?)	165k	+	+

5. p125 is present in both cytopathic and noncytopathic biotypes of BVDV isolates. It is highly immunogenic and can be readily detected.

6. Of the two cleavage products of p125 in cytopathic BVDV isolates, p80 was easily identified while p54 was not detected perhaps because of its comigration with gp53. In retrospect, the species that is resistant to endo-glycosidase digestion and tunicamycin inhibition shifts may represent p54.

7. 37k is detected in both biotypes by either method. Not all bovine sera used in RIP assays contain antibodies to this polypeptide.

8. 165k is readily detectable and may represent a precursor of p133 and 37k.

9. p133 is detected with difficulty in HIB and is absent in RIP. It is not glycosylated and is the precursor of p58 and p75.

10. p58 is likely to be masked by gp53, we did not detect it.

11. p75 could be masked by p80, since we did not detect it.

Antigenic analysis of BVDV glycoproteins with monoclonal antibodies

To carry out antigenic analyses and to obtain functional information about the viral polypeptides we set out to produce a panel of hybridomas secreting monoclonal antibodies (MABs). For this, mice were immunized with puri-fied virions and their splenocytes fused to a murine myeloma cell line. Screening of the hybrid cell lines was carried out by indirect immuno-fluorescence (IFA) and virus neutralization. Fifteen hybridomas were fully characterized and used for the analyses discussed below [11, 16].

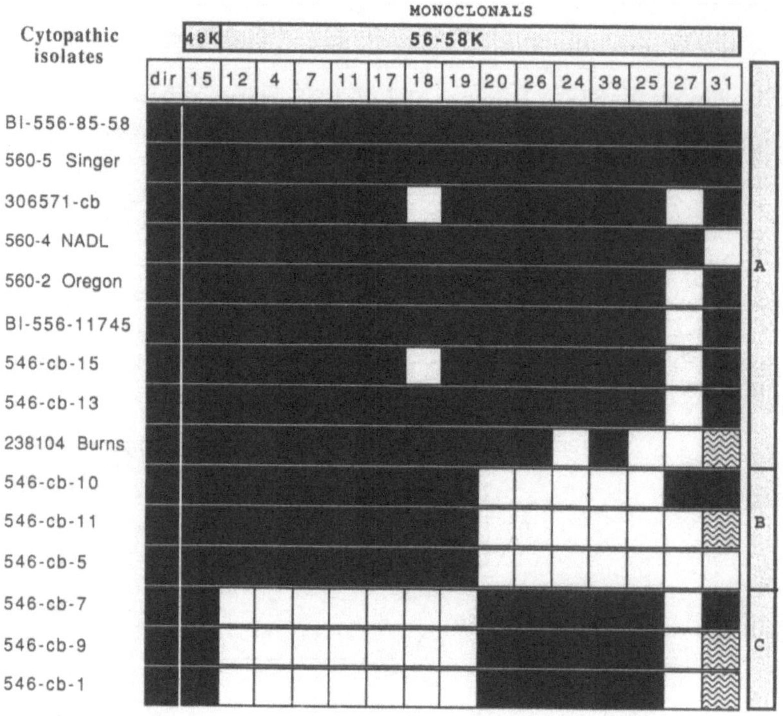
MONOCLONALS
Cytopathic isolates
48K
56-58K
dir 15 12 4 7 11 17 18 19 20 26 24 38 25 27 31
BI-556-85-58
560-5 Singer
306571-cb
560-4 NADL
560-2 Oregon
BI-556-11745
546-cb-15
546-cb-13
238104 Burns
546-cb-10
546-cb-11
546-cb-5
546-cb-7
546-cb-9
546-cb-1
A
B
C

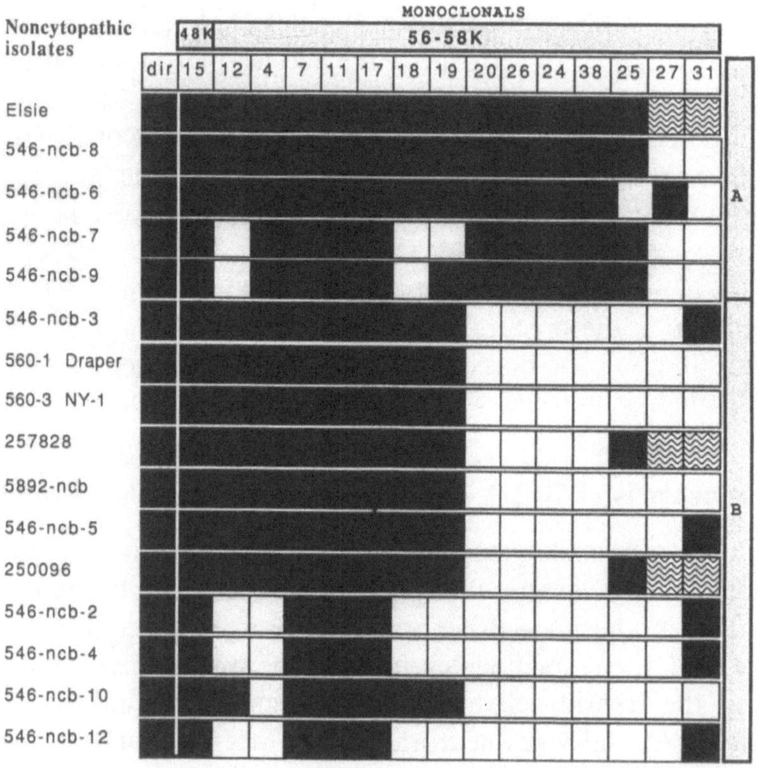
MONOCLONALS
Noncytopathic isolates
48K
56-58K
dir 15 12 4 7 11 17 18 19 20 26 24 38 25 27 31
Elsie
546-ncb-8
546-ncb-6
546-ncb-7
546-ncb-9
546-ncb-3
560-1 Draper
560-3 NY-1
257828
5892-ncb
546-ncb-5
250096
546-ncb-2
546-ncb-4
546-ncb-10
546-ncb-12
A
B

Virus neutralization

Approximately half of the MABs possessed virus neutralization titers > 256. Approximately 25% of the MABs had low neutralization titers (32–64). The other 25% lacked virus neutralizing activity. All the neutralizing MABs were either of the IgG_1 or IgG_2 subclass.

Molecular specificity of the MABs

To identify the viral polypeptide recognized by each MAB we carried out RIP analyses. The results indicated that all MABs except one bound to the 56–58k glycoprotein. Only one MAB bound to the 48k glycoprotein. All neutralizing MABs bound to the 56–58k polypeptide, providing the first direct functional evidence that this glycoprotein is present in virions and contains neutralization epitope(s). It is still unknown if the 48k is present in virions since the MAB that recognizes it does so in infected cells, but fails to bind to virions [16]. The exact location of the epitopes within 56–58k awaits the sequencing of neutralization escape mutants. However, the majority of the MABs of the panel bind to a vaccinia virus recombinant expressing the N-terminal 177 amino acids of the 56–58k glycoprotein (data not shown) [20]. Thus, we predict that this glycoprotein is a typical class I transmembrane glycoprotein, with an N-terminal immunodominant domain and a C-terminal anchor. Antiserum against the vaccinia virus recombinant expressing the N-terminal half of the 56–58k glycoprotein neutralizes BVDV infectivity, in agreement with the predictions resulting from the MAB data [20].

Antigenic analyses of BVDV isolates

The panel of MABs against the viral glycoproteins was used to carry out a preliminary antigenic analysis of BVDV field isolates. The reactivity of MABs with a given isolate was determined by the IFA assay on infected bovine testicle cell monolayers. A typical result of these analyses is shown in Fig. 1a and b. Some generalizations that can be derived from these reactivity patterns include:

1) There are multiple patterns of reactivity of the viral isolates with the panel of MABs, suggesting extensive antigenic diversity among field isolates of BVDV.

Fig. 1. Indirect immunofluorescence reactivity patterns of cytopathic (1.a) and non-cytopathic BVDV (1.b) isolates (rows) with a selected panel of MABs numbered at the top of the figure with roman numerals. Control reactivity with a polyclonal antiserum is shown in "dir" column. The protein specificity of the MABs is indicated at the top: 48k and 56–58k. Letters (A, B, etc.) inside bars indicate major antigenic grouping of the isolates. Blank squares indicate no reaction, boxes with wavy lines a weak reaction. A complete reaction is shown as a black box

2) There are multiple patterns of reactivity among the MABs with the virus isolates, suggesting that most MABs recognize distinct epitopes (an exception is represented by MABs 7, 11 and 17 which appear to have identical patterns) (Fig. 1).

3) MAB 15, specific for the 48k polypeptide, appears to react with a highly conserved epitope: all isolates tested to date react with this MAB.

MAB-Neutralization escape mutants

The diversity of antigenic patterns (as defined with this panel of MABs) among different isolates of BVDV is likely to be the consequence of the typically high error rate of viral RNA replicases. As a result, multiple lineages of BVDV may have evolved and continue to circulate in the cattle population. To quantitate the error rate of the viral replicase we used MABs: the frequency of MAB neutralization resistant mutants present in cloned viral stocks reveals, albeit indirectly, this rate. The frequency of neutralization escape mutants in low multiplicity-passage of cloned Singer BVDV stocks is $10^{-2.47}$ (R. Donis, unpublished). This rate is similar to that of other RNA viruses known for their antigenic variability such as influenza and foot and mouth disease viruses [5, 38]. Natural populations of BVDV consist of multitudes of nonidentical genomes. The concept that genetic elements that replicate with limited fidelity will give rise to a set of related nonidentical genomes was introduced by Eigen et al. [21, 22]. They used the term "quasispecies" to refer to the extremely heterogeneous nature of viral RNA populations. RNA viruses have as a result rapidly evolving genomes with a high degree of adaptability and phenotypic flexibility [1, 3, 12, 13, 23, 39, 40, 41, 47].

Control strategies and antigenic variation

The availability of large number of susceptible cattle born every year throughout the world ensures continual BVDV replication. Thus, high BVDV genomic RNA mutation rates coupled with short generation intervals will necessarily result in high rates of genetic evolution. Two implications for the field of vaccinology can be derived from the previous discussion: 1) The extent of antigenic variability may result in incomplete protection afforded by monovalent vaccines, raising the issue of the possible need to include a few antigenically divergent isolates in new oligovalent vaccines; 2) The high mutation rates of BVDV could enable the virus to survive even after the widespread use of vaccines. Vaccination could drive the evolution of BVDV through immune selective pressure. Immunity to the circulating strains of human influenza virus has been demonstrated to play a critical role in driving the evolution of the hemagglutinin gene of the virus [15, 36, 45]. Therefore, continued surveillance of the antigenic structure of field

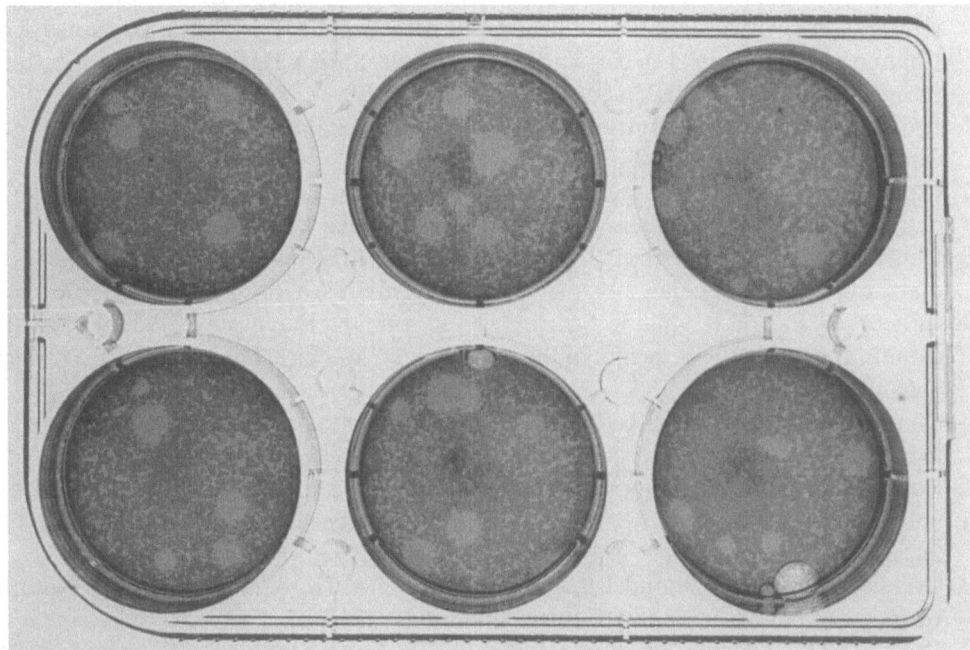

Fig. 2. Plaque assay of a cloned stock of Singer BVDV isolate on bovine testicle cell monolayer overlayed with 1/30 dilution of MAB 19 in agarose. Stained and photographed three days after inoculation. Wild-type virus plaques (approximately 1,500 in each well) appear as pinpoint unstained areas. In contrast, neutralization escape mutants form plaques 3 mm in diameter

isolates would be warranted to monitor the appearance of "escape" strains which elude immunity induced by the vaccines. The MAB panel we have developed should be of use in surveillance to immunophenotype field isolates of BVDV and assess the need for changes in the antigenic composition of vaccines [11, 16].

Acknowledgements

The support of Dr. J. Schmitz and the Dept. of Veterinary Science is gratefully acknowledged. We thank Jerry Phillips for his excellent technical assistance.

References

1. Almond JW, Stanway G, Cann AJ, Westrop GD, Evans DM, Ferguson M, Minor PD, Schild GC (1984) New poliovirus vaccines: a molecular approach. Vaccine 2: 177–82
2. Baker JC (1987) Bovine viral diarrhea virus: a review. J Am Vet Med Assoc 190: 1449–58
3. Birrer MJ, Udem S, Nathenson S, Bloom BR (1981) Antigenic variants of measles virus. Nature 293: 67
4. Bolin SR, McClurkin AW, Cutlip RC, Coria MF (1985) Severe clinical disease induced in cattle persistently infected with noncytopathic bovine viral diarrhea virus by superinfection with cytopathic bovine viral diarrhea virus. Am J Vet Res 46: 573–576

5. Bolwell C, Parry NR, Rowlands DJ, Brown F, Ouldridge EJ (1986) Foot and mouth disease virus variants with broad and narrow antigenic spectra. In: Lerner RA, Channock RN, Brown F (eds), Vaccines 86. Cold Spring Harbor Laboratory, Cold Spring Harbor, New York, p 51

6. Brownlie J, Clarke MC, Howard CJ (1984) Experimental production of fatal mucosal disease in cattle. Vet Rec 114: 535–537

7. Collett MS, Anderson DK, Retzel E (1988) Comparisons of the pestivirus bovine viral diarrhoea virus with members of the flaviviridae. J Gen Virol 69: 2637–43

8. Collett MS, Larson R, Belzer SK, Retzel E (1988) Proteins encoded by bovine viral diarrhea virus: the genomic organization of a pestivirus. Virology 165: 200–208

9. Collett MS, Larson R, Gold C, Strick D, Anderson DK, Purchio AF (1988) Molecular cloning and nucleotide sequence of the pestivirus Bovine Viral Diarrhea virus. Virology 165: 191–199

10. Commerce Dpt, Bureau of the Census (1984) 1982 Census of Agriculture. US Government Publications

11. Corapi WV, Donis RO, Dubovi EJ (1988) Monoclonal antibody analyses of cytopathic and noncytopathic viruses from fatal bovine viral diarrhea virus infections. J Virol 62: 2823–2827

12. Crainic R, Couillin P, Blondel B, Cabau N, Bouse A, Horodniceau F (1983) Natural variation of poliovirus epitopes. Infect Immun 41: 1217–23

13. Diamond DC, Jameson BA, Bonin J, Kohara M, Abe S, Crainic R, Wimmer E (1985) Antigenic variation and resistance to neutralization in poliovirus type 1. Science 229: 1090–93

14. Done JT, Terlecki S, Richardson C, Harkness JW, Sands JJ, Patterson DSP, Sweasey D, Shaw IG (1980) Bovine virus diarrhea-mucosal disease virus: pathogenicity for the fetal calf following maternal infection. Vet Rec 108: 473–479

15. Donis RO, Bean WJ, Webster RG (1989) Distinct lineages of influenza virus H4 hemagglutinin genes in different regions of the world. Virology 169: 408–417

16. Donis RO, Corapi WV, Dubovi EJ (1988) Neutralizing monoclonal antibodies to BVD virus bind to the 56–58k polypeptide. J Gen Virol 69: 77–86

17. Donis RO, Dubovi EJ (1987) Glycoproteins of bovine viral diarrhea-mucosal disease virus in infected bovine cells. J Gen Virol 68: 1607–1616

18. Donis RO, Dubovi EJ (1987) Differences in virus-induced polypeptides in cells infected by cytopathic and noncytopathic biotypes of bovine virus diarrhea-mucosal disease virus. Virology 58: 168–173

19. Donis RO, Dubovi J (1987) Characterization of bovine viral diarrhea-mucosal disease virus-specific proteins in bovine cells. J Gen Virol 68: 1597–1605

20. Dubovi EJ, White T, Cordell B, Dales B, Donis RO (1987) Expression of a glycoprotein antigen of bovine viral diarrhea virus that elicits neutralizing antibodies in mice. In: Chanock M, Lerner R (eds) Modern approaches to new vaccines. Cold Spring Harbor Laboratory, New York, pp 435–439

21. Eigen M (1971) Self-organization of matter and the evolution of biological macro-molecules. Naturwissenschaften 58: 465–482

22. Eigen M, Schuster P (1979) The hypercycle: a principle of natural self-organization. Springer, Berlin Heidelberg New York Tokyo

23. Emini EA, Jameson BA, Lewis AJ, Larsen GR, Wimmer E (1982) Poliovirus neutral-ization epitopes: analysis and localization with neutralizing monoclonal antibodies. J Virol 43: 997–1002

24. Gillespie JH, Coggins L, Thompson J, Baker JA (1961) Comparison by neutralization tests of strains of virus isolated from virus diarrhea and mucosal disease. Cornell Vet 51: 155–159

25. Gillespie JH, Madin SH, Darby NB (1962) Cellular resistance in tissue culture, induced by noncytopathogenic strains, to a cytopathogenic strain of virus diarrhea virus of cattle. Proc Soc Exp Biol Med 110: 248–250

26. Goudswaard J, Van Der Donk JA, Noordzij A (1978) Protein A reactivity of various mammalian immunoglobulins. Scand J Immunol 8: 21–28

27. Harkness JW (1987) The control of bovine viral diarrhoea virus infection. Ann Rech Vet 18: 167–74

28. Harkness JW, Roeder PL, Drew T, Wood L, Jeffrey M (1985) The efficacy of an experimental inactivated BVD-MD vaccine. CEC seminar: Pestivirus infection of ruminants

29. Kessler SW (1975) Rapid isolation of antigens from cells with a staphylococcal protein A-antibody adsorbent: parameters of the interaction of antibody-antigen complexes with protein A. J Immunol 115: 1617–1624

30. Kniazeff AJ, Huck RA, Jarrett WFH, Prichard WR, Ramsey FK, Schipper IA, Stober M, Liess B (1961) Antigenic relationship of some isolates of bovine viral diarrhea-mucosal disease viruses from the United States, Great Britain and West Germany. Vet Rec 73: 768–769

31. Kniazeff AJ, Pritchard WR (1960) Antigenic relationships in the bovine viral diarrhea-mucosal disease complex. Proceedings U.S.A.H.A. 64: 345–350

32. Lee KM, Gillespie JH (1957) Propagation of virus diarrhea virus of cattle in tissue culture. Am J Vet Res 18: 952–953

33. Lobmann M, Charlier P, Florent G, Zygraich N (1984) Clinical evaluation of a temperature sensitive bovine viral diarrhea vaccine strain. Am J Vet Res 45: 2498–2504

34. Nuss DL, Oppermann H, Koch G (1975) Selective inhibition of host protein synthesis in RNA-virus-infected cells. Proc Nat Acad Sci 72: 1258–1262

35. Office for Statistics, European Community (1989) Agricultural Statistical Yearbook 1989. Brussels-Luxembourg CECA-CEE-CEEA

36. Palese P, Young JF (1982) Variation of influenza A, B, and C viruses. Science 215: 1468–74

37. Peters AR (1987) Vaccines for respiratory disease in cattle. Vaccine 5: 164–69

38. Portner A, Webster RG, Bean WJ (1980) Similar frequencies of antigenic variants in Sendai, vesicular stomatitis and influenza A viruses. Virology 104: 235–41

39. Prabhakar BS, Haspel MD, McClintock PR, Notkins AL (1982) High frequency of antigenic variants among naturally occurring human Coxsackie B4 virus isolates identified by monoclonal antibodies. Nature 300: 374–77

40. Prabhakar BS, Menegus MA, Notkins AL (1985) Detection of conserved and non-conserved epitopes on Coxsackie virus B4: frequency of antigenic change. Virology 146: 302–7

41. Sheshberadaran H, Norrby E (1986) Characterization of epitopes on the measles virus hemagglutinin. Virology 152: 58–67

42. Svitkin YV, Lyapustin VN, Lashkevich VA, Agol VI (1984) Differences between translation products of tick-borne encephalitis virus RNA in cell-free systems from Krebs-2 cells and rabbit reticulocytes: involvement of membranes in the processing of nascen precursors of flavivirus structural proteins. Virology 135: 536–41

43. Underdahl NR, Grace OD, Hoerlein AB (1957) Cultivation in tissue culture of cytopathogenic agent from bovine mucosal disease. Proc Soc Exp Biol Med 94: 795–797

44. Vonderfecht HE (1980) Why I recommend vaccinating cattle against bovine virus diarrhea (BVD). VMSAC Vet Med Small Anim Clin 75: 853–858

45. Webster RG, Laver WG, Air GM, Schild GC (1982) Molecular mechanisms of variation in influenza viruses. Nature 296: 115–21

46. Wellink J, Van-Kammen A (1988) Proteases involved in the processing of viral polyproteins. Brief review. Arch Virol 98: 1–26
47. Wiktor TJ, Koprowski H (1980) Antigenic variants of rabies virus. J Exp Med 152:
99–16

Author's address: R. O. Donis, Department of Veterinary Science, University of
Nebraska, Lincoln, NE 68583-0905, U.S.A.

Arch Virol (1991) [Suppl 3]: 41–46

A "zinc finger-like" domain in the 54 KDA protein
of several pestiviruses

L. De Moerlooze[1], A. Renard[2], C. Lecomte[2], and J. A. Martial[1]

[1] Laboratoire Central de Génie Génétique, Université de Liège, Institut de Chimie, B6
Belgium and
[2] Eurogentec s.a. Campus du Sart Tilman, Belgium

Accepted March 14, 1991

Summary. We sequenced cDNAs, amplified by the Polymerase Chain Reaction (PCR) which correspond to the carboxy-terminal portion of the 54 KDa protein of various cytopathic (cp) or non-cytopathic (ncp) pestiviral strains. Except for the previously described insertions in two cp strains of the Bovine Viral Diarrhea virus (BVDV), we did not find comparable insertions in this gene. The predicted amino acid sequences of this 54 KDa protein portion contain a conserved cysteine-rich stretch remarkably similar to a "zinc finger"-type binding domain found in many gene-regulatory proteins. Thus, this protein may be involved in the binding to nucleic acids. As the 54 KDa protein is released from its 125 KDa precursor only in the cp strains, we propose the cytopathology may be a consequence of this event.

A major difference between cp and ncp biotypes is in the cleavage of the 125 KDa protein to 80 KDa and 54 KDa subunits in only cp biotypes. The 80 KDa protein which is abundantly produced by cp biotypes is not observed in ncp viruses [17]. Moreover the 54 KDa protein was positioned in the predicted amino acid sequence of NADL by radioimmunoprecipitation with anti-fusion protein sera [5] and by comparison with Flaviviruses [4]. The nomenclature used was proposed by Collett et al. [5].

When the complete nucleotide sequences of NADL and Osloss strains are compared, an insertion is noticed (Fig. 1). This insertion is located in the genomic region coding for the 54 KDa protein of the cp-BVDV Osloss strain and it corresponds to an ubiquitin-like coding sequence [18]. However, a different insertion has been found in this same region in the cp-BVDV NADL genome [3]. This insertion is of 270 nucleotides, it has 99% homology with a host cellular mRNA [14] and it also begins with an ATG codon. No comparable insertion can be found in the complete nucleotide sequence of the ncp-HCV Alfort strain [15]. As both insertions were in cp isolates, it

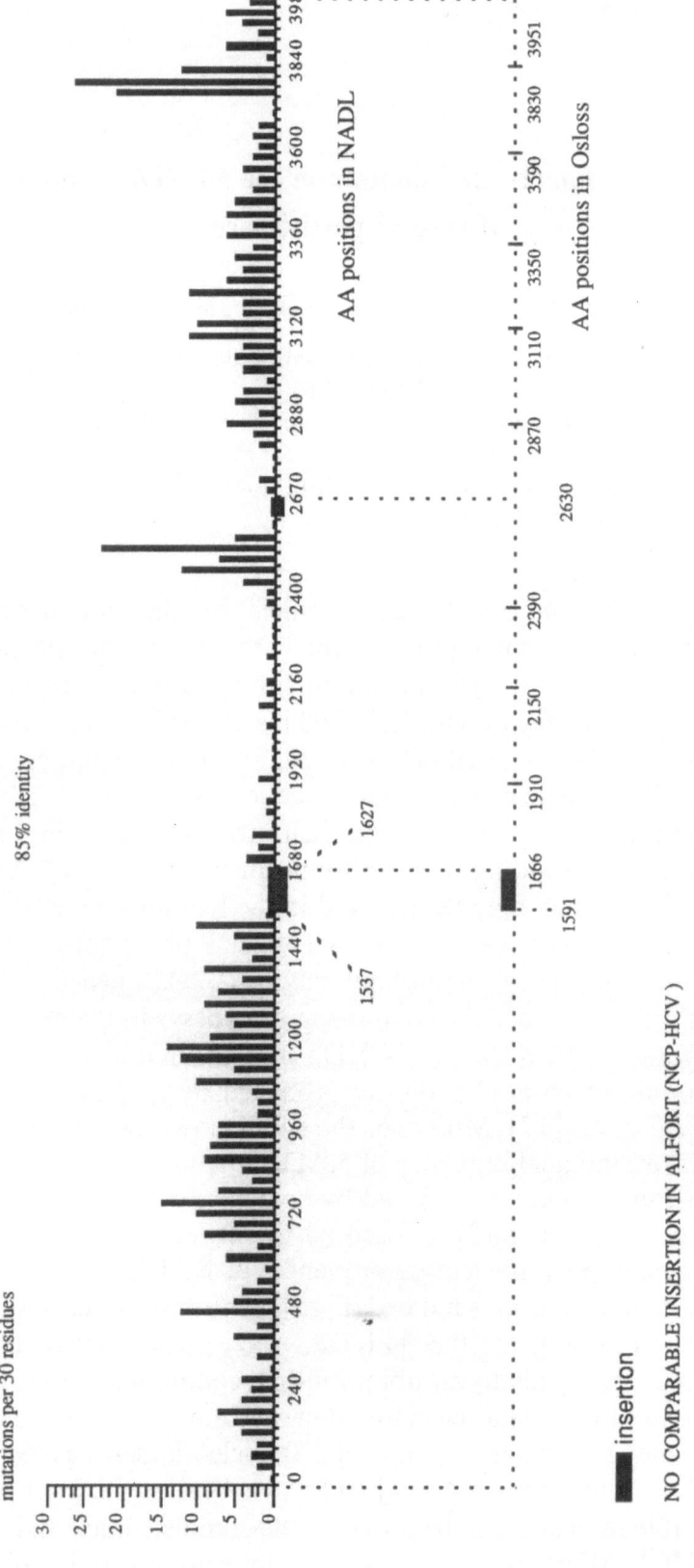

Fig. 1. Amino acid comparison between NADL and Osloss BVDV strains

```
                     4630
A:    5'- GGAATTCAGCACTCATAGAGCTAAACTGGT -3'
          Eco RI                    *

                     5590
B:    5'- GGGATCCAGTCCCCTCCTCACTTTTAG -3'
          Bam HI  *
```

Fig. 2. Sequences of the two amplimers used in the PCRs. Restriction enzyme sites used for subcloning the PCR fragments are underlined and the numbers above the amplimers position them in the NADL genomic sequence [3]. Stars indicate mismatches in the NADL sequence in comparison to Osloss [18] from which the amplimers were chosen: the A is replaced by a G and the T by A in NADL

has been suggested as a predictive means of differentiating the two biotypes [13]. In order to evaluate this possibility, we sequenced the region including the insertion sites in a number of cp and ncp pestiviral strains.

Using the polymerase chain reaction (PCR) we amplified this region corresponding to the insertions (Fig. 1) in three ncp-BVDV strains called Osloss ncp [10], New-York (from ATCC) and Pe515 ncp [1], five cp-BVDV strains called NADL [9], Singer [16], Lamspringe [12], Pe515 cp [1] and Osloss cp [10] and one cp-BDV strain called BD. All the viruses were received from G. Chappuis (Rhône-Mérieux, Lyon, France) except for Pe515 received from J. Brownlie (AFRC Institute for animal health, Compton, UK). We chose amplimer sequences from regions of good homology between the NADL and Osloss strains. The sequences and positions of these oligonucleotides are shown in Fig. 2.

To investigate the presence of insertions in the cp strains, we analysed the PCR products by agarose gel electrophoresis. The results showed that there is a detectable insertion in the NADL and Osloss strains only. The nucleotide sequences of the PCR products were obtained by the dideoxy method either by direct sequencing or after subcloning into plasmid pGEM 3zf (−) (Promega) as described elsewhere [7].

The deduced amino acid sequences of the different strains are presented on Fig. 3. We found only one new insertion in this region consisting of a sequence of four amino acids (YPSE) in the ncp BVDV New-York strain. Unlike the NADL and Osloss insertions this small insertion does not begin with a Met residue and is therefore not an indication of the biotype. No consistent difference between cp and ncp strains was found in these ten sequences. It is nevertheless possible that consensus mutations or insertions in another part of the 125 KDa protein may led to a conformational change affecting its potential cleavage into the 54 KDa and 80 KDa proteins. The region we sequenced encodes a conserved cysteine-rich stretch bearing a remarkable similarity to the "Zinc-finger" sequences found in several nucleic acid-binding proteins [11]. Such "Zinc finger-like" sequences were also described for proteins interacting with RNA: in the gag region of reverse

```
               1420      1430      1440      1450      1460
NADL cp    ALIELNWSMEEEESKGLKKFYLLSGRLRNLIIKHKVRNETVASWYGEEE
Sg cp      ................................................
Lam cp     .........G.................................DA..R.....
BD cp      .........G.................................DA..R.....
Pe cp      .........G..................................D..R.....
Os cp      ..................FI.....KA..........Q...........
Os ncp     ..................FI.....KA..........Q...........
NY ncp     ..........N......FI........L.........Q...........
Pe ncp     .........G.................................DA..R.....
Alfort ncp ....V..AFDN..V......F...S.VKE...........V.VR.F.D..
   (MEYERS, G. ET AL)
               1470      1480      1490      1500      1510
NADL cp    VYGMPKIMTIIKASTLSKSRHCIICTVCEGREWKGGTCPKCGRHGKPITC
Sg cp      ..................K..M......S...................M.
Lam cp     ...............N.NW...........K..................
BD cp      ...............N.N............K..................
Pe cp      ...............N.N............K..................
Os cp      ......VV...R.CS.N.NK......AKK....N...............
Os ncp     ......VV...R.CS.N.NK......AKK....N...............
NY ncp     ......V....R.C..N.NK......A......N...............
Pe ncp     ...............N.N............K..................
Alfort ncp I.....LIGLV..A...RNK..ML.....D.D.R.E......F.P.VV.
               1520      1530      1630      1640      1650
NADL cp    GMSLADFEERHYKRIFIREGNFEGPFRQEYNGFVQYTARGQLFLRNLPVL
                                  INS
Sg cp      .........................................LV.....
Lam cp     .......................D..........I.....E.......I.
BD cp      .......................D..........I.............I.
Pe cp      .......................D..........I.............I.
Os cp      ..T..............T......HS.....................I.
Os ncp     ..T..............T......HS.....................I.
NY ncp     ..T...........................E..S.............I.
                                           YPSE
Pe ncp     .......................D..........I.............I.
Alfort ncp ..T......K.........DQSG..L.E.HA.YL..K.............
               1660      1670      1680      1690      1700
NADL cp    ATKVKMLMVGNLGEEIGNLEHLGWILRGPAVCKKITEHEKCHINILDKLT
Sg cp      ...................D.....................R..V......
Lam cp     ...................D.....................R..V......
BD cp      ...................D.....................R..V......
Pe cp      ...................D.....................R..V......
Os cp      .................V...D........K..............V.....
                                     UBQ
Os ncp     ...................D........................V.....
NY ncp     ...................D........................V.....
Pe ncp     ...................D.....................R..V......
Alfort ncp .......L.....T...D......V.........V....R.TTS.M....
               1710      1720      1730
NADL cp    AFFGIMPRGTTPRAPVRFPTSLLKVRRGL
Sg cp      ....
Lam cp     .............................
BD cp      .............................
Pe cp      ....I........................
Os cp      ....V.............A..........
Os ncp     ....V.............A..........
NY ncp     ....V.............A..........
Pe ncp     ....I........................
Alfort ncp ....V................I....
```

Fig. 3. Alignment of the deduced amino acid sequences of similar portion of various pestiviral strains. Sequences corresponding to the PCR amplimers are underlined, the one which conforms to a "zinc finger" binding domain is doubly underlined. The numbers above the sequences indicate the amino acids positions in the NADL genomic sequence. Abbreviations used for BVDV strains: Sg, Singer; Lam, Lamspringe; BD, Border Disease; Os, Osloss; NY, New-York; Pe, Pe515. Other abbreviations: INS, 90 amino acids insertion in the NADL sequence; UBQ, 75 amino acids insertion corresponding to an ubiquitin-like protein in the Osloss sequence

transcribing elements [6], in the P gene of paramyxoviruses [2, 20] and in non-structural proteins of plus-stranded RNA plant viruses. A report on the role of the coat protein of Tobacco streak virus and Alfalfa mosaic virus in RNA replication proposes a model of protein-RNA interactions where the protein is able to activate the infectivity of the RNA [19]. The 54 KDa protein where the Cysteine-rich stretch was found is only present in the cp strains (shown in NADL [5]). Helical prediction [8] shows a conserved helix following the "Zinc finger-like" sequence consistent with the model of protein-nucleic acid interaction. It might thus be that this protein is related to the activation of viral RNA for infection and to the regulation of its replication. It is also possible that the presence of this "Zinc-finger" domain, in the 54 KDa protein is similarly associated with cytopathology. The uncleaved 125 KDa protein would not mediate the same biological effect because of a distinct conformation of its potential finger region.

The remarkable homology shown in these sequences of both biotypes of pestiviruses (Fig. 3), must indicate that other regions of the genome require examination before the molecular basis of biotypic differences can be explained.

Key words: Cysteine-rich stretch, cytopathic and non-cytopathic strains, insertions, PCR (Polymerase Chain Reaction), nucleic acid-binding protein.

References

1. Brownlie J, Clarke MC, Howard CJ (1984) Experimental production of fatal mucosal disease in cattle. Vet Rec 114: 535–536
2. Cattaneo R, Kaelin K, Baczko K, Billeter MA (1989) Measles virus editing provides an additional cysteine-rich protein. Cell 56: 759–764
3. Collet MS, Larson R, Gold C, Strick D, Anderson DK, Purchio AF (1988) Molecular cloning and nucleotide sequence of the Pestivirus bovine viral diarrhea virus. Virology 165: 191–199
4. Collett MS, Anderson DK, Retzel E (1988) Comparison of the Pestivirus bovine viral diarrhoea virus with members of the Flaviviridae. J Gen Virol 69: 2637–2643
5. Collet MS, Larson R, Belzer SK, Retzel E (1988) Proteins encoded by bovine viral diarrhea virus: the genomic organization of a Pestivirus. Virology 165: 220–208
6. Covey SN (1985) Amino acid sequence homology in gag region of reverse transcribing elements and the coat protein gene of cauliflower mosaic virus. Nucl Acids Res 14: 623–632
7. De Moerlooze L, Desport M, Renard A, Lecomte C, Brownlie J, Martial JA (1990) The coding region for the 54 KDa protein of several Pestiviruses lacks host insertions but reveals a "zinc finger-like" domain. Virology 177: 812–815
8. Garnier J, Osguthorpe DJ, Robson B (1978) Analysis of the accuracy and implications of simple methods for predicting the secondary structure of globular proteins. J Mol Biol 120: 97–120
9. Gutekunst DE, Malmquist WA (1963) Separation of a soluble antigen and infectious particles of Bovine Virus Diarrhea virus and their relationship to hog cholera. Can J Comp Med Vet Sc 27, n°5.
10. Hafez SM (1967) Charakterisierung und Darstellung des Virus der "Virusdiarrhoe-Mucosal Disease" des Rindes (VD-virus). Thesis, Hannover

11. Klug A, Rhodes D (1987) 'Zinc fingers': a novel protein motif for nucleic acid recognition. TIBS 12: 464–469
12. Liess B (1967) Die ätiologische Abgrenzung selbständiger Virusinfektionen, insbesondere der Virusdiarrhoe-Mucosal Disease im sogenannten "Mucosal-Disease-Komplex" bei Rindern. Rindern Dtsch Tierärztl Wochenschr 74: 46–49
13. Meyers G, Rumenapf T, Thiel H-J (1989) Ubiquitin in a togavirus. Nature 341: 491
14. Meyers G, Rumenapf T, Thiel H-J (1989) Insertion of host cell derived sequences identified in the RNA genome of a Togavirus. Proceedings of 11th International Symposium of the World Association of Veterinary Microbiologists, Immunologists and Specialists in Infections Diseases, 239
15. Meyers G, Rumenapf T, Thiel H-J (1989) Molecular cloning and nucleotide sequence of the genome of hog cholera virus. Virology 171: 555–567
16. McClurkin AW, Coria MF (1978) Selected isolates of bovine viral diarrhea (BVD) virus propagated on bovine turbinate cells: virus titre and soluble antigen production as factors in immunogenicity of killed BVD virus. Arch Virol 58: 119–128
17. Pocock DH, Howard CJ, Clarke MC, Brownlie J (1987) Molecular variation between BVD virus isolates. Seminar in the CEC programme on Pestivirus infections of ruminents (Harkness JW (ed), UK) pp 43–51
18. Renard A, Dino D, Martial JA (1987) Vaccines and diagnostics derived from bovine viral diarrhea virus. European Patent Appl. n°86870095.6. Publ. n°0208672
19. Sehnke PC, Mason AM, Hood SJ, Lister RM, Johnson JE (1989) A "Zinc-Finger"-type binding domain in tobacco streak virus coat protein. Virology 168: 48–56
20. Thomas SM, Lamb RA, Paterson GR (1988) Two mRNAs that differ by two non-templated nucleotides encode the amino coterminal proteins P and V of the paramyxovirus SV5. Cell 54: 891–902

Authors' address: L. de Moerlooze, Laboratoire central de Génie génétique, Université de Liège, Institut de Chimie Batiment B6, B-4000 Sart-Tilman, Belgium.

Arch Virol (1991) [Suppl 3]: 47–54

BVD monoclonal antibodies: relationship between viral protein specificity and viral strain specificity

D. J. Paton, J. J. Sands, and **P. M. Roehe**

Central Veterinary Laboratory, U.K.

Accepted March 14, 1991

Summary. Seventeen monoclonal antibodies raised against bovine viral diarrhoea virus were divided into three groups on the basis of radio-immunoprecipitation results. Seven monoclonal antibodies precipitated a polypeptide of 80kD and defined four domains, all of which showed considerable conservation amongst the 180 pestivirus strains and isolates examined. Nine monoclonal antibodies, including six with virus neutralizing activity, precipitated a 53kD polypeptide and all appeared to be directed towards a single domain of clustered epitopes. Several of these epitopes were present in many ruminant virus strains and isolates, but not in hog cholera viruses. A single monoclonal antibody precipitated a 48kD polypeptide, defining an epitope that was also present on many ruminant viruses, but not hog cholera viruses. Most pestiviruses from cattle and some from sheep shared a number of epitopes located on three different proteins.

Key words: Pestivirus, BVD monoclonal antibodies, epitope conservation.

Introduction

The pestivirus genus comprises a heterogeneous but overlapping family of economically significant viruses including bovine viral diarrhoea virus (BVDV), hog cholera virus (HCV) and border disease virus of sheep (BDV). Antigenic similarities and differences both between and amongst these three viral-species types, as well as cross host-species infections, make differentiation and classification of these viruses difficult. Monoclonal antibodies (Mabs) are well suited to analysing differences between related viruses. This paper attempts to define some sites of antigenic conservation and divergence amongst pestiviruses by comparing viral strain specificity with viral protein specificity for a group of Mabs.

Materials and methods

Monoclonal antibodies and viruses

Preparation of 17 Mabs to the cytopathogenic BVDV strains Oregon C24V and NADL has been described [6]. The viral strain specificity of the Mabs was determined by a peroxidase linked assay (PLA) [9] testing each Mab at a single predetermined dilution against diverse laboratory strains and field isolates of pestivirus grown in PK15 cells (HCV) or bovine turbinate cells (BVDV and BDV). The method employed, along with results, and details of origins for many of the viruses has already been reported [6, 7]. All the viruses in this study are named according to the host of origin. Pestiviruses from pigs that were not recognised by HCV specific Mabs have been excluded.

Radioimmunoprecipitation

Radiolabelled cell lysates were prepared by a method similar to one previously described [13]. BVDV strains Oregon C24V or NADL were used to infect calf testis cells at a multiplicity of infection of ten. Radiolabelling was for 4 hours, beginning 18 hours post infection and using L-[^{35}S]-methionine or D-[^{3}H]-glucosamine at 100 μCi/ml. Cells were disrupted with a lysis buffer containing 0.15 M NaCl, 1% sodium deoxycholate, 1% Triton x-100, 0.1% SDS, 0.02 M Tris HCl pH 7.6 and 1 mM phenyl methyl sulphonyl fluoride.

Oregon C24V infected cell lysate was used for most of the radioimmunoprecipitations (RIPs). Mabs which did not react with this virus in PLA or RIP were tested against NADL. RIPs were performed using Protein G Sepharose (Pharmacia) as immunosorbent and lysis buffer as diluent and wash fluid. Immunosorbent was incubated with ascites or serum, washed and then used to capture antigens from the cell lysates. After four washes, the immunosorbent complex was treated with sample buffer (0.06 M Tris HCl pH 6.8, 10% glycerol, 2% SDS, 5% mercaptoethanol, 0.005% w/v bromophenol blue) and immersed in boiling water for four minutes. The fluid phase was immediately loaded for polyacrylamide gel electrophoresis (PAGE), which employed a 4% stacking gel and a 10% separating gel with a discontinuous buffer system [10]. ^{14}C labelled molecular weight markers (Amersham UK) were in the range 14.3kD to 200kD. After electrophoresis, gels were fixed and dried before autoradiography using Hyperfilm Bmax (Amersham UK) and 3–5 day exposures.

Competition assay

Competition was assessed between biotinylated and unbiotinylated Mabs for binding to BVDV infected bovine turbinate cells grown in microtitre plates. The majority of the Mabs were tested against Oregon C24V infected cells whilst those that were unreactive to Oregon C24V were tested against NADL. Mabs were biotinylated as previously described [5]. Equal volumes of unbiotinylated and biotinylated Mabs were added to the plates concomitantly and binding was detected by a streptavidin-biotinylated horse radish peroxidase complex (Amersham UK). Each biotinylated Mab was titrated and then used in competition studies at twice its optimal endpoint concentration. Unbiotinylated mabs were in the form of ascites at doubling dilutions starting from 1/10 or 1/50, depending on availability.

Results

Radioimmunoprecipitation

All of the Mabs produced multiple bands on PAGE following RIP although in most cases one product clearly predominated. Similar bands were present

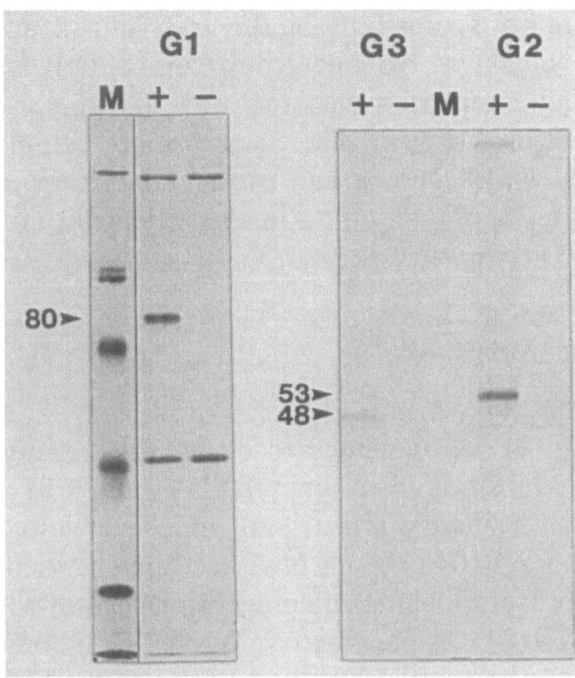

Fig. 1. Radioimmunoprecipitation patterns G1 = Mab 112, G2 = Mab 163, G3 = Mab 210 + = Oregon C24V infected cell lysate − = mock infected cell lysate M = molecular weight markers Arrowed numbers indicate apparent molecular weights ($\times 10^3$) of virus associated polypeptides

Table 1. Classification of monoclonal antibodies by parent virus and protein specificity

		C24V Mabs	NADL Mabs
G1	80kD	103, 105, 106, 109, 112	160, 212
G2	53kD	115, *162*, 165, *215*, *163*	*158*, *166*, *214*, 170
G3	48kD	210	
		Mabs = neutralizing	

with Oregon C24V and NADL infected cell lysates. Three basic patterns of reactivity were observed with L-[^{35}S]-methionine labelling (Fig. 1) and the Mabs were grouped on this basis (Table 1). Group 1 (G1) Mabs precipitated an 80kD polypeptide, group 2 (G2) Mabs a 53kD polypeptide and a single group 3 (G3) Mab precipitated a 48kD polypeptide. The 53kD polypeptide could sometimes be resolved into a doublet. Additional virus associated bands were present to variable extents at 115kD, 76kD, 62kD and 25kD depending in part on the ratio of cell lysate to Mab. All of these bands were evident in RIPs performed with an antiserum from a BVDV immunised calf. With D-[^{3}H]-glucosamine labelling, only the 115kD, 62kD, 53kD and 48kD bands were evident.

Competition data

Mabs WB166 and WB170 were not successfully biotinylated and so were only tested as unbiotinylated blockers. No intergroup competition was observed. Amongst G1 Mabs all competition was two way, allowing four competition subgroups to be defined (Fig. 2). There was a more complex relationship between G2 Mabs, with both one and two way blocking. All Mabs showed at least some degree of one way blocking activity against one or more other Mabs within the group.

Viral strain specificity

The viral strain specificity of the Mabs arranged by RIP and competition groups is shown in Fig. 3. The G1 Mabs tended to be broadly reactive, whilst most of the G2 and G3 Mabs bound to many BVDVs and a lesser proportion of BDVs but not HCVs. There was a striking difference between the binding of Mabs WB160 and WB109 in terms of viral strain specificity, even though they showed reciprocal blocking in competition testing with Oregon C24V. Whereas Mab WB109 was panreactive, Mab WB160 bound to just one HCV and only a proportion of BVDVs and BDVs. When HCV strain Baker A was used as antigen instead of Oregon C24V in competition testing, then Mab WB160 no longer blocked the binding of Mab WB109.

Results of an attempt to subdivide the ruminant pestiviruses on the basis of their recognition by some of the more discriminating Mabs is shown in Fig. 4.

Discussion

Current theory on the genomic organisation of BVDV [2] suggests that the virus encoded proteins are all derived from a single polyprotein by successive cleavages, producing "families" of polypeptides, each with a common intermediate precursor. One such family comprises a group of structural glycoproteins which include gp53 and gp48, both believed to be envelope proteins. Another family of polypeptides is the non-structural group: p125/p80/p54. Non-cytopathogenic BVDVs produce uncleaved p125, but in cytopathogenic biotypes, this polypeptide is cleaved so that p80 predominates [13]. These viral polypeptides have for the most part been demonstrated by RIPs using polyclonal antisera [3]. Mabs against gp53, gp48 and p125/80 have been described [4, 12] and all neutralizing Mabs so far described have been directed against gp53.

Precipitation of more than one BVDV polypeptide by individual Mabs has been attributed to coprecipitation [4] and this probably accounts for the multiple bands obtained in many RIPs in this study, although product–precursor relationships might be involved in some cases. For the G1 Mabs,

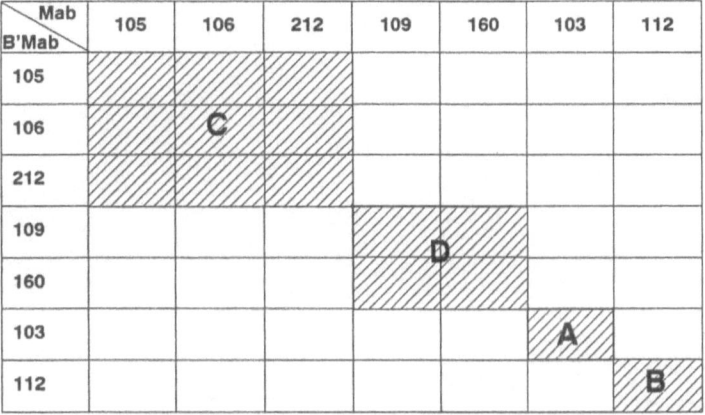

B' = Biotinylated

= Blocking activity (A-D = competition group)

Fig. 2. Competitive binding: G1 Mabs

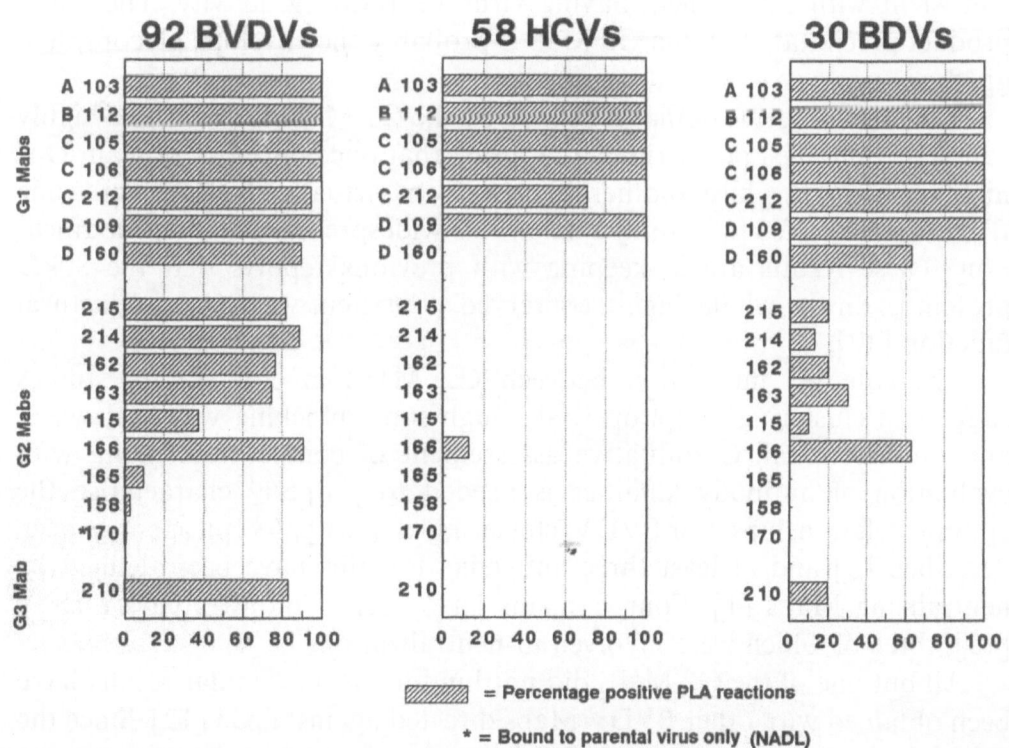

= Percentage positive PLA reactions

* = Bound to parental virus only (NADL)

Fig. 3. Viral strain specificity of Mabs

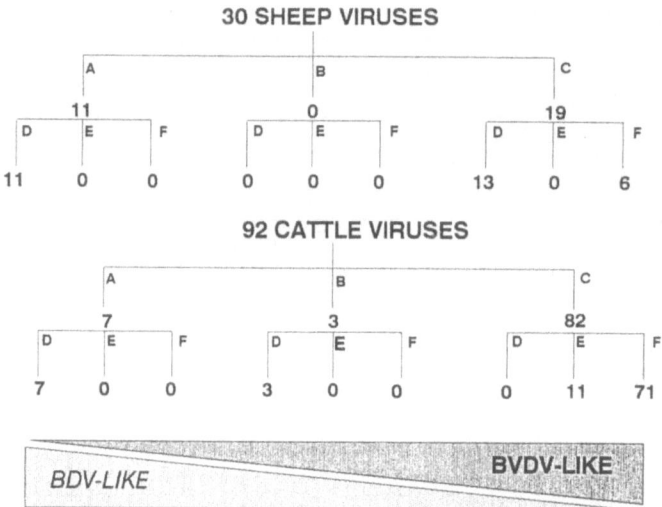

Fig. 4. Subdivision of ruminant pestiviruses A–F indicate PLA reaction with Mabs 160 and 166 (A, B, C) or Mabs 215 and 210 (D, E, F). A, D = negative to both Mabs, B, E = negative to one Mab, C, F = positive to both Mabs

an 80kD polypeptide was always the predominant product and this probably represents the non-structural viral protein associated with biotype specificity. Characterization of the eight G2 Mabs as anti gp53 would be consistent with six of them having virus neutralizing activity. The 48kD product precipitated by the G3 Mab is probably the envelope glycoprotein gp48.

The four antigenic domains defined by the G1 Mabs appear to be highly conserved amongst pestiviruses. It appears that one epitope in domain D is highly conserved, whilst another is absent from virtually all HCVs and some BVDVs and BDVs indicating a small but widespread structural modification. These results are in keeping with previous reports that the 80kD protein is, on the whole, highly conserved, consistent with a non-structural function [12].

The complex interaction between G2 Mabs in competition studies suggests a clustering of epitopes, although some blocking was both weak and unidirectional. Quantitative assessement of competition, along with evaluation of antibody affinities is needed to properly characterise the epitopic relationships. For BVDV, clustering of most gp53 epitopes has been described [3] and at least three antigenic domains have been defined by neutralizing Mabs [1]. Four domains have been demonstrated for HCV [15], three of which were involved in neutralization.

All but one of the G2 Mabs did not bind to HCVs. Similar results have been obtained with other BVDV Mabs directed against gp53 [12]. Since the G3 Mab also failed to recognise HCVs it appears that the antigenic epitopes on the envelope proteins of HCV and BVDV may be relatively distinct. This however conflicts with the known ability of some polyclonal antibodies to cross-neutralize HCVs and BVDVs [11]. This is particularly enigmatic, since neutralizing Mabs seem to have an even narrower spectrum of viral recognition by neutralization than by binding [3]. Differentiation of HCVs from other pestiviruses is feasible, using panels of Mabs raised against either BVDV or HCV [7, 8, 14]. Separation of BVDVs from BDVs is more problematical [6]. The majority (71/92) of BVDVs were homologous with respect to binding by Mabs WB160, WB166, WB215 and WB210 indicating a high degree of structural conservation at the level of three different proteins. Six of the 30 BDV strains and isolates showed the same reactivity suggesting that all 77 of these ruminant pestiviruses are very closely related. By contrast, 7 BVDVs and 11 BDVs did not bind any of these Mabs suggesting a fairly distant relationship with the previous 77. In between were 11 BVDVs that reacted with Mabs WB160 and WB166 but only one or other of WB215 and WB210; 13 BDVs that reacted with Mabs WB160 and WB166 but neither Mab 215 nor Mab 210 and 3 BVDVs that reacted with only one of Mabs WB160 and WB166 but neither Mab WB215 nor Mab WB210. The division of ruminant pestiviruses into BVDV-like and BVDV-unlike (?BDV) groups independent of the host of origin has been reported previously [6]. However, the knowledge that some of these homologies and differences exist at the level of three different proteins makes this classification much more compelling. The evidence that the majority of cattle isolates appear antigenically similar and share at least one major neutralizing domain has significance for vaccine development, as does the fact that some isolates are different. How much diversity exists amongst these "different" strains remains an important question to be answered.

Acknowledgements

This work was carried out while PMR was in receipt of grants from The British Council and The Brazilian National Research Board (CNPq).

References

1. Bolin S, Moennig V, Kelso Gourley NE, Ridpath J (1988) Monoclonal antibodies with neutralizing activity segregate isolates of bovine viral diarrhoea virus into groups. Arch Virol 99: 117–123
2. Collet MS, Larson R, Belzer SK, Retzel E (1988) Proteins encoded by bovine viral diarrhea virus; the genomic organization of a pestivirus. Virology 165: 200–208
3. Collet MS, Moennig V, Horzinek MC (1989) Recent advances in pestivirus research. J Gen Virol 70: 253–266
4. Donis RO, Corapi W, Dubovi EJ (1988) Neutralizing monoclonal antibodies to bovine viral diarrhea virus bind to the 56k to 58k glycoprotein. J Gen Virol 69: 77–86
5. Edwards S, Gitao GC (1987) Highly sensitive antigen detection procedures for the diagnosis of infectious bovine rhinotracheitis: amplified ELISA and reverse passive haemagglutination. Vet Microbiol 13: 135–141

6. Edwards S, Sands JJ, Harkness JW (1988) The application of monoclonal antibody panels to characterize pestivirus isolates from ruminants in Great Britain. Arch Virol 102: 197–206

7. Edwards S, Sands JJ (1990) Antigenic comparisons of hog cholera virus isolates from Europe, America and Asia using monoclonal antibodies. Dtsch Tieraerztl Wochenschr 97: 79–81

8. Hess RG, Coulibaly COZ, Greiser-Wilke I, Moennig V, Liess B (1988) Identification of hog cholera viral isolates by use of monoclonal antibodies to pestiviruses. Vet Microbiol 16: 315–321

9. Holm Jensen M (1981) Detection of antibodies against hog cholera virus and bovine viral diarrhoea virus in porcine serum. A comparative examination using CF, PLA and NPLA assays. Acta Vet Scand 22: 85–98

10. Laemmli UK (1970) Cleavage of structural proteins during the assembly of the head of bacteriophage T4. Nature 227: 680–685

11. Liess B, Frey HR, Prager D (1977) Antibody response of pigs following experimental infections with strains of hog cholera and bovine viral diarrhoea virus. CEC seminar on hog cholera/classical swine fever, and African swine fever, CEC Agriculture Series EUR 5904, Brussels, pp 200–213

12. Moennig V, Bolin SR, Coulibaly COZ, Kelso Gourley NE, Liess B, Mateo A, Peters W, Greiser-Wilke I (1987) Studies into the antigenic structure of pestiviruses using monoclonal antibodies. Dtsch Tieraerztl Wochenschr 94: 572–576

13. Pocock DH, Howard CJ, Clarke MC, Brownlie J (1987) Variation in the intracellular polypeptide profiles from different isolates of bovine virus diarrhoea virus. Arch Virol 94: 43–53

14. Wensvoort G, Terpstra C, Boonstra J, Bloemraad M, Van Zaane D (1986) Production of monoclonal antibodies against swine fever virus and their use in laboratory diagnosis. Vet Microbiol 12: 101–108

15. Wensvoort G (1989) Topographical and functional mapping of epitopes on hog cholera virus with monoclonal antibodies. J Gen Virol 70: 2865–2876

Author's address: D. J. Paton, Central Veterinary Laboratory, Weybridge, Surrey KT15 3NB, U.K.

Arch Virol (1991) [Suppl 3]: 55–65

Correlation of bovine viral diarrhoea virus induced cytopathic effects with expression of a biotype-specific marker

I. Greiser-Wilke[1], B. Liess[1], J. Schepers[1], C. Stahl-Hennig[2], and V. Moennig[1]

[1] Institute of Virology, Hannover Veterinary School, Hannover, and
[2] German Primate Centre, Göttingen, Federal Republic of Germany

Accepted March 14, 1991

Summary. The purpose of this study was the identification of antigenic differences between cytopathic (cp) and noncytopathic (ncp) bovine viral diarrhoea viruses (BVDV). Cells infected with 19 strains of each viral biotype were analyzed for reactivity with the monoclonal antibody (mab) BVD/C38. Reactivity was examined using an enzyme immunoassay on fixed infected monolayers of fetal calf kidney cells. In the majority of cases, the mab discriminated between cells infected with each of the two viral biotypes. Three reactivity patterns could be distinguished. Most cpBVDV strains yielded monolayers where 80–100% of infected cells reacted with the mab. Most of the ncpBVDV infected cells showed either no reaction, or only single cells or foci were stained. However, about one third of either cp- or ncpBVDV strains tested yielded infected monolayers where 30–50% of the cells reacted with the antibody. Cell damage other than the typical cytopathic effect might be responsible for the BVD/C38 reactivity of cells infected with BVDV.

In addition, it was analyzed whether the antigenic marker associated with cpBVDV was expressed in cells infected with viral isolates from 21 animals with clinical mucosal disease. In 14 cases cpBVDV was isolated and the antigenic marker was found throughout. In seven cases ncpBVDV was cultivated and the antigenic marker was detected in four isolates.

Key words: Bovine viral diarrhoea virus, monoclonal antibodies, cytopathogenicity, marker.

Introduction

In cultured bovine cells two biotypes of bovine viral diarrhoea virus (BVDV) can be distinguished, namely the noncytopathic (ncp) and the cytopathic (cp)

[5, 11]. The cytopathic effect (cpe) producing biotype is held responsible for the induction of fatal mucosal disease (MD). It can be isolated from cases of MD and—under certain conditions—the disease can be precipitated by superinfection of persistently infected animals with cpBVDV [1, 2]. An important molecular marker of cpBVDV is the presence of an additional nonstructural protein (p80) presumably derived by proteolytic cleavage from the viral p125 [4, 19]. In addition, it has been shown that the gene coding for this protein in cells infected with the cpBVDV strains Osloss or NADL carry an insert of non-viral RNA [16]. However, no broadly reactive probes specific for cpBVDV are known. As antigenic markers, monoclonal antibodies have been described that discriminate between cells infected with either of the biotypes [18]. In this study we show that one of the monoclonal antibodies (mab) which is directed against the minor envelope glycoprotein gp48 reacts predominantly—but not exclusively—with cultured cells infected with cpBVDV. In addition, we analyzed field isolates for expression of the antigenic marker. For this purpose 100 patients of the local cattle clinic were screened for BVDV. The animals displayed enteric, respiratory, central nervous or growth retardation signs. In 25 cases isolation attempts were successful. The relation between apparent MD, cytopathic effects of isolated virus in tissue culture and reactivity of the isolates with the above mab was analyzed.

Materials and methods

Virus strains

The BVDV strains originated from the National Animal Disease Center, Ames, or from the Institute of Virology, Hannover (Table 1).

Stock viruses were propagated in fetal calf kidney cells (FCKC) and stored at $-80\,^{\circ}$C. Virus titres were determined in microassays [8, 10].

Primary fetal calf kidney cells

FCKC monolayer cultures were prepared as described [17]. Secondary cultures were tested for the absence of BVDV antigen by immunofluorescence [10]. FCKC were subcultured by seeding 8×10^4 cells suspended in 1 ml culture medium (Eagle's MEM supplemented with 5% fetal calf serum, 100 units of penicillin-G-Na and 0.1 µg of dihydrostreptomycin per ml) into macrotitration plates (COSTAR, USA), or 1.2×10^5 cells in 1 ml into plastic culture tubes (Greiner, FRG). Fetal calf sera were determined to be free of BVDV contamination.

Hybridomas

Isolation and initial characterization of mabs BVD/C16 and BVD/C38 have been described [18]. The homologous antigen for the mabs was the cpBVDV strain NADL. Hybridomas were propagated in Dulbecco's modification of Eagle's Medium supplemented with 10% donor horse serum (Biochrom, FRG).

Table 1. Source and cytopathogenicity of strains and isolates of bovine viral diarrhoea virus (BVDV)

Source: NADC[a] strains/isolates	Source: Hannover[b] strains/isolates
Cytopathic:	
NADL	NADL
Oregon C24V	Oregon C24V
Singer	Singer
Burtenshire	MD-1
New Zealand-cp	TMV-2
TGAC	Osloss 2482
Illinois	A 1138/69
Sturgis	Lamspringe/738
Indiana	DII/265
5960	
Noncytopathic:	
7443	New York-1
New Zealand-ncp	0321/80
McCann	0710/80
Auburn	0712/80
New York-1	0715/80
Nebraska	2204/82
TGAN	22146/81
Sanders	R56/74
639	DII/1102
9762	

[a] Strains maintained in the National Animal Disease Centre, Ames, Iowa.
[b] Strains maintained in the Institute for Virology, Hannover Veterinary School.

Infected monolayer enzyme immunoassay

The reactivity of the mabs against different BVDV strains was tested in an infected monolayer enzyme immunoassay (IM-EIA) [8], as described previously [6]. Briefly, macro-titre plates were seeded with 8×10^4 FCKC per well and simultaneously infected with 10^3 tissue culture infectious doses/50 (TCID/50) of BVDV. After 48 hours the cells were fixed (2 hours at 80 °C). Incubation with mabs, followed by biotinylated anti-mouse IgG and Streptavidin-biotinylated peroxidase (PO) complex (Amersham Buchler, FRG) were for 1 hour at room temperature. Washes between incubations were performed with PBS/0.03% Tween-20. As a substrate for PO, the precipitating substrate 3-amino-9 ethylcarbazol (AEC) (Sigma, FRG) was used. The substrate was freshly prepared as follows: 2 mg AEC were dissolved in 0.3 ml dimethylformamide per 5 ml of 50 mM citrate buffer, pH 5, and 0.03% H_2O_2. Microtitre plates were read using a light microscope.

Selection of cattle

Blood and faeces were collected from animals hospitalized in the cattle clinic of the Hannover Veterinary School. The cattle displayed clinical signs associated with the enteric, respiratory and/or central nervous system. Some animals had retarded growth. Samples were collected over a period of 8 months from a total of 100 animals.

Sampling and processing of specimens

For isolation of BVDV from leucocyte fractions (buffy coats), samples of peripheral blood were collected in polystyrene tubes coated with EDTA (Greiner, FRG). The blood samples were processed as described [13]. The leucocyte fractions were used directly for inoculation of FCKC cultures in culture tubes, or were stored at $-80\,^\circ$C after addition of 10% dimethyl sulfoxide.

For isolation of virus from serum, blood was collected in kaolin-coated polystyrene tubes. After low speed centrifugation ($1200 \times$ g, 10 minutes, $20\,^\circ$C), the serum was used for inoculation of FCKC cultures in culture tubes or macrotitre plates.

Faecal samples were obtained and diluted with 9 volumes of Earle's balanced salt solution supplemented with 0.5% lactalbumin hydrolysate and antibiotics (100 units penicillin-G-Na and 0.1 μg dihydro-streptomycin per ml). After low speed centrifugation, the supernatant from the diluted faecal samples were incubated for 1 hour at room temperature. Samples (2 ml) were either directly used for inoculation of FCKC cultures in culture tubes or frozen at $-80\,^\circ$C.

Virus screening and isolation

Screening for BVDV antigen was performed either by direct immunofluorescence or in a direct IM-EIA [10]. Virus isolation was performed as described [14, 17].

Cloning of virus

The cp strain Oregon was cloned in a plaque assay [7]. The ncp isolate DII/1102 was cloned by limiting dilution in microtitre plates. For this, microtitre wells containing 1×10^4 FCKC were inoculated with $\log_{10}$ dilutions of the virus. After 48 hours of growth, the supernatants were transferred to new sterile microtitre plates and frozen at $-80\,^\circ$C until use. The cells were fixed and the presence of virus was determined in an IM-EIA as described above. Virus from the last positive culture was propagated once on FCKC. It was then again diluted and used for inoculation of FCKC in microtitre wells. The cloning was performed four times.

Results

Reactivity of BVD/C38 with cytopathic and noncytopathic BVDV strains and isolates

For the analysis of the reactivity of the monoclonal antibody with the two biotypes of BVDV, FCKC monolayers were infected for 48 hours with 19 cytopathic (cp) and 19 noncytopathic (ncp) viruses (Table 1). After heat fixation, the indirect IM-EIA using hybridoma supernatant was performed. As a control for the infection of cells, the panpestivirus specific mab BVD/C16 was used. With this antibody, all infected cells (100%) in the

Table 2. Reactivity of BVD/C38 with cytopathic (cp) and noncytopathic (ncp) BVDV strains and isolates in an indirect enzyme immunoassay (IM-EIA) on infected monolayers[a]

Staining pattern	cp BVDV n = 19	ncp BVDV n = 19
No reaction	0	3
Foci or single cells	0	9
30–50% stained cells	6	7
80–100% stained cells	13	0

[a] In all tests, 100% of the cells were stained when the panpestivirus specific mab BVD/C16 was used for EIA on parallel cultures.

monolayer showed intense cytoplasmic staining (Fig. 1 and 2, top rows), showing that all the cells in the monolayer were virus-infected. With mab BVD/C38, cp- and ncpBVDV strains/isolates reacted differently, as shown in Table 2.

All cpBVDV infected monolayers reacted positively and two different patterns were observed. In most cases (13 strains/isolates) between 80 and 100% of the cells had intensely stained cytoplasms (Fig. 1, bottom left). Six strains/isolates had a lower fraction of positive cells (30–50%; Fig. 1, bottom right).

Reactivity of ncpBVDV with the mab BVD/C38 was clearly different. In most cases (9 strains/isolates), only foci or single cells were intensely stained (Fig. 2, bottom right). With seven of the strains/isolates, about 30 to 50% of the infected cells showed either a strong perinuclear staining reaching into the cytoplasm or a weak cytoplasmic staining (Fig. 2, bottom left). Three ncpBVDV strains (New York-1 passaged in Hannover; TGAN; isolate R56/74) showed no reaction with the mab. While New York-1 (Hannover) was negative, the same strain from the NADC showed the foci-single cell reactivity pattern. Staining of 80–100% of the infected cells was never observed with ncpBVDV.

Interestingly, two of the strains passaged in different laboratories (NADC and Hannover) showed different reactivity patterns. While the mab reacted with 80–100% of cells infected with NADL (NADC) or Oregon (Hannover), only about 50% of cells infected with the corresponding strains—NADL (Hannover) or Oregon (NADC)—gave positive reactions. In contrast, Singer from both laboratories displayed identical reaction patterns (80–100% positive cells).

To assess whether the reactivity patterns observed with BVD/C38 were dependent on the time after infection (p.i.) the monolayers were tested, cells

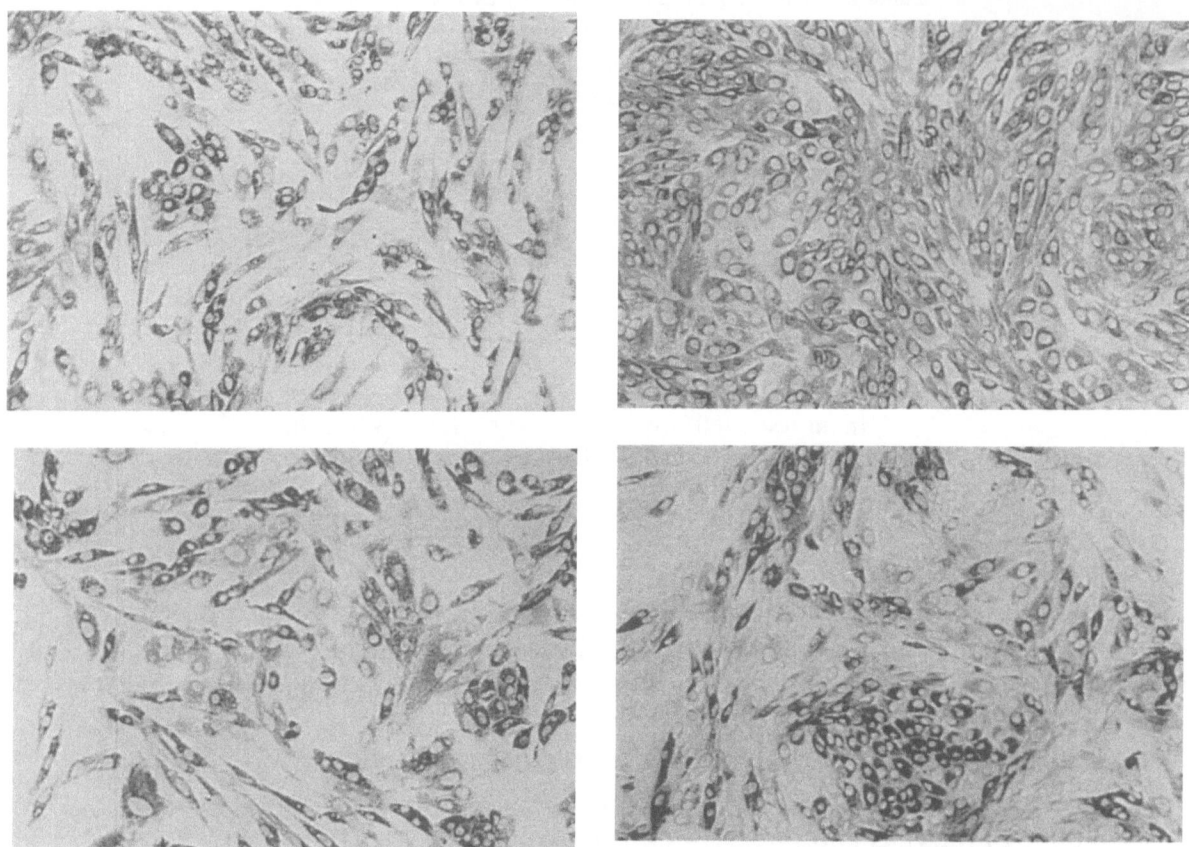

Fig. 1. Reactivity of the monoclonal antibodies BVD/C16 (top row) and BVD/C38 (bottom row) in IM-EIA on fetal calf kidney cells infected with two cytopathic BVDV strains.

Left side: FCKC infected with NADL (NADC)
Right side: FCKC infected with Illinois

The monolayers were fixed 48 hours p.i. and IM-EIA was performed as described in Materials and Methods. Magnification: 400 ×

were infected with the ncp strains Auburn (30–50% BVD/C38 reactive cells 48 hours p.i.), 7443 (single cells or foci stained 48 hours p.i.), and with the cpBVDV strains NADL(NADC) (80–100% reactive cells at 48 hours p.i.) and NADL (Hannover) (30–50% reactive cells at 48 hours p.i.). Vacuolization and advance of cpe was evaluated microscopically every 6 hours p.i.. The monolayers were fixed and tested for reactivity with mabs BVD/C16 and BVD/C38 after 24, 48 and 72 hours p.i. .

Reactivity in the IM-EIA with the mab BVD/C16 showed that after 24 hours about 50–80% of the cells in each culture were virus-infected. After 48 hours p.i., infection was complete, i.e., 100% of the cells were intensely stained with BVD/C16. At 72 hours p.i., staining of the monolayers infected with the two ncpBVDV strains was weaker but still complete. Staining

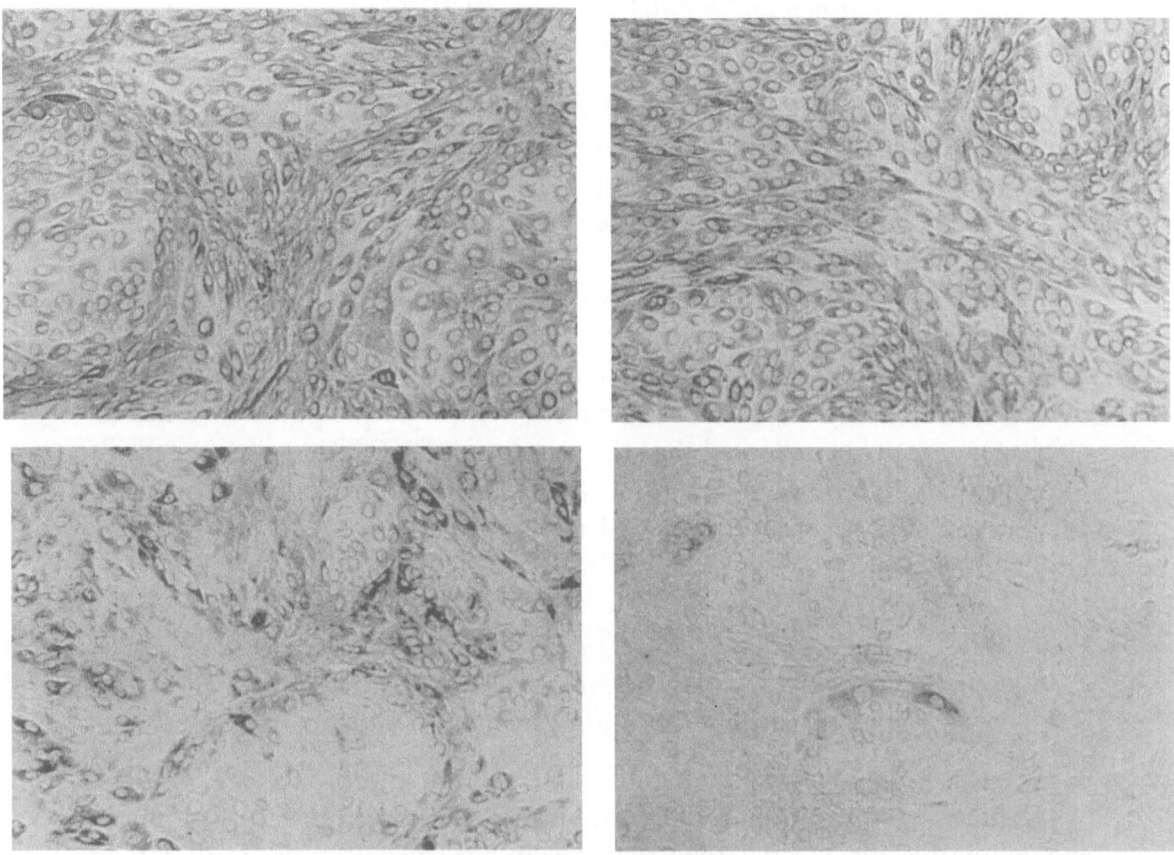

Fig. 2. Reactivity of the monoclonal antibodies BVD/C16 (top row) and BVD/C38 (bottom row) in IM-EIA on fetal calf kidney cells infected with two noncytopathic BVDV strains.

Left side: FCKC infected with Auburn
Right side: FCKC infected with 7443

The monolayers were fixed 48 hours p.i. and IM-EIA was performed as described in Materials and Methods. Magnification: 400×

of these cultures with BVD/C38 after 48 and 72 hours showed that the characteristic staining pattern of each of the strains was maintained.

In the cultures infected with cpBVDV NADL (NADC), the first vacuoles appeared in the cell cytoplasm about 36 hours p.i.. After 48 hours, the cells were highly vacuolized and the cpe was already advanced (Fig. 1, left). After 72 hours, the cpe had completely destroyed the monolayer. In contrast, cells infected with NADL (Hannover) were only slightly vacuolized and almost intact 48 hours p.i.. An advanced cpe was observed with a delay of 12–16 hours, and at 72 hours p.i. the cpe had only destroyed about 30% of the monolayer. The remaining highly vacuolized cells gave a weaker but complete staining with BVD/C16, while with BVD/C38 only a fraction of about 50% of the cells was reactive. Complete detachment of the monolayers was achieved after 96 hours, i.e. with a 24 hours delay.

Reactivity of a ncp and a cpBVDV with BVD/C38 before and after cloning of the viruses

The assess whether the reactivity of the mab with certain fractions of an infected monolayer only is due to contamination with cp and ncp viruses, respectively, two BVDV strains were cloned and analyzed in the indirect IM-EIA. The ncp isolate DII/1102 was cloned four times by limiting dilution, and the cp strain Oregon was cloned two times in plaque assays. Both the clones and the original viruses were tested on FCKC monolayers for reactivity with the mab. While the ncp isolate and the four clones showed identical staining patterns (foci, single cells), the fraction of BVD/C38 reactive cells was clearly increased after infection with plaque purified clones of strain Oregon. This indicated that the fraction of BVD/C38 reactive cells in monolayers infected with ncpBVDV cannot be altered by cloning the virus. In contrast, the reactive fraction of cells infected with cpBVDV can be enriched by cloning the virus.

Reactivity of mab BVD/C38 with isolates from cattle with and without classical mucosal disease clinical signs

BVDV was isolated from serum, buffy coat fraction and/or faeces of 25 cattle with enteric, respiratory and/or central nervous clinical signs. For monitoring cytopathic effects (cpe), the isolates were passaged 10 times in FCKC in cell culture tubes. The isolates were tested before being passaged and after appearance of cpe for reactivity with BVD/C38; the isolates that stayed ncp were tested after the fourth and after the last passage (Table 3). Only 7 isolates displayed a cpe in the first passage, and six of them reacted with the mab. After 10 passages in cell culture, 8 more isolates displayed cytopathic effects, and all of them reacted with BVD/C38. The mab gave positive reactions with 5% (1 isolate), with 30–50% (7 isolates) and with 80–100% (7 isolates) of the infected cells, respectively. From the 10 isolates that did not develop cpe, five did not react with the mab. The remaining 5 isolates showed the characteristic perinuclear staining in addition to the weak cytoplasmic staining observed with the above ncpBVDV strains. About 5–40% of the cells in the infected monolayer were stained.

Only 21 of the 25 animals from which BVDV was isolated developed clinical signs of Mucosal Disease (MD). From these, 14 were cpBVDV; one animal from which cpBVDV was isolated did not develop MD signs in vivo. In addition, seven of the 10 animals from which ncpBVDV was isolated also succumbed with MD clinical signs.

The relation between development of MD, cpe and BVD/C38 reactivity in vitro, is summarized in Table 4. From the 21 isolates that originated from animals that developed MD, 18 (14 cp and 4 ncp) reacted with the mab. Three ncp isolates lacked BVD/C38 reactivity. One animal from which

Table 3. Passage of BVDV isolates on fetal calf kidney cells. Correlation between appearance of cytopathic effect and reactivity with mab BVD/C38 in the indirect IM-EIA

	Passage number	Number of isolates	Reactivity with BVD/C38
Development of cpe (n = 15)	1	7	6/7[a]
	2	4	
	3	2	
	4		13/13
	5		
	6		
	7		
	8	1	
	9	1	15/15
	10		
no cpe	10	10	5

[a] Isolates reacting with BVD/C38/Total cp isolates.

Table 4. Reactivity with mab BVD/C38 of BVDV isolates from cattle with and without symptoms typical for MD

Symptoms	Biotype of BVDV	Total	Reactivity with BVD/C38[a]	
			positive	negative
MD; n = 21	cytopathic	14	14	0
	noncytopathic	7	4	3
others; n = 4	cytopathic	1	1	0
	noncytopathic	3	1	2

[a] Reactivity was assayed in an EIA on infected monolayers.

cpBVDV was isolated did not develop MD but showed BVD/C38 reactivity. From the three ncp isolates from this group, only one reacted with the mab.

Discussion

When initially characterized, the mab BVD/C38 was described as reacting specifically with cpBVDV strains/isolates only [18]. The results described here show that this statement has to be modified and extended. The analysis of FCKC monolayers where 100% of the cells were infected with 19 cp- and 19 ncpBVDV allowed us to distinguish several reactivity patterns.

The majority of cpBVDV strains, i.e., those which produced a visible cpe, yielded infected cells where 80–100% reacted with the mab. In one third of the cases the proportion of reactive cells was lower. In contrast, with ncpBVDV, i.e., strains not producing a visible cpe, a reverse gradient of reactive infected cells was found. With most ncp viruses either no cells, single cells or foci showed cytoplasmic staining. About one third of the viruses with no visible cpe (ncpBVDV) had proportions of reactive cells between 30–50%. The staining pattern for the individual BVDV strains were independent from the TCID/50 used for infecting the monolayers (data not shown), and independent from the time p.i. chosen to perform the IM-EIA.

The differential staining patterns obtained with mab BVD/C38 indicate that the BVDV stocks represent a consensus of closely related viruses, but are not pure clones. This can be attributed to the high mutation rates predicted for all positive strand RNA viruses [21]. The failure to eliminate the BVD/C38-reactive component from ncpBVDV indicates that this fraction might emerge "de novo" at a constant rate in infected cells. A genetic "flip-flop" mechanism might be discussed, although the cause for this effect is unknown. We succeeded in enriching the BVD/C38-reactive component of cpBVDV by repeated cloning using the plaque assay (Oregon-NADC). However, it was observed that the reactive fraction decreased again after several passages (data not shown). These results suggest that the expression of the epitope for BVD/C38 is a relatively unstable genetic trait yet characteristic for each virus strain. It may be altered during different passage histories, as illustrated by the observed differences in BVD/C38 reactivity between batches of identical strains maintained in different laboratories, e.g., strains New York-1, Oregon and NADL.

A consistent feature of cpBVDV is the presence of the nonstructural protein p80 in infected cells. This polypeptide is derived from p125 [4, 19]. This shows that in cpBVDV infected cells the processing of viral proteins may be altered. Likewise the structure of gp48 might be affected by altered processing. Reactivity with BVD/C38 may be a marker for this event; appearance of the cpe might depend on the accumulation of BVD/C38 reactive protein. This would explain the 24 hours delay of the cpe in infected cultures where only 30–50% of the infected cells react with the mab. Nevertheless, it cannot be excluded that this delay is due to a mixed infection with newly emerged mutants in the stocks.

After the above "in vitro" studies we analyzed BVDV isolates from 21 clinical cases of MD [5, 12; for review see 20], to find whether reactivity with the mab BVD/C38 correlated with cytopathogenicity of the corresponding viral isolates. The viruses were passaged 10 times in cell culture before testing. In this limited study there was a high (100%) correlation between cytopathogenicity, strong BVD/C38 reactivity (>50% of cells) and clinical appearance of MD (14 out of 14 animals). In seven cases, the cattle succumbed to clinical MD although only ncpBVDV was isolated. Three of these isolates showed no reaction with the mab BVD/C38. Here, the correlation between MD "in vivo" and expression of the BVD/C38 epitope dropped to about 57%.

It has been suggested that the two biotypes of BVDV display a tissue tropism, whereby cp virus is mainly isolated from intestinal samples and ncp virus from the respiratory tract, blood and blood associated organs [3]. In the cases described here, there was no opportunity to isolate the virus from the gut. This would possibly have displayed the cp biotype, since there is evidence that both biotypes can be isolated from animals suffering from MD [2, 9, 15].

The mab BVD/C38 defined an epitope on the minor glycoprotein gp48 of BVDV which is closely associated with the cp biotype of BVDV. The reaction with this mab clearly discriminated between cells infected with either of the BVDV biotypes. However, about 30% of isolates of either biotype yielded an overlapping reactivity pattern with BVD/C38 in infected cells. This might indicate that the conventional cytopathic effect with detachment of the cells from the monolayers and lysis [5, 12] might not be the only marker for cytopathogenicity. A more subtle analysis of virus induced cell damage could lead to a better understanding of this biologically important phenomenon related to or responsible for the pathogenesis of fatal BVDV infections.

Acknowledgements

The support of the director of the Cattle Clinic, Hannover Veterinary School, Professor Dr. med. vet DDr. h.c. M. Stöber is gratefully acknowledged. The authors thank M. Kaps and G. Müller for expert technical assistance.

References

1. Bolin SR, McClurkin AW, Cutlip RC, Coria MF (1985) Severe clinical disease induced in cattle persistently infected with noncytopathic bovine viral diarrhea virus by superinfection with cytopathic bovine viral diarrhea virus. Am J Vet Res 48: 573–576
2. Brownlie J, Clarke MC, Howard CJ (1984) Experimental production of fatal mucosal disease in cattle. Vet Rec 114: 535–536
3. Clarke MC, Brownlie J, Howard CJ (1987) Isolation of cytopathic and non-cytopathic bovine viral diarrhoea virus from tissues of infected animals. In: Harkness JW (ed) Agriculture. Pestivirus infections of ruminants. Commission of the European Communities, Brussels, publication CD-NA-10238-EN-C, pp 3–12
4. Donis RO, Dubovi EJ (1987) Differences in virus-induced polypeptides in cells infected by cytopathic and noncytopathic biotypes of bovine virus diarrhea–mucosal disease virus. Virology 158: 168–173
5. Gillespie JH, Baker JA, McEntee K (1960) A cytopathogenic strain of virus diarrhea virus. Cornell Vet 50: 73–79
6. Greiser-Wilke I, Moennig V, Coulibaly COZ, Dahle J, Leder L, Liess B (1990) Identification of conserved epitopes on a hog cholera virus protein. Arch Virol 111: 213–225

7. Hafez SM, Liess B (1972) Studies on bovine viral diarrhea–mucosal disease virus. I. Cultural behaviour and antigenic relationship of some strains. Acta Virol 16: 388–398

8. Holm-Jensen M (1981) Detection of antibodies against hog cholera virus and bovine viral diarrhea virus in porcine serum. A comparative examination using CF, PLA and NPLA assays. Acta Vet Scand 22: 85–98

9. Howard CJ, Brownlie J, Clarke MC (1987) Comparison by the neutralization assay of pairs of non-cytopathogenic and cytopathogenic strains of bovine virus diarrhoea virus isolated from cases of mucosal disease. Vet Microbiol 13: 361–369

10. Hyera JMK, Dahle J, Liess B, Moennig V, Frey H-R (1987) Production of potent antisera raised in pigs by anamnestic response and use for direct immunofluorescent and immunoperoxidase techniques. In: Harkness JW (ed) Agriculture. Pestivirus infections of ruminants. Commission of the European Communities, Brussels, publication CD-NA-10238-EN-C, pp 87–102

11. Lee KM, Gillespie JH (1957) Propagation of virus diarrhea virus of cattle in tissue culture. Am J Vet Res 18: 952–953

12. Liess B (1967) Die ätiologische Abgrenzung selbständiger Virusinfektionen, insbesondere der Virusdiarrhoe-Mucosal Disease im sogenannten "Mucosal Disease-Komplex" bei Rindern. Dtsch Tierärztl Wochenschr 74: 46–49

13. Liess B, Plowright W (1964) Studies on the pathogenesis of rinderpest in experimental cattle. I. Correlation of clinical signs, viremia and virus excretion by various routes. J Hyg (Camb) 62: 80–100

14. Liess B, Orban S, Frey H-R, Trautwein G, Wiefel H, Blindow H (1984) Studies on transplacental transmissibility of a bovine virus diarrhoea (BVD) vaccine virus in cattle. II. Inoculation of pregnant cows without detectable neutralizing antibodies to BVD-virus 90–229 days before parturition (51st to 1090th days of gestation). J Vet Med B 31: 669–681

15. McClurkin AW, Bolin SR, Coria MF (1985) Isolation of cytopathic and noncytopathic bovine viral diarrhea virus from the spleen of cattle acutely and chronically affected with bovine viral diarrhea. JAVMA 186: 568–569

16. Meyers G, Ruemenapf T, Thiel H-J (1989) Ubiquitin in a togavirus. Nature 341: 491

17. Orban S, Liess B, Hafez SM, Frey H-R, Blindow H, Sasse-Patzer B (1983) Studies on transplacental transmissibility of a bovine virus diarrhoea (BVD) vaccine virus. I. Inoculation of pregnant cows 15 to 90 days before parturition (190th to 265th day of gestation). J Vet Med B 30: 619–634

18. Peters W, Greiser-Wilke I, Moennig V, Liess B (1986) Preliminary serological characterization of bovine viral diarrhoea virus strains using monoclonal antibodies. Vet Microbiol 12: 195–200

19. Purchio AF, Larson R, Collett MS (1984) Characterization of bovine viral diarrhea virus proteins. J Virol 50: 666–669

20. Radostits OM, Littlejohns IR (1988) New concepts in the pathogenesis, diagnosis and control of diseases caused by the bovine viral diarrhea virus. Can J Vet 29: 513–528

21. Steinhauer DA, Holland JJ (1987) Rapid evolution of RNA viruses. Ann Rev Microb 41: 409–433

Author's address: V. Moennig, Institute of Virology, Hannover Veterinary School, Bischofsholer Damm 15, D-W-3000 Hannover, Federal Republic of Germany.

Arch Virol (1991) [Suppl 3]: 67–70

Cytopathogenicity of pestiviruses isolated post mortem from cattle

Brief Report

T. Løken

National Veterinary Institute, Oslo, Norway

Accepted March 14, 1991

Summary. Cytopathic pestivirus was isolated from different tissues of only eight of 23 cattle with mucosal disease. Three persistently infected cows were healthy until slaughter after death of all their seven offspring, out of which one of four examined demonstrated cytopathic pestivirus.

Key words: Mucosal disease, pestiviruses, persistent infection, cytopathogenicity.

*

Infection with pestivirus in early foetal life in cattle induces persistent infection. In such infected cattle, mucosal disease (MD) may develop, mostly between 4 and 24 months of age [6]. Both characteristic MD and chronic, atypical forms of MD have been described [1, 4]. According to a hypothesis favoured by several authors, MD is induced in cattle persistently infected with non-cytopathic pestivirus by superinfection with antigenically homologous cytopathic MD virus [2]. Also superinfection with heterologous cytopathic strains has been shown to induce MD [7]. It seems reasonable to suppose that other inducing mechanisms or factors may also be involved in the pathogenesis of MD.

From 1987 and onwards, pestivirus has been isolated from different organs of 23 cattle necropsied at the National Veterinary Institute in Oslo. All these animals were Norwegian Red Cattle, aged between 6 months and 2 years. Six of the animals showed both signs and lesions characteristic of MD, and five others such lesions. The other cases had varying unspecific signs and lesions of unthriftiness and enteritis, and several had bacterial lung

Table 1. Isolation of cytopathic pestivirus (cpMDV) from tissue of 23 cattle aged between 6 and 24 months, by inoculation onto calf kidney cell cultures. Non-cytopathic pestivirus was isolated from all tissues

Tissue	Number examined[1]			Number positive for cpMDV[1]					
				primary passage			second passage		
	A	CS	CL	A	CS	CL	A	CS	CL
Peyer's patches	9	2	3	3	1	1	1		1
Tonsil	5	2	2	0			0		
Spleen	15	5	8	2			2		1
Myocard/lung/liver/ lymph nodes	19(29)	6(8)	11(18)	2(2)	1(1)	1(1)	4(9)		3(5)
Totals	23(58)	6(17)	11(31)	3(7)	1(1)	1(1)	5(12)		3(7)

[1] Animals; tissues in brackets. A = All animals.
CS = Animals with characteristic signs of mucosal disease
CL = Animals with characteristic lesions of mucosal disease

infections. The present material included four animals which were offspring of three cows in different herds, shown to be persistently infected with pestivirus.

Totally 58 tissue samples of the 23 cattle (see Table 1), including Peyer's patches and/or tonsil from 10 animals and spleen from 15, were inoculated onto primary calf kidney (CK) cell cultures and incubated at 37 °C for 8 to 10 days. All samples were passaged twice. In 13 of the animals, the same material including 32 tissues was also inoculated and similarly passaged in a bovine turbinate (BT) cell line. Non-cytopathic pestivirus was demonstrated immunoenzymatically [5]. Cells and foetal calf serum used in the tests were similarly examined for absence of non-cytopathic pestivirus.

Pestivirus was demonstrated immunoenzymatically in the primary passage from all 58 examined tissues. Cytopathic strains were isolated from 19 samples from eight of the cattle, which were from 1 to 2 years of age (Table 1). Seven tissues from three of the animals induced cytopathic effect in the primary passage, and from further nine tissues of three animals in the second passage, this being the case for all the two to four organs examined from each of the animals. Three of five tissues of two other animals showed cytopathic effect in the second passage. In four of the animals, cytopathic strains were isolated from eight tissues in CK cells but from none of them in the BT cells. Only one animal which had shown characteristic signs and lesions of MD demonstrated cytopathic virus, which was isolated from both examined organs (Peyer's patches and liver). Three other animals which harboured cytopathic strains had lesions indicative of MD, while four such animals had exhibited signs and lesions only weakly indicative of this disease.

All seven offspring from different pregnancies in the three persistently infected cows developed chronic disease and died before reaching 2 years of age. However, the three dams were still healthy when slaughtered 1 to 2 months after the death of their offspring. Though cytopathic pestivirus was isolated from spleen, lymph node and Peyer's patches from one of these offspring, only non-cytopathic strains were isolated from different organs from three offspring of the other two cows.

In the present material, cytopathic strains were isolated from only 1/3 of the tissues and animals. In tissues from which cytopathic strains were not isolated, there may well have been such virus which could have been demonstrated in other systems of cell culture. Longer incubation periods and additional passages in cell cultures might have demonstrated cytopathic strains from more tissues. Cytopathic pestiviruses have been readily isolated from tissues of MD cases by two passages in bovine cell cultures [3]. It should be borne in mind that cytopathogenicity is not necessarily a constant characteristic of a virus, but a property which may vary dependant both on genetic changes of the virus and different cultural conditions. Furthermore, the pathogenicity of pestivirus in persistently infected cattle, i.e. the ability of the virus to induce MD, has not been shown to be directly linked genetically to cytopathogenicity in cell cultures. MD may therefore also be induced by superinfection with certain strains of non-cytopathic pestivirus. It is also possible that development of MD, either in the characteristic or chronic form, may be influenced by factors which are as yet unknown, and that superinfection with pestivirus is not always involved in the pathogenesis.

References

1. Baker JC (1987) Bovine viral diarrhea virus: a review. J Am Vet Med Assoc 190: 1449–1458
2. Brownlie J, Clarke MC, Howard CJ (1987) Clinical and experimental mucosal disease —defining a hypothesis for pathogenesis. In: Harkness (ed) Agriculture. Pestivirus infections of ruminants. A seminar in the CEC programme of coordination of research on animal husbandry, Brussels 1985. Office for Official Publications of the European Communities, Luxembourg, pp 147–156
3. Clarke MC, Brownlie J, Howard CJ (1987) Isolation of cytopathic and non-cytopathic bovine viral diarrhoea virus from tissues of infected animals. In: Harkness (ed) Agriculture. Pestivirus infections of ruminants. A seminar in the CEC programme of coordination of research on animal husbandry, Brussels 1985. Office for Official Publications of the European Communities, Luxembourg, pp 3–10
4. Grønstøl H, Berge GE, Løken T (1988) Clinical observations in chronic bovine virus diarrhoea. In: Proceedings of the 15th World Buiatrics Congress, Mallorca 1988, pp 890–895
5. Meyling A (1983) An immunoperoxidase (PO) technique for detection of BVD virus in serum of clinically and subclinically infected cattle. In: Proceedings of the 3rd International Symposium of Veterinary Laboratory Diagnosticians, Ames 1983, pp 179–184
6. Roeder PL (1982) Bovine mucosal disease—a persistent viral infection. In: Proceedings

of the 7th International Symposium of the World Association of Veterinary Microbiologists, Immunologists and Specialists in Infectious Diseases of animals, Barcelona
1982
7. Westenbrink F, Straver PJ, Kimman TG, De Leeuw PW (1989) Development of
   a neutralising antibody response to an inoculated cytopathic strain of bovine virus
   diarrhoea virus. Vet Rec 125: 262–265

Author's address: T. Løken, National Veterinary Institute, N-0033 Oslo 1, Norway.

Arch Virol (1991) [Suppl 3]: 71–78

Diaplacental infections with ruminant pestiviruses

B. I. Osburn[1] and **G. Castrucci**[2]

[1] School of Veterinary Medicine, University of California, Davis, California and
[2] Università di Perugia, Facoltà di Veterinaria, Perugia, Italy

Accepted March 14, 1991

Summary. Pestiviruses are capable of causing diaplacental infections. Maternal viremias are important for localizing virus in the ruminant placentome. Placental lesions occur with cytopathic BVDV and noncytopathic BDV. The ruminant fetus is very susceptible to pestivirus infections once the virus crosses the placenta because the fetus is 1) agammaglobulinemic, 2) immunologically immature, and 3) it has many immature organ systems with undifferentiated cells. Cytopathic BVDV (NADL) in calves and noncytopathic BDV (BD-31) in lambs cause a variety of clinical syndromes including early embryonic death, abortion, stillbirth, malformed fetuses, and/or low birth weight with viral persistence and immunological tolerance. The cytopathic BVDV (NADL) reviewed herein caused pulmonary, placental and dermal lesions when infection occurred at 80–90 days gestation. In contrast, infection at 140–150 days resulted in retinal dysplasia and cerebellar hypoplasia. The lesions were attributed to direct viral cytopathology. Noncytopathic BDV (BD-31) in lambs caused weak lambs, with hairy fleece and tonic-clonic tremors. The lambs were of low birth weight, persistently viremic and immunologically tolerant. The lambs are hypothyroid and had severe hypomyelination. It is hypothesized that the central lesion leading to many of the neural, skeletal and dermal lesions was the endocrine dysfunction leading to hypothyroidism.

Key words: Pestivirus, congenital disease, cerebellar hypoplasia, immune tolerance, fetal infections, border disease, bovine virus diarrhea.

Introduction

Pestiviruses are well known causes of diaplacental infections in domestic ruminants. These infections are best described in cattle and sheep where the

infections are associated with early embryonic deaths, abortions, malformations and unthrifty newborns. In this paper, the pathogenesis of diaplacental
pestivirus infections will be reviewed.

Status of fetus

The ungulate fetus has three factors which favor diaplacental infections.
These factors include 1) agammaglobulinemia 2) immature immune system,
and 3) undifferentiated organ systems [21]. The fetal ungulate is isolated
from the insults of microbial agents by the maternal immune and innate
immune systems. The syndesmochorial ungulate placenta is incapable of
transferring immunoglobulins from the maternal circulation to the fetus. As
a result, the fetus lacks maternal resistance during the entire gestation. It is
not until after birth that the newborn ungulate receives maternal immunity
in the colostrum.

Like all of the other organ systems of the fetus, the lymphoid and
accessory immune systems continue to develop and mature throughout
gestation and into the neonatal period. The lymphoid system in the fetal
lamb and fetal calf develop in a sequential manner [21]. Stem cells are
present in the yolk sac and later appear in the liver. Lymphocytes appear in
blood by 30–32 days in the fetal lamb. In the developing fetus, the thymus is
the first lymphoid organ populated by lymphocytes at 40 days followed by
lymph nodes at 45 days, the spleen at 56 days and the Peyers patches and
related lymphoid tissues in the intestinal tract at 75 days. Although lymphocytes are present in the organs by midgestation, the lymph nodes and spleen
are sparsely populated until 120 days. Immune responses measured by
specific antibody activity occurs in the fetal lamb by 40 days gestation.
Specific antibody activity to × 174 bacteriophage has been observed by
40 days; however, the activity to more complex antigens occurs later and in
a somewhat scheduled pattern. For instance, graft rejection does not occur
until 75 days or midgestation in the fetal lamb. Other antigens such as BCG
will not evoke a specific antibody response until 4 to 6 weeks after birth.

The accessory immune system develops gradually during fetal life and in
many instances the components are not functionally active until late in
gestation [21]. Neutrophils are not present in peripheral blood until 56–60
days gestation. Even at that age, the numbers are few and the cells not fully
mature or differentiated. The numbers of neutrophils in peripheral blood
remain low until 130 days of gestation. It is not until shortly before birth that
the numbers approach or exceed the number found in postnatal and adult
animals. Both monocytes and neutrophils of fetal lambs lack lipases and
esterases until the time of birth. Apparently neutrophils will rapidly differentiate if the appropriate inflammatory stimulant is administered by midgestation. Another accessory system that is not functional is the complement
system. Various components appear at different times during ontogeny with

C_1 appearing by 39 days gestation. However, full hemolytic activity is not present before 120 days gestation. The level of hemolytic activity at 120 days is low and it is not until the postnatal period before adult levels are reached. Some factors, such as the interferons, are present early; however, these interferons are ineffective in limiting viral infections of the fetus. Other factors develop gradually as well. As a result, the fetus responds poorly and inadequately to injurious agents such as teratogenic viruses. Studies on inflammatory responses in fetal rhesus monkeys clearly demonstrated the inadequacy of responses to various irritants. The patterns of inflammation changes as the fetus matures [26]. It is not until late in gestation or the early neonatal period that inflammatory responses occur in a timely and fully complementary manner to that observed in the adult. It is clear that the fetal host defense system is inadequate to cope with major insults such as viral infections. This inadequacy permits viral agents to replicate with minimal interference.

Another important factor favoring fetal infection is the undifferentiated cells in organ systems. A list of these are included in Table 1. Each of the systems undergo cellular differentiation and maturation during ontogeny. All organs are morphologically identifiable at the time the embryo becomes a fetus. The lamb embryo becomes a fetus at 36 days gestation and the bovine at 45 days gestation. The organs will vary in maturation. The organs listed in Table 1 have cells which undergo differentiation at predetermined times during fetal ontogeny. Interference with cellular differentiation at the specified time often leads to maturation arrest with subsequent morphological and/or functional deficits. The character of the lesions depend upon the precise time during ontogeny when the insult occurs.

The pathogenesis of viral teratogenic lesions in the fetus during ontogeny are due to 1) destruction of undifferentiated dividing cells, 2) cytolytic effects on cells by the virus, 3) vascular compromise, and 4) indirectly through interference with endocrine function. Destruction of undifferentiated dividing cells is the mechanism in parvovirus infections [16]. These viruses cause cerebellar hypoplasia by destroying the external granular layer. Viruses

Table 1. Organ systems with undifferentiated cells

Nervous
Pulmonary
Integumentary
Skeletal
Urinary
Immune (lymphoid)
Endocrine
Muscular

causing cytolytic effects include bluetongue [20], epizootic hemorrhagic disease virus [18], Akabane [23], and Cache Valley [14]. The consequences of these viruses may be extensive destruction leading to hydranencephaly, porencephaly, cerebellar hypoplasia, retinal dysplasia, hydrocephalus and/or arthrogryposis. Vascular compromise has been suggested as a cause of lesions with some virus infections. Presently there is no known virus that causes this; however, the presence of cerebral cysts or porencephaly in some fetuses with congenital viral infections suggests that this is a possible mechanism. The fourth mechanism involves indirect effects on other organ systems resulting from a primary functional deficit in an endocrine organ. This mechanism has been proposed for lymphocytic choriomeningitis virus infection in mice [19].

Pestiviruses

Pestiviruses are nonarboviruses in the Togaviridae. Pestiviruses which cause major disease and economic losses in domestic animals include bovine virus diarrhea virus (BVDV), border disease virus (BDV) and the hog cholera or swine fever virus. Two biotypes (cytopathic and noncytopathic) of bovine virus diarrhea and border disease virus types have been described. Bovine virus diarrhea has been associated with diaplacental disease as well as disease in adult cattle. Border disease is associated only with diaplacental infection and congenital diseases. The pattern of lesions associated with the 2 BVD biotypes in the bovine fetuses are different. Earlier findings suggest that cytopathic strains of BVD are pathogenic resulting in diaplacental infections with subsequent embryonic mortalities, abortion, malformation or subclinical disease [3, 11, 13]. In contrast, the noncytopathic strains are associated with birth of calves which are persistently infected and immunologically tolerant to BVD virus [6, 8, 24]. With one exception, border disease virus isolates are noncytopathic.

Ruminant pestiviruses have been reported to cause a wide array of diaplacental syndromes including early embryonic death, abortion, mummification, a variety of congenital defects, runting, persistent viral infection with immune tolerance, or birth of normal calves and lambs immune to BVDV or BDV. Results of a number of studies indicated that cytopathic biotypes of BVDV are associated with teratogenic lesions in fetal calves whereas the noncytopathic biotypes are associated with persistent infections and immune tolerance which continues throughout life [6, 8]. BDV causes a number of teratogenic lesions, viral persistence and immune tolerance. The pattern of teratogenic lesions are dependent upon 1) the timing of fetal infection and 2) the biotype of virus causing infection [22]. For instance, infection of the bovine fetus between 75 and 120 days gestation with noncytopathic strains of BVDV leads to persistent infection with immune tolerance [8]. Infection with the NADL cytopathic strain of BVDV at 75 to

90 days gestation results in hypotrichosis and pulmonary lesions whereas infection at 140 to 150 days gestation results in cerebellar hypoplasia [11, 13]. These calves clear virus by 190–200 days gestation and have anti BVDV antibodies [7]. For BDV to cause teratogenic lesions, viral persistence and immune tolerance, viral infection occurs between 40 and 80 days gestation.

Pathogenesis of ruminant pestivirus infection

An example of a diaplacental infection with a cytopathic strain of BVDV in cattle and a noncytopathic BDV in sheep are described.

Maternal viremia occurs leading to placental infection in the ruminant placentome. Focal villous necrosis has been reported in both noncytopathic BDV infection in sheep and cytopathic BVDV infection in cattle [5, 11]. Vasculitis with endarteritis and placentitis has been reported with BVDV. The lesions are localized in nature. It has been suggested that the lesions contributed to the fetal changes by interfering with fetal nutrition [5]. It was suggested that the placental insufficiency contributed to the low birth weight of infected fetuses.

Cytopathic BVDV (NADL strain) in bovine fetus

Infection at 80–90 days gestation

Infection of the bovine fetus with the NADL-cytopathic strain of BVDV at 80–90 days of gestation leads to widespread lesions [11]. The virus has an affinity for vascular endothelium and for epithelial tissues. Necrotizing and proliferative changes occur in the endothelium of some arteries. The changes in the lungs consist of a necrotizing bronchiolitis associated with mononuclear infiltrates in the mesenchymal lung stroma. These types of lesions may lead to pulmonary hypoplasia since the precursor bronchiolar epithelium which is necessary for alveolar development are destroyed. Local necrotizing lesions with underlying inflammatory responses can be observed in the epidermis. Destruction of the adnexa in these areas leads to permanent damage of the epidermis. At birth, the manifestation is hypotrichosis resulting from a loss of hair follicles and associated adnexa [11].

Infection at 140–150 days gestation

Cytopathic BVDV infection at 140–150 days gestation is associated with the virus targeting different organ systems than those affected at 80–90 days gestation [3, 11]. The principal organ system affected at this stage of fetal development is the cerebellum and retina. A necrotizing retinitis occurs in the undifferentiated retina. The result is a blind newborn calf because of retinal dysplasia. These calves are also uncoordinated and often they are unable to stand. These clinical signs are caused by severe cerebellar hypoplasia. This hypoplasia is the result of destruction of the external granular layer in the undifferentiated cerebellum at 150 days gestation. The external granular layer at that stage of gestation contains most of the precursor neurones which migrate from this area into the internal granular cell layer of the cerebellar folia. The result is a failure of the cerebellar folia to develop morphologically and functionally.

Regardless of when fetal infection occurs with the cytopathic strain of BVDV, immune responses occur at 180 days gestation [5]. In most instances, this immune response participates in immune clearance of the virus. As a result, it is not possible to recover virus in most of the affected newborn calves. Instead, immunoglobulins with specific neutralizing antibody are present and these represent the principal means of diagnosing the cause of the congenital lesions.

Noncytopathic BDV (BD-31) infection in sheep

Infection at 45–80 days gestation

Infection of pregnant ewes during this period leads to a variety of morphological and functional lesions in the fetal lamb [1, 2, 15]. Apparent morphologic changes during the fetal period are inapparent. The clinical and pathologic changes are most apparent at birth. The clinical appearance of newborn lambs includes low birth weight which averages 1.5 kg less than control or normal lambs; a 1 cm average shorter crown-rump length and a 1.3 cm shorter tibia/radius length [1, 2]. Most of the lambs have a hairy rather than wool birthcoat. This is due to the almost exclusive predominance of primary medullated hairs rather than secondary hairs. Abnormal pigmentation often occurs. There are facial abnormalities often characterized by a slightly domed appearance to the skull. Radiographs reveal arrested growth plates in the long bones. Most affected lambs are depressed and many have varying degrees of tonic-clonic tremors. In some cases, the tonic-clonic tremors are severe and the lambs are unable to stand. Most infected lambs die in the early neonatal period.

Other morphologic and histologic changes are subtle. Rarely, hydranencephaly and cerebellar changes are observed [12]. More commonly, the neurological lesions are detected by resorting to special myelin stains [4, 12]. There is a severe hypo- or dysmyelination. Immunologically, the lambs are compromised or tolerant [25]. There are no neutralizing antibodies and the lambs are persistently viremic. Reports of perturbations in lymphocyte subpopulations with a preponderance of B-cells have been reported [10, 17]. Also there is evidence of impaired functional responses and increased susceptibility to secondary infections. Lambs that live remain viremic for life.

Virus is associated with lymphocytes. Viral shedding has been reported in the saliva, urine, feces and placental tissues and fluids.

Fluorescent antibody and immunocytochemical staining of tissue indicates a wide distribution of virus in tissues. It occurs in blood, lymphoid, nervous, thymus, and endocrine tissues. The virus is found in most cells in the nervous system and in the thyroid gland [2, 27]. In the latter, virus is found in follicular epithelium and occasionally cells in the interstitium. The rather extensive infiltration in the thyroid is associated with a significant reduction in T_3 and T_4 levels. These represent levels of 60% and 73%, respectively, of normal [2]. The hypothyroidism may play a central role in the pathogenesis of the lesions. Apparently the principal reason for the hypothyroidism can be attributed to follicular dysfunction since the pituitary secretes high levels of TSH.

It has been hypothesized that hypothyroidism plays the critical role in the pathogenesis of the series of lesions. There is evidence that T_3 and T_4 are critical for normal myelin formation [2]. These hormones are also important for maintaining basal metabolism, normal dermal function and bone growth. Although neurons, astrocytes and oligodendroglial cells may be infected with virus, there is no direct evidence that these cells are dysfunctional.

References

1. Anderson CA, Higgins RJ, Smith ME, Osburn BI (1987) Border disease: virus-induced decrease in thyroid hormone levels with associated hypomyelination. Lab Invest 57: 168–175
2. Anderson CA, Higgins RJ, Smith ME, Sawyer MM, Osburn BI (1988) Hypomyelination in border disease: the roles of thyroid hormones, growth hormone and virus infection of the CNS. In: Scarpelli E, Magaki G (eds) Transplacental effect on fetal health. Prog in Clinical and Biol Res. A. R. Liss, New York, 281: 59–70
3. Badman RT (1981) Association of bovine viral diarrhoea virus infection to hydranencephaly and other central nervous system lesions in perinatal calves. Aust Vet J 57: 306–307
4. Barlow RM, Storey IJ (1977) Myelination of the ovine CNS with special reference to border disease. In: Qualitative aspects. Neuropathol Appl Neurobiol 3: 225–265
5. Barlow RM (1980) Morphogenesis of hydranencephaly and other intracranial malformations in progeny of ewes infected with pestiviruses. J Comp Pathol 90: 87–98
6. Bolin SR, McClurkin AW, Cutlip RC, Coria MF (1985) Severe clinical disease produced in cattle persistently infected with noncytopathic bovine viral diarrhea virus by super-infection with cytopathic bovine viral diarrhea virus. Am J Vet Res 46: 573–576
7. Braun RK, Osburn BI, Kendrick JW (1973) Immunologic response of bovine fetus to bovine virus diarrhea virus. Am J Vet Res 34: 1127–1132
8. Brownlie J, Clarke MC, Howard CJ (1984) Experimental production of fatal mucosal disease in cattle. Vet Rec 114: 535–536
9. Brownlie J (1989) Bovine virus diarrhea: molecular studies. Proc 11th Int Symp, Wld Assn Vet Microbial, Immunol Spec Inf Dis, pp 253–256
10. Burrells C, Nettleton PF, Reid HW, Miller HR, Hopkins J, McConnell I, Gorrell MD,

Brandon MR (1989) Lymphocyte subpopulations in the blood of sheep persistently infected with border disease virus. Clin Exp Immunol 76: 446-451

11. Casaro A (1969) The experimental pathology of bovine virus diarrhea infection in fetal calves. Thesis, University of California, Davis

12. Clarke GL, Osburn BI (1975) Border disease-like syndrome in California lambs. Proc. 18th Annual Mtg Am Assn Vet Lab Diagnosticians, pp 303–325

13. Done JT, Terlecki S, Richardson C, Harkness JW, Sands JJ, Patterson DSP, Sweasey D, Shaw IG, Winkler CE, Duffel LJ (1980) Bovine virus diarrhea-mucosal disease virus: pathogenicity for the fetal calf following maternal infection. Vet Rec 106: 473–479

14. Edwards JF, Livingston CW, Chung SI, Collisson EC (1989) Ovine arthrogryposis and central nervous system malformations associated with in utero cache valley virus infection. Spontaneous Disease 26: 33–40

15. Hughes LE, Kershaw GF, Shaw IG (1959) "B" or border disease: an undescribed disease of sheep. Vet Rec 71: 313–317

16. Johnson RT (1982) Viral infection of the central nervous system. Raven Press, New York

17. Lamontagne L, LaFontaine P, Fournel M (1989) Modulation of the cellular immune responses to T-cell dependent and T-cell independent antigens in lambs with induced bovine viral diarrhea virus infection. Am J Vet Res 50: 1604–1608

18. MacLachlan NJ, Osburn BI, Stott JL, Ghalib HW (1985) Orbivirus infection of the bovine fetus. Barber TL, Jochim MM (eds) Bluetongue and related orbiviruses. A. R. Liss, New York, pp 79–86

19. Oldstone MBA, Sinha YN, Blount P, Tishon A, Rodreguez M, von Wedel R, Lampert PW (1982) Virus-induced alterations in homeostasis: atterations in differentiated functions in infected cells in vivo. Science 218: 1125–1126

20. Osburn BI, Johnson RT, Silverstein AM, Prendergast RA, Jochim MM, Levy SE (1971) Experimental viral-induced congenital encephalopathies. II. The Pathogenesis of Bluetongue Virus Infection of Fetal Lambs. Lab Invest 25: 206–210

21. Osburn BI (1988) Ontogeny of host defense systems and congenital infections. In: Transplacental effects on fetal health. A. R. Liss, New York (Prog Clin Biol Res 281: 15–32)

22. Osburn BI (1989) Bovine virus diarrhea: molecular studies. Proc. 11th Int. Symp. Wld Assn Vet Microbiol, Immunol Spec Inf Dis, pp 253–256

23. Parsonson IM, Della-Porta AJ, Selleck PW (1985) Pathology of akabane virus in the ovine foetus: immuno-fluorescence and histological findings. In: Della-Porta AJ (ed) Vet viral disease. Acad Press, Sydney, pp 439–442

24. Pohlenz JFL, Trautwein G, Bolin SR, Moennig V (1988) Congenital pestivirus disease (Bovine Virus Diarrhea) of Cattle. In: Scarpelli D, Wigoki G (eds) Transplacental effects on fetal health. A. R. Liss, New York (Prog Clin Biol Res 281: 49–58)

25. Sawyer MM, Schore CE, Menzies PI, Osburn BI (1986) Border disease in a flock of sheep: epidemiologic, laboratory and clinical findings. J Am Vet Med Assoc 189: 61–65

26. Schwartz LW, Osburn BI (1974) An ontogenic study of the acute inflammatory reaction in the fetal rhesus monkey. I. Cellular response to bacterial and nonbacterial Irritants. Lab Invest 31: 441–453

27. Terpstra C (1978) Detection of border disease antigen in tissues of affected sheep and in cell culture by immunofluorescence. Res Vet Sci 25: 350–355

Author's address: B. I. Osburn, School of Veterinary Medicine, University of California, Davis, California 95616, U.S.A.

Arch Virol (1991) [Suppl 3]: 79–96

The pathways for bovine virus diarrhoea virus biotypes in the pathogenesis of disease

J. Brownlie

AFRC, Institute for Animal Health, Compton Laboratory, Compton, Newbury, Berkshire, RG16 ONN, U.K.

Accepted March 14, 1991

Summary. BVDV infections of cattle ranges from the transient acute infections, which may be inapparent or mild, to mucosal disease which is inevitably fatal. On occasions the acute infections can lead to clinical episodes of diarrhoea and agalactia but as these syndromes cannot be reproduced experimentally, the pathogenesis remains unclear. The immuno-suppressive effect of acute BVDV infections can enhance the clinical disease of other pathogens and this may be an important part of the calf respiratory disease complex. Although BVDV antigen has been demonstrated within the lymphoid tissues, for prolonged periods, the evidence for viral latency remains to be proven. Venereal infection is shown to be important in the transfer of virus to the foetus and congenital infections can cause abortions, malformations and the development of persistently viraemic calves.

The two biotypes of BVDV, non-cytopathogenic and cytopathogenic, are described. Their sequential role in the pathogenesis of mucosal disease arises from the initial foetal infection with the non-cytopathogenic virus and the subsequent production of persistently viraemic calves. These calves may later develop mucosal disease as a result of superinfection with a "homologous" cytopathogenic virus and the possible origin of this biotype by mutation is discussed. Chronic disease is defined as a progressive wasting and usually diarrhoeic condition; it is suggested that this may develop following superinfection of persistently viraemic cattle with a "heterologous" cytopathogenic biotype.

Key words: Pestivirus, BVDV, pathogenesis, mucosal disease, biotypes, acute disease. chronic disease.

Introduction

Bovine virus diarrhoea virus (BVDV) is a major pathogen of cattle and contributes to the genesis of a wide variety of pathology. The clinical aspects of disease following BVDV infection reflect this pathology and range from inapparent or mild to the inevitably fatal syndrome of mucosal disease.

In the last few years, the proposal that the two biotypes of BVDV play different and, at times, interacting roles in the pathogenesis of disease has gained credence [9, 13]. These two biotypes of the virus, one of which is non-cytopathogenic [2] and the other cytopathogenic [32, 73], can be distinguished by their cytopathology in bovine cell culture. A detailed description of the pathways taken by these two viruses has recently been reviewed [14] and is now given with proposals for certain immunological mechanisms involved in the pathogenesis.

Non-cytopathogenic BVDV infections

It has on occasions been stated that the non-cytopathogenic BVD virus is less pathogenic than the cytopathogenic biotype. An examination of the epidemiology and the pathogenesis will soon reveal that the natural history and most of the pathology caused by BVDV is associated with the non-cytopathogenic virus. It is this biotype that is responsible for both the majority of in utero damage and also the maintenance of BVDV within the cattle population.

Acute disease

The majority of cattle, about 70%, have seroconverted to BVDV by 4 years of age, with evidence of only mild if not inapparent infection [36]. Detailed clinical examination may reveal a limited period of pyrexia, leucopenia and occasionally a nasal discharge following initial infection [22, 36]. The virus isolated from these acute infections is non-cytopathogenic and is usually only recoverable from blood and nasal secretions during the first 3–10 days. The antibody response arises slowly but there is, however, a slow and prolonged increase for 10–12 weeks post-infection [14].

The pathway taken by the virus during acute infection is not understood. It is likely that the initial site of replication is within the oronasal mucosa, particularly the palatine tonsil (Brownlie & Clarke, unpublished results) and from there the spread is systemic. The rapid growth of virus within epithelial cells of the oronasal mucosa may be responsible for mucosal ulceration [3] and for any subsequent salivation or nasal discharge. The systemic spread could be either from free virus in serum or from virus associated with circulating leucocytes.

Systemic spread of infection may occur as virus free in serum or as virus associated with the cells within the buffy coat fraction of blood; the lymphocytes and monocytes are generally regarded as being particularly sensitive to BVDV infection [72]. A decreased in the B-lymphocytes following acute BVDV infection has been reported [52] and a decreased T-cell responsiveness to mitogens [58]. However, these findings were not confirmed in subsequent work on a group of 4 to 6-month old calves [20]. Changes in the lymphocyte populations have also been examined by the use of fluorescent antibody assays [8]. This study has shown that during the leucopaenia following BVDV infection, there is a transient decrease in both the B and T-cell lymphocytes but there was effective recovery by 11–17 days post-infection. It would appear that such reductions in lymphocytes do not prejudice the ability of the normal young calf to recover from infection. BVDV will also multiply in non-lymphoid cells, e.g. calf kidney and calf testes cell cultures, under laboratory conditions, but there is limited information about the relevance of this for the pathogenesis of acute disease.

Although most acute infections with BVDV are mild, and this is particularly true of experimental infections, there is sufficient evidence from field reports that more severe outbreaks of disease can occur. Therefore, it would be simplistic to discount the pathogenic effect of BVDV during the course of all acute infections. It is clearly evident from clinical observations that BVDV can, under certain circumstances, cause disease. Episodes of agalactia and diarrhoea have been recorded in adult cattle from which BVDV has been identified as the causal organism [56]. The original description of disease was of a transmissible diarrhoea in adult cattle [53] and in many of the early studies severe lesions were produced [22, 72]. The reason for the severe experimental disease of yesteryear and the failure of present day researchers to reproduce it, is not clear. It may, in part, be the result of a confusion with the acute disease that leads to a clinical manifestation and mucosal disease; recently, the pathogenesis of mucosal disease was clarified [9, 15] and will be explained in detail below. The virus may also have altered in virulence over the years but this is not easy to quantify. Many of the early BVDV isolates have become laboratory-adapted (i.e. NADL) and there is recent evidence that these have become altered in antigenicity [23], perhaps by host nucleic acid insertion into the viral genome [48]. There is also the possibility of differences in tissue tropisms between field isolates, some of which may be better adapted to multiply in either the respiratory mucosa or the general systemic tissues. It is interesting to note that in a study of 21 animals, aged between 53 and 440 days, the titre of virus recovered from nasal secretions was age-related and highest in the younger animals whereas the differences in viraemia were not (Clarke, Howard and Brownlie; unpublished observations). This may help explain the variation in experimental disease when animals of different age are used. However, the pathogenesis of the severe clinical disease following acute infection will remain unresolved until it can be reproduced experimentally. The pathology of acute infections does not appear to have received attention and this may reflect both the lack of interest in examining mild infections and the failure to reproduce the severe acute disease.

An exception to this premise is a fatal condition that can follow experimental infection with the Cornell BVDV isolate CD87 [27]. This isolate was originally recovered from an outbreak of disease in a milking herd of 100 Holsteins in New York State [63]. The clinical signs included fever, agalactia and diarrhoea. Experimental infections with the CD87 BVDV isolate produced severe disease with profound thrombocytopaenia, internal haemorrhages and death in young calves. This isolate is non-cytopathogenic and could be recovered from blood for 16–40 days post-infection. The most noticeable change, that of platelet depletion, would account for the gross pathology; petechial and ecchymotic haemorrhages are visible on many of the body's mucosal surfaces [63]. It would also explain the bloody diarrhoea and the deaths in veal calves reported during outbreaks of disease with this virus in New York State. It is both a fascinating and worrying observation that this BVD virus, which normally gives mild clinical signs, can change to be so severe a pathogen. The molecular basis for this change have yet to be reported.

Congenital infections

The premature exposure of the bovine foetus to non-cytopathogenic BVDV illustrates two of the tenets required for viral persistence [51]; firstly that infection of a foetus before it is immunocompetent can lead to an immuno-tolerance which allows the virus to avoid all subsequent immune-detection and secondly that the most likely biotype to persist is non-cytopathogenic in nature. Both these factors are central to BVDV pathogenesis.

BVDV rarely infects the foetuses of sero-positive cattle. Maternal antibodies appear well able to prevent the access of virus through the placentome. Whether maternal antibody prevents the virus becoming viraemic has not been determined. It appears that the problem of in utero and congenital infections is restricted to the BVDV sero-negative dam. In these animals, foetal infection can follow from either acute or persistent viraemias.

During acute infection the virus invades the placentome, replicates and may cross to the foetus without producing lesions [24]. In sheep, BVDV has been shown to damage the maternal vascular endothelium within 10 days of infection and the resulting cellular debris is ingested by the foetal trophoblast [4]. This could be a mechanism of virus transfer from dam to offspring but may also account for the placentitis that leads to the high level of abortion following BVDV infection. It is well recorded that early embryonic death, infertility and "repeat breeder" cows are often the sequel to pestivirus infection during pregnancy [76]. In a herd infected with BVDV, the conception rates were reduced from 78.6% in the immune cows to 22.2% in infected cattle [77].

In cattle that are persistently viraemic, there is less certainty about the pathway and timing of foetal invasion because all tissues, including the uterus, are continually infected. However most, if not all, foetuses born of viraemic dams become likewise persistently viraemic. Whether infection of these foetuses occurs at the level of the germ cell or subsequent to the rupture of the zona pellucida upon implantation is still to be clarified. It has been reported that Border disease virus antigen can be found in the germinal cells of the sheep ovary [31] but reports in the cow demonstrate the failure of cytopathogenic BVDV to infect the early embryo before rupture of the zona pellucida [68]. However, this biotype fails to infect the early foetus in utero [18] and further studies with the non-cytopathic biotype are required.

Whether, following acute or persistent infection, the virus infects the foetus by either direct cell to cell transmission or systemic spread is not clear. The time taken for the passage of virus from dam to foetus is variable but it has been recorded that abortions due to BVDV can occur within 10–18 days after intramuscular infection [77]. Our own experience has shown that abortions can take place several months after foetal infection.

The outcome of foetal infection is dependent on two main variables; the age of the foetus at the time of infection and the biotype of the infecting virus. There is uncertainty about the pathogenesis of infection during the first 30 days of pregnancy. There is good evidence that BVDV will reduce the conception rate during this period [77] and that the virus will replicate freely in the maternal placenta [54]. However, there is also the view that limited transplacental infection occurs during this early stage [78] because the contact between maternal epithelium and foetal trophoblast is not sufficiently intimate for vertical transmission until the "bridge" formation at around 30 days [4, 43]. This has implications for the use of infected semen or even during embryo transfer [50].

There is little doubt that foetal infection will occur after this 30-day period and the outcome depends on whether the virus establishes during the first (up to about 110–120 days), the second (to about 180–200 days), or third trimester (to full term, about 280 days). Infection during the first two trimesters can result in abortions [33] whereas infection during the first trimester can also produce calves that remain persistently viraemic for life (see below). Calves infected during the last trimester are able to mount an active immune response [12].

The outcome of infection with the non-cytopathogenic biotype during the first and second trimesters is frequently death, abortion or mummification of the foetus [24, 39, 42]. Foetal death can follow directly from viral invasion but damage of the maternal placenta may contribute by disrupting its vascular supply of nutrients. Experimental infections during this period have shown that more than 30% of foetuses are aborted [19] but recovery of virus from aborted tissues is poor. However, experimental infection of cattle during the first trimester of pregnancy with the cytopathogenic biotype does not give abortions and there is some doubt whether this biotype can even establish in the early foetus [18].

Teratogenesis

Viruses that establish in the early foetus during organogenesis can have the distinction of causing bizarre malformations that permanently affect the animal. BVDV has a well documented teratogenic effect, in common with other nonarbo togaviruses [74]. When the lesions induced by BVDV infection are particularly severe, the foetus will die and be aborted. However, it is evident that the non-cytopathogenic biotype can replicate in the early foetus, often causing damage to selected tissues but not sufficient to cause death. Such calves are born with a variety of clinical signs that range from apparently normal to the weakly, unthrifty calf or occasionally brain-damaged calf.

The pathogenesis of this wide range of lesions is unlikely to be due to a single defect. The virus appears catholic in its choice of cell in which to replicate. It has a preference for mitotically active cells, particularly those of the central nervous system (CNS) and lymphoid tissues [6, 11, 28, 30]. Whether the pathogenic event is an inhibition of normal cell division and differentiation or due to a direct lytic action of the virus is difficult to determine. Certainly, BVDV causes significant intrauterine growth retardation in many tissues of the foetus, particularly in the CNS and the thymus [28] and a direct cytolytic effect has been suggested for the hypoplasia in the germinal layer of the cerebellum [11, 67] and other tissues [22]. Hypomyelination of the CNS, which is often associated with thymic hypoplasia, has also been observed [1, 7]. A further consistent finding within the pestiviruses is the localisation of the virus in the vascular endothelium and from the resulting vasculitis, there can be inflammation, oedema, hypoxia and cellular degeneration [74].

Persistent viraemia

Another outcome of foetal infection during the first trimester is the establishment of a viraemia that persists for life [24, 41]. The basis for this persistence is that the bovine immune system which, before 110–120 days, has not developed sufficient immunocompetence to recognise the BVDV within the foetus as foreign. When "self" antigens are recognised, soon after this 110–120 day period, the virus is accepted as a "self" tissue and there is immunotolerance. It is this immunotolerance, reflected by the lack of specific antibody to the persisting virus, that allows the virus to persist in the blood and tissues for the lifetime of the animal. It is worthy of mention that in all the recorded field and experimental data there is no evidence for persistence with the cytopathogenic biotypes; only with the non-cytopathogenic [18].

There is considerable variation in the signs and pathology described for these persistently viraemic cattle. Their identity is based on the recovery of non-cytopathogenic virus in high titre on successive occasions and the lack of antibody to the persisting virus. Their clinical appearance can range from normal to the grossly abnormal. Why some are more damaged than others can, at present, only be a speculation about the age, size, and timing of viral challenge for the early foetus. The pathogenesis of the grossly abnormal calf reflects the viral tropism for the CNS, lymphoid and epithelial cells. Within the CNS, the predilection sites for viral persistence are the cerebral cortex and the hippocampus [30]. Lesions in such tissues are often more severe when the foetus is infected during the second trimester [7, 67] and account for the depression and incoordination seen in some new-born calves. Frequently these calves fail to survive and grossly abnormal brain lesions, such as cerebellar hypoplasia [11, 28], can be seen at post-mortem.

Lesions within the lymphoid tissues, apart from the reduced size of organ, such as the thymus [28], are not so evident. The gross changes, seen in the Peyer's patches of the small intestine during mucosal disease, are not observed [15]. However, there are cellular changes that are said to account for the immunosuppression seen in persistently viraemic animals. There is a reduction in the recirculating B-cells [52] and also in T-cells [71]. There are preliminary data to show that the recirculating gamma/delta T-cells are also depressed (Howard, Clarke and Brownlie, unpublished data). It has been estimated that 4.4% of blood leukocytes, 5.4% of T-cells and 2.1% of B-cells are infected with virus [10]. Interestingly, in sheep persistently infected with Border disease virus, it was demonstrated that B-cells were significantly increased whereas the T-cells and lymphocytes expressing class I MHC antigen were decreased [21].

Several epithelial tissues sustain BVDV replication. BVDV antigen can be demonstrated within the keratinocytes of the tongue, skin and labia [5] and this may account for the erosive oral lesions which characterise clinical disease.

Venereal infections

The major interest in any BVDV invasion of the uro-genital tract is the possibility of subsequent congenital infection; this risk is greater with the persistently viraemic animal. Acute infections of the uro-genital tract of sero-negative cattle with BVDV can produce clinical disease and may be a greater cause of loss to the national herd than results from the persistently viraemic animal.

The virus can infect both ovarian and testicular tissues and can be recovered from semen of acutely infected bulls [55, 78]. The semen is often of poor quality [78] and has the potential to spread infection to sero-negative heifers [49]. However, the pathogenesis of uro-genital infection during acute disease is poorly described.

Mixed BVDV infections

A further complication of acute infections occurs when there is invasion of BVDV along with another pathogen. It has been well documented that a mixed infection of BVDV with infectious bovine rhinotracheitis virus [35, 59], bovine respiratory syncytial virus (Dr. EJ Stott, pers. comm.) or *Pasteurella haemolytica* [60] produces a more severe disease than with either pathogen alone. It is particularly interesting to note that all those dual infections mentioned above are respiratory and, therefore, it is not surprising that field surveys have implicated BVDV as a causal agent in the calf respiratory disease complex [70]. Furthermore, mixing BVDV with an enteric pathogen, such as rotavirus and coronavirus [75] or Salmonella sp. [79], has been demonstrated to exacerbate enteric disease.

The basis for the pathogenesis of mixed infections would appear to be the immunosuppression consequent on the transient leukopenia (see above) and possibly on a neutrophil dysfunction [66] following acute BVDV infections. There is also the suggestion that BVDV may stimulate the release of prostaglandins from blood mononuclear cells and that the prostaglandins in turn would depress lymphocyte blastogenesis [47]. Unfortunately, there has been, as yet, no documented research into the effect of BVDV infections on the local immune response. This would have particular relevance for respiratory and enteric infections which are essentially mucosal invasions. Epithelial cells appear to be affected during the acute phase and this may permit the establishment of these surface pathogens, thereby promoting bacteraemias following acute BVDV infections [64].

The outcome of infecting cattle, that are persistently viraemic with BVDV, with other pathogens has been reported. Infecting such animals with bovine leucosis virus reduced the ability of 4 out of 6 to make strong antibody responses as measured by immunodiffusion [65]. The pathology of these mixed infections is highly dependent on the nature of the second pathogen. In the case of *Pasteurella haemolytica*, there is a fibrinopurulent pneumonia and pleuritis [60] but from the other reports, there is a lack of descriptive pathology.

BVDV latency

An aspect of BVDV pathogenesis that has received little attention is the possibility of viral latency. Under normal conditions, the progress towards recovery following acute infections is the development of a specific neutralising antibody response and possibly a cell-mediated response. However, what is striking about this response is the slow rise in antibody over the first 10–12 weeks after infection and yet the failure to recover virus after the first 3–10 days (Fig. 1) [38].

The failure to isolate virus from either nasal swabs or blood after about day 10 is perplexing for two reasons. Firstly, there is often undetectable antibody by this stage and secondly the antibody response continues in the apparent absence of virus for a further 10–11 weeks. Explanations for this observation may be either that non-infectious virus (i.e. viral antigen) is being continually presented to the immune cells during these 10–11 weeks or that infectious virus is sequestered in lymphoid tissues. It has been shown that viral antigen is present in the macrophages within the lymphoid tissues of foetuses following experimental infection [5] and this may be a mechanism for continual stimulation of the antibody response. The persistence of viral antigen has been suggested to occur also in the ovaries of cattle for at least 60 days after intramuscular inoculation [69]. However, it is uncommon to recover infectious virus from tissues later than 14–21 days post-infection. In earlier years BVDV was reported to be present in various tissues after prolonged periods following infection; virus was isolated from mesenteric and bronchial lymph nodes on days 39 and 56 post-inoculation [50] and from blood and the nares on days 72 and 102 after infection [44]. In recent times, this has not been observed and may reflect the closer attention now paid to adventitious virus infecting experimental calves or cell cultures. Experiments to attempt the recrudescence of virus from convalescent animals have not been reported but a summary of our present knowledge suggests there is little evidence for BVDV latency.

Cytopathogenic BVDV infections

The distinction of these two biotypes of the same virus is made more relevant and more curious by their separate pathways in the pathogenesis of disease.

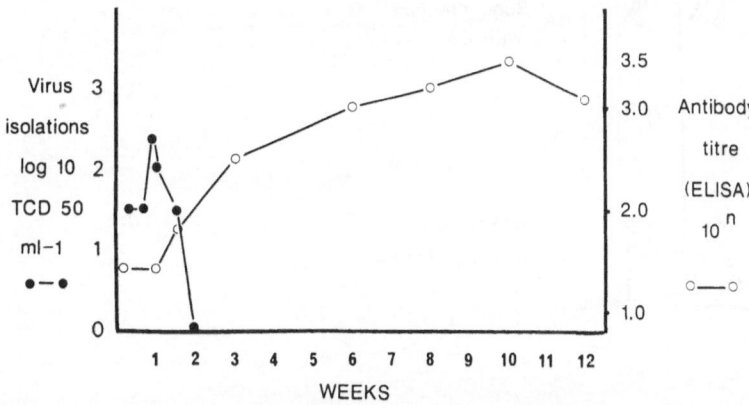

Fig. 1. Typical virus detection and specific antibody response following acute BVDV infection of calves

In fact, the syndrome first associated with the cytopathogenic virus was considered to be an entirely new and different disease from bovine virus diarrhoea and accorded a new name—*Mucosal disease* [61].

Mucosal disease

Mucosal disease was first reported in 1953 and described as a fatal condition of cattle, characterised by severe erosive lesions in the intestinal mucosa [61]. The virus isolated from this condition was understandably called mucosal disease virus (MDV) but reinoculation into cattle did not reproduce the fatal mucosal disease. Later, it was demonstrated that BVDV and MDV were serologically similar and gave the same mild illness in response to acute infection [34]. Over the next thirty years a series of observations were made about these viruses, reviewed by Brownlie [13] and most of which have been identified above. Finally, two significant findings led to the proposal for the aetiology of mucosal disease. The first was that only persistently viraemic animals succumbed to mucosal disease [45]. Secondly, that persistently viraemic animals had only the non-cytopathogenic virus whereas those that died of mucosal disease were infected with both biotypes, non-cytopathogenic and cytopathogenic [15].

These observations were refined into a hypothesis [15], Fig. 2, and requires that cattle, sero-negative to the virus, become infected during early pregnancy with the non-cytopathogenic biotype. The virus, infecting the dam, transfers across the placenta to the foetus and can result in the birth of a persistently viraemic calf, due to the immunotolerance described above.

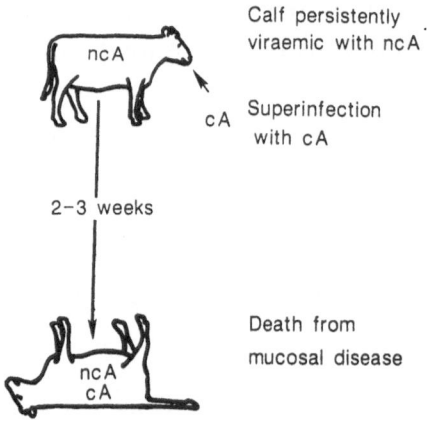

Calf persistently
viraemic with ncA

Superinfection
with cA

2–3 weeks

Death from
mucosal disease

nc = non-cytopathogenic BVDV

c = cytopathogenic BVDV

Fig. 2. Pathogenesis of mucosal disease

Some time after birth, usually when the animal is 6–18 months of age, superinfection of these viraemic animals with the cytopathogenic biotype may occur. This results in the rapid development of fatal mucosal disease.

The truth of this hypothesis was demonstrated by the experimental production of mucosal disease in exactly the manner predicted [9, 15]. Subsequently, the need for antigenic "homology" between the persisting and superinfecting biotypes has been recognised as crucial to the development of mucosal disease and has illustrated the precision of the immunotolerance [16]. The origin of the cytopathogenic virus continues to fascinate us; it has been suggested that this may arise by a molecular event, such as mutation, from the persisting non-cytopathogenic virus [16, 37].

Mucosal disease has been described as a sequel to vaccination with modified-live virus [46]. These vaccines are mostly derived from cyto-pathogenic virus strains and our hypothesis would suggest that such vaccines may be acting as the superinfecting challenge.

The pathology of mucosal disease has been described, particularly the characteristic erosive lesions seen in the gut lymphoid tissues [40]. It has now been demonstrated that the cytopathogenic biotype has a particular tropism for the gut-associated lymphoid tissue [25] and that, in the sequential development of lesions, this biotype rapidly homes to the Peyer's patch tissue [17]. Although, the reason for the gross lesions visible over the Peyer's patches is likely to be a result of the direct lytic action of the virus, there is now preliminary evidence that a synergism between the two biotypes is required for the full expression of mucosal disease [26].

Prominent in mucosal disease is the lesion that develops after the destruction of the lymphoid tissue in the Peyer's patches and the collapse of its overlying intestinal mucosa. This gives the characteristic erosions along the small intestine. The ultimate question for pathologists, the cause of death, has a less certain answer. In studies on the sequential development of mucosal disease it was evident that these Peyer's patch lesions occurred both late and rapidly in the course of superinfection with cytopathogenic virus and coincided with clinical disease [17]. However, there were animals that died without clinical diarrhoea and this would suggest that the diarrhoea and any resulting dehydration was not an essential part of the syndrome. Whether in the destruction of the lymphoid tissue, there is a release of toxic factors or an excess of inflammatory products has not been assessed. The cause of death still remains an enigma.

Chronic BVDV disease

The aspect of BVDV pathogenesis, above all others, that gives rise to confusion is the "chronic" disease. It is used to describe not only animals that are continually or "chronically" infected with BVDV but also animals that are clinically ill for prolonged periods. A clearer definition is required nowadays in order to make progress in its understanding. There is no doubt that a clinical entity exists where cattle become progressively cachectic following long lasting bouts of diarrhoea. The course of disease may be several weeks or months and ultimately the animal dies or it may apparently recover. There is little disagreement that this could be described as chronic disease. From some field cases of this syndrome, non-cytopathogenic BVDV has been isolated. Should this biotype be isolated on repeated occasions, in absence of specific antibody to the virus, then the animal would accurately be described as persistently viraemic. The problem of definition arises from the knowledge that all persistently viraemic animals do not have a chronic clinical disease and that the syndrome has not been experimentally reproduced.

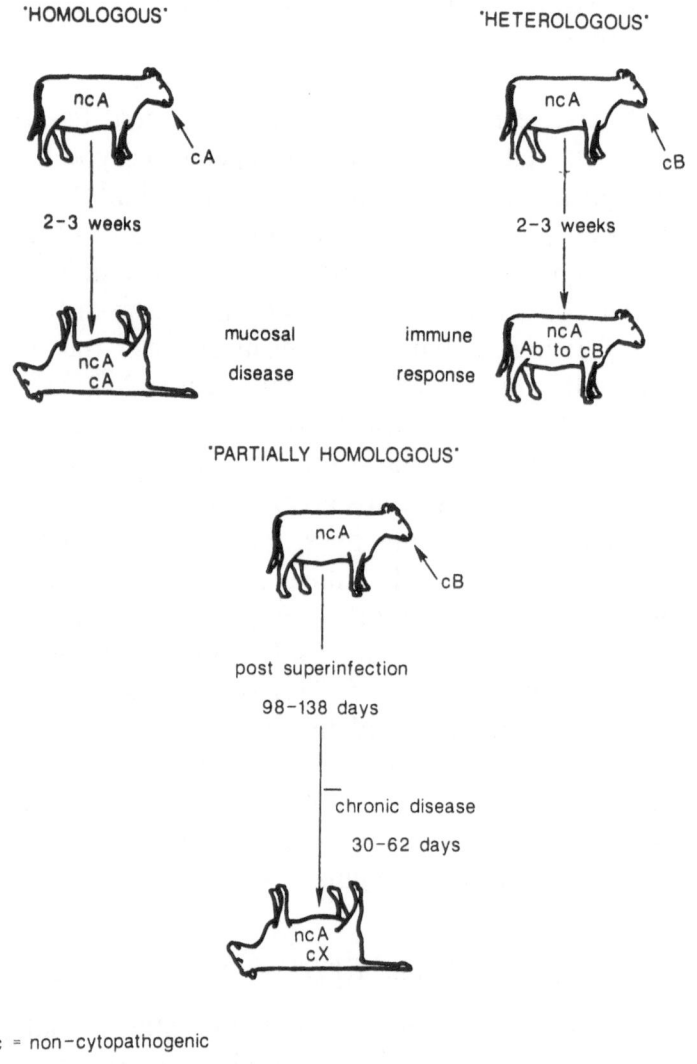

nc = non-cytopathogenic

c = cytopathogenic

Fig. 3. Pathogenesis of chronic disease

In 1985, we suggested that chronic disease may occur as a result of superinfection of persistently viraemic animals with cytopathogenic isolates that have partial antigenic "homology" with the persisting virus [16]. Our understanding of partial "homology" was not defined but was intended to indicate an antigenic position between "homology" and "heterology", as shown in Fig. 3. "Homology" was conferred by the immunotolerance which allows the non-cytopathogenic virus to persist and also enables the super-infecting cytopathogenic virus to survive. The failure of the immune system to recognise the superinfecting cytopathogenic virus was taken to mean antigenic homology. It is now understood that the two biotypes isolated from cases of fatal mucosal disease are antigenically similar [37, 57]. Furthermore, it has been shown that those cytopathogenic viruses that are antigenically different, i.e. against which antibody is made, do not cause mucosal disease. These can be said to be "heterologous".

Our proposal for chronic disease suggested that immunotolerant animals were superinfected with cytopathogenic isolates that were "hetero-logous" but only partially recognised by the immune system and therefore not eliminated. Recently, we have superinfected a series of persistently viraemic cattle with 'heterologous' cytopathogenic virus [19]. In two animals there was, several months after superinfection, a slow onset of disease with bouts of diarrhoea and progressive wasting. During this period, both animals preferred roughage and refused 'concentrate' food. They had periods of several days when their condition appeared to ameliorate, only to relapse into further clinical disease. The animals were finally killed, and at post-mortem, gut erosive lesions typical of mucosal disease were revealed. These preliminary experiments may be the first to reproduce chronic disease. The pathology of field and experimental chronic disease has not been described.

Conclusions

Few viruses appear to have so diverse and complex a pathogenesis as the ruminant pestiviruses. Our recent understanding only serves to highlight the ingenious ways in which the virus subverts the hosts defences. Like certain other viruses that are able to persist for long periods, it invades the lymphoid tissue and disables the recognition and effector functions of immunity. In his book, Mims examines the mechanisms of viral persistence [51] and, from this, it is evident that BVDV is the pathogen par excellence for persistence [17]. The clear definition we are now able to give to the biological roles of the two biotypes has opened up new areas in our immunological and molecular biology research. It is possible that the apparent unique pathogenesis of mucosal disease will later be shown to be in the vanguard of new understanding of other infections.

The emergence of the separate biological roles for the biotypes has now given greater impetus to understanding anew the molecular basis of this virus [27, 29] and its pathogenesis.

References

1. Anderson CA, Higgins RJ, Smith ME, Osburn BI (1987) Border disease Virus-induced decrease in thyroid hormone levels with associated hypomyelination. Lab Invest 57: 168–175
2. Baker JA, York CJ, Gillespie JH, Mitchell GB (1954) Viral diarrhea in cattle. Vet Res 15: 525–531
3. Baker JC (1987) Bovine viral diarrhea virus: a review. J Am Vet Med Assoc 190: 1449–1458
4. Barlow RM (1972) Experiments in border disease IV. Pathological changes in ewes. J Comp Pathol 72: 151–157
5. Bielefeldt Ohmann H (1983) Pathogenesis of bovine viral diarrhoea–mucosal disease: distribution and significance of BVDV antigen in diseased calves. Res Vet Sci 34: 5–10
6. Bielefeldt Ohmann H (1988) BVD virus antigens in tissues of persistently viraemic, clinically normal cattle: implications for the pathogenesis of clinically fatal disease. Acta Vet Scand 29: 77–84
7. Binkhorst GJ, Journee DLH, Wouda W, Straver PJ, Vos JH (1983) Neurological disorders, virus persistence and hypomyelination in calves due to intra-uterine infections with bovine virus diarrhoea virus. Vet Q 5: 145–155
8. Bolin SR, McClurkin AW, Coria MF (1985) Effects of bovine viral diarrhoea virus on the percentages and absolute numbers of circulating B and T lymphocytes in cattle. Am J Vet Res 46: 884–886
9. Bolin SR, McClurkin AW, Cutlip RC, Coria MF (1985) Severe clinical disease induced in cattle persistently infected with noncytopathic bovine viral diarrhea virus by super-infection with cytopathic bovine viral diarrhea virus. Am J Vet Res 46: 573–576
10. Bolin SR, Sacks JM, Crowder SV (1987) Frequency of association of non-cytopathic bovine viral diarrhea virus with mononuclear leukocytes from persistently infected cattle. Am J Vet Res 48: 1441–1445
11. Brown TT, de Lahunta A, Scott FW, Kahr RF, McKentee K, Gillespie JH (1973) Virus induced congenital anomalies of the bovine fetus. II Histopathology of cerebellar degeneration (hypoplasia) induced by the virus of the bovine viral diarrhea–mucosal disease. Cornell Vet 63: 561–578
12. Brown TT, Schultz AD, Duncan JR, Bistner SI (1979) Serological response of the bovine fetus to bovine viral diarrhea virus. Infect Immun 25: 93–97
13. Brownlie J (1990) Pathogenesis of mucosal disease and molecular aspects of bovine virus diarrhoea virus. Vet Microbiol 23: 371–382
14. Brownlie J (1990) The pathogenesis of bovine virus diarrhoea virus infections. Rev Sci Tech Off Jnt Epiz 9: 43–59
15. Brownlie J, Clarke MC, Howard CJ (1984) Experimental production of fatal mucosal disease in cattle. Vet Rec 114: 535–536
16. Brownlie J, Clarke MC, Howard CJ (1987) Clinical and experimental mucosal disease–defining a hypothesis for pathogenesis. In: Harkness JW (ed) Pestivirus infections of ruminants. CEC Seminar, Brussels, September 1985, pp 147–157
17. Brownlie J, Clarke MC, Howard CJ (1988) Mucosal disease—sequential studies on the

infectivity of bovine virus diarrhoea (BVD) virus on the gut associated lymphoid tissue. 15th World Buiatrics Congress, Palma, Spain. Proceedings II, pp 899–904

18. Brownlie J, Clarke MC, Howard CJ (1989) Experimental infection of cattle in early pregnancy with a cytopathic strain of bovine virus diarrhoea virus. Res Vet Sci 46: 307–311

19. Brownlie J, Clarke MC, Howard CJ, Pocock DH (1986) Mucosal disease: the dilemma of experimental disease. 14th World Congress of Cattle Diseases, Dublin, Ireland, 1: 199–203

20. Brownlie J, Nuttall PA, Stott EJ, Taylor G, Thomas LH (1980) Experimental infection of calves with two strains of bovine virus diarrhoea virus: certain immunological reactions. Vet Immunol Immunopathol 1: 371–378

21. Burrells C, Nettleton PF, Reid HW, Miller HRP, Hopkins J, McConnell I, Gorrell MD, Brandon MR (1989) Lymphocyte subpopulations in the blood of sheep persistently infected with Border disease virus. Clin Exp Immunol 76: 446–451

22. Carlson RG, Pritchard WR, Doyle LP (1957) The pathology of the virus diarrhea of cattle in Indiana. Am J Vet Res 18: 560–568

23. Cay B, Chappuis G, Coulibaly C, Dinter Z, Edwards S, Greiserwilke I, Gunn M, Have P, Hess G, Juntti N, Liess B, Mateo A, McHugh P, Moennig V, Nettleton P, Wensvoort G (1989) Comparative analysis of monoclonal antibodies against pestiviruses: Report of an international workshop. Vet Microbiol 20: 123–129

24. Casaro APE, Kendrick JW, Kennedy JW (1971) Response of the bovine fetus to bovine viral diarrhea–mucosal disease virus. Am J Vet Res 32: 1543–1562

25. Clarke MC, Brownlie J, Howard CJ (1987) Isolation of cytopathic and non-cytopathic bovine viral diarrhoea virus from the tissues of infected animals. In: Harkness JW (ed) Pestivirus infections of ruminants. CEC Seminar, Brussels, September 1985, pp 3–12

26. Clarke MC, Brownlie J, Howard CJ (1989) The effect of immunosuppression with corticosteroid on the infection of calves with bovine virus diarrhoea virus. Immunobiol 4: 151–152

27. Collett MS, Moennig V, Horzinek MC (1989) Recent advances in Pestivirus research. J Gen Virol 70: 253–266

28. Done JT, Terlecki S, Richardson C, Harkness JW, Sands JJ, Patterson DSP, Sweasey D, Shaw IG, Winkler CE, Duffell SJ (1980) Bovine virus diarrhoea–mucosal disease virus: pathogenicity for the fetal calf following maternal infection. Vet Rec 106: 473–479

29. Donis RO, Dubovi EJ (1987) Characterization of bovine viral diarrhoea–mucosal disease virus-specific proteins in bovine cells. J Gen Virol 68: 1597–1605

30. Fernandez A, Hewicker M, Trautwein G, Pohlenz J, Liess B (1989) Viral antigen distribution in the central nervous system of cattle persistently infected with bovine viral diarrhoea virus. Vet Pathol 26: 26–32

31. Gardiner AC (1980) The distribution and significance of border disease viral antigen in infected lambs and foetuses. J Comp Pathol 91: 467–470

32. Gillespie JH, Baker JA, McEntee K (1960) A cytopathogenic strain of virus diarrhoea virus. Cornell Vet 50: 73–79

33. Gillespie J, Bartholomew P, Thomson R, McEntee K (1967) The isolation of non-cytopathic virus diarrhoea virus from two aborted fetuses. Cornell Vet 57: 564–571

34. Gillespie J, Coggins L, Thompson J, Baker JA (1961) Comparison by neutralisation tests of strains of virus isolated from virus diarrhoea and mucosal disease. Cornell Vet 51: 155–159

35. Grieg A, Gibson IR, Nettleton PF, Herring JA (1981) Disease outbreak in calves caused by a mixed infection with infectious bovine rhinotracheitis virus and bovine virus diarrhoea virus. Vet Rec 108: 480

36. Harkness JW, Sands JJ, Richards MS (1978) Serological studies of mucosal disease in England and Wales. Res Vet Sci 24: 98–103

37. Howard CJ, Clarke MC, Brownlie J (1987) Comparisons by neutralisation assays of pairs of non-cytopathogenic and cytopathogenic strains of bovine virus diarrhoea virus isolated from cases of mucosal disease. Vet Microbiol 13: 361–369

38. Howard CJ, Clarke MC, Brownlie J (1989) Protection against respiratory infections with bovine virus diarrhoea virus by passively acquired antibody. Vet Microbiol 19: 195–203

39. Huck RA (1957) A mucosal disease of cattle. Vet Rec 69: 1207–1215

40. Jubb KVF, Kennedy PC, Palmer N (1985) Pathology of domestic animals. 3rd ed, 2: 95–100

41. Kahrs RF (1973) Effects of bovine viral diarrhea on the developing fetus. J Am Vet Med Assoc 163: 877–878

42. Kendrick JW (1971) Bovine viral diarrhea–mucosal disease virus infection in pregnant cows. Am J Vet Res 32: 533–544

43. Kendrick JW (1976) Bovine viral diarrhea virus induced abortion. Therio 5: 91–93

44. Lambert G, Fernelius AL (1968) Bovine viral diarrhea virus and *Escherichia coli* in neonatal calf enteritis. Can J Comp Med 32: 440–446

45. Liess B, Frey H-R, Kittsteiner H, Baumann F, Neumann N (1974) Beobachtungen und Untersuchungen über die "Mucosal disease" des Rindes. Dtsch Tierärztl Wochenschr 81: 477–500

46. Liess B, Orban S, Frey H-R, Trautwein G, Wiefel A, Bindow H (1984) Studies on transplacental transmissibility of a bovine virus diarrhea (BVD) vaccine in cattle. II. Inoculation of pregnant cows without detectable neutralising antibodies to BVD virus 90–229 days before parturition (51st to 190th day of gestation). Zentralbl Veterinarmed (B) 31: 669–681

47. Markham RJF, Ramnaraine ML (1985) Release of immunosuppressive substances from tissue culture cells infected with bovine viral diarrhea virus. Am J Vet Res 46: 879–883

48. Meyers G, Rumenapf T, Thiel H-J (1989) Ubiquitin in a togavirus. Nature 341: 491

49. Meyling A, Jensen AM (1988) Transmission of bovine virus diarrhoea virus (BVDV) by artificial insemination (AI) with semen from a persistently-infected bull. Vet Microbiol 17: 97–105

50. Mills JHL, Luginbuhl RE (1968) Distribution and persistence of mucosal disease virus in experimentally exposed calves. Am J Vet Res 29: 1367–1375

51. Mims C (1987) The pathogenesis of infectious disease. Academic Press

52. Muscoplat CC, Johnson DW, Teuscher E (1973) Surface immunoglobulin of circulating lymphocytes in chronic bovine diarrhea of cattle: abnormalities in cell populations and cell function. Am J Vet Res 34: 1101–1104

53. Olafson P, MacCallum AD, Fox FH (1946) An apparently new transmissible disease of cattle. Cornell Vet 36: 205–213

54. Parsonson IM, O'Halloran ML, Zee YC, Snowdon WA (1979) The effect of bovine viral diarrhoea–mucosal disease (BVD) virus on the ovine foetus. Vet Microbiol 4: 279–292

55. Paton DJ, Goodey R, Brockman S, Wood L (1989) Evaluation of the quality and virologic status of the semen from bulls acutely infected with BVDV. Vet Rec 124: 63

56. Perdrizet JA, Rebhun WC, Dubovi EJ, Donis RO (1987) Bovine virus diarrhea–clinical syndromes in dairy herds. Cornell Vet 77: 46–74

57. Pocock DH, Howard CJ, Clarke MC, Brownlie J (1987) Variation in the intracellular polypeptide profiles from different isolates of bovine virus diarrhoea virus. Arch Virol 49: 43–53

58. Pospisil Z, Machatkova M, Mensik J, Rodak I, Muller G (1975) Decline in the phytohaemagglutinin responsiveness of lymphocytes of calves infected experimentally

with bovine virus diarrhoea–mucosal diarrhoea virus and parainfluenza 3 virus. Acta Vet Brno 44: 369–375

59. Potgieter LND, McCracken MD, Hopkins FM, Walker RD (1984) Effect of bovine viral diarrhea virus infection on the distribution of infectious bovine rhinotracheitis-virus in calves. Am J Vet Res 45: 687–690

60. Potgieter LND, McCracken MD, Hopkins FM, Walker RD, Guy JS (1984) Experimental production of bovine respiratory disease with bovine diarrhea virus. Am J Vet Res 45: 1582–1585

61. Ramsey FK, Chivers WH (1953) Mucosal disease of cattle. North Am Vet 34: 629–633

62. Ramsey FK, Chivers WH (1957) Symposium on the mucosal disease complex 11 Pathology of a mucosal disease of cattle. J Am Vet Med Assoc 130: 381–383

63. Rebhun WC, French TW, Perdrizet JA, Dubovi EJ, Dill SG, Karcher LF (1989) Thrombocytopenia associated with acute bovine virus diarrhea infection in cattle. J Vet Intern Med 3: 42–46

64. Reggiardo C, Kaeberle ML (1981) Detection of bacteremia in cattle inoculated with bovine viral diarrhea virus. Am J Vet Res 42: 218–221

65. Roberts DH, Lucas MH, Wibberley G, Westcott D (1988) Response of cattle persistently infected with bovine virus diarrhoea virus to bovine leukosis virus. Vet Rec 122: 293–296

66. Roth JA, Kaeberle ML, Griffith RW (1981) Effects of BVDV on bovine polymorpho-nuclear leukocyte function. Am J Vet Res 42: 24–250

67. Scott FW, Kahrs RF, de Lahunta A, Brown TT, McEntee K, Gillespie JH (1973) Virus induced congenital abnormalities of the bovine foetus I cerebellar degeneration (hypoplasia), ocular lesions and fetal mummifications following experimental infection with bovine viral diarrhea–mucosal disease. Cornell Vet 63: 536–5

68. Singh EL, Eaglesome MD, Thomas FC, Papp-Vid G, Hare WCD (1982) Embryo transfer as a means of controlling the transmission of viral infections. 1. The in vitro exposure of the preimplantation bovine embryo to Akabane, bluetongue, and bovine viral diarrhea viruses. Therio 17: 437–444

69. Ssentonga YK, Johnson RH, Smith JR (1980) Association of bovine viral diarrhoea-mucosal disease virus with ovaritis in cattle. Aust Vet J 56: 272–273

70. Stott EJ, Thomas LH, Collins AP, Jebett NJ, Smith GS, Luther PD, Caswell R (1980) A survey of virus infections of the respiratory tract of cattle and their association with disease. J Hyg 85: 257–270

71. Truitt RL, Schchmeister IL (1973) The replication of the bovine viral diarrhea–Mucosal Disease virus in bovine leukocytes in vitro. Arch Ges Virusforsch 42: 78–87

72. Tyler DE, Ramsey FK (1965) Comparative pathologic, immunologic and clinical responses produced by selected agents of the bovine mucosal disease virus diarrhea complex. Am J Vet Res 26: 903–913

73. Underdahl NR, Grace OD, Hoerlein AB (1957) Cultivation in tissue-culture of cytopathogenic agent from bovine mucosal disease. Proc Soc Exp Biol Med 94: 795–797

74. Van Oirschot JT (1983) Congenital infections with nonarbo togaviruses. Vet Microbiol 12: 14–18

75. Van Openbosch E, Wellemans G, Oudewater J (1981) Interaction of BVDV, corona and rotavirus in neonatal calf diarrhea: experimental infections in newborn calves. Vlaams Diergeneesk. Tijdschr 50: 163–173

76. Virakul P, Fahning ML, Joo HS, Zemjanis R (1988) Fertility of cows challenged with a cytopathic strain of bovine virus diarrhoea virus during an outbreak of spontaneous infection with a noncytopathic strain. Therio 29: 441–449

77. Whitmore HL, Zemjanis R, Olson J (1981) Effect of bovine viral diarrhea virus on conception in cattle. J Am Vet Med Assoc 178: 1065–1067
78. Whitmore HL, Gustafsson BK, Hauareshti P, Duchateau AB, Mather EC (1978) Inoculation of bulls with bovine virus diarrhea virus: excretion of virus in semen and effects on semen quality. Therio 9: 153–169
79. Wray C, Roeder PL (1987) Effect of bovine virus diarrhoea–mucosal disease infection on salmonella infection in calves. Res Vet Sci 42: 213–218

Author's address: J. Brownlie, AFRC, Institute for Animal Health, Compton Laboratory, Compton, Newbury, Berkshire, RG16 ONN, U.K.

Arch Virol (1991) [Suppl 3]: 97–100

Border disease of sheep—
Aspects for diagnostic and epidemiologic consideration

M. M. Sawyer, C. E. Schore, and **B. I. Osburn**

Department of Veterinary Pathology, University of California, Davis, USA

Accepted March 14, 1991

Summary. Border Disease (BD) is a condition of newborn sheep that results from congenital infection by a non-cytopathic pestivirus, occurring during the first half of gestation. The variations in expression of the virus directly relate to the age of the fetus at the time of infection. There are four distinct disease syndromes: (1) early embryonic death, (2) abortion and stillbirth, (3) birth of lambs with malformations, and (4) birth of small, weak lambs, lacking characteristic clinical signs, but bearing features of immunosuppression. In the newborn, the BD virus may be recovered from all tissues and teratogenic lesions are found in the endocrine, nervous, skeletal, integumentary and immune systems. These effects of virus infection are manifest in the clinical signs characteristic of the disease, such as tremors, ataxia, hairy birthcoat, low birth weight, facial bone malformations, short-boxy stature, and eye abnormalities. The consequences of the BD compromised immune system is an increased susceptibility to infection, a failure to produce specific antibody to BD virus, and an inability to clear the virus; features characteristic of the immuno-tolerant state.

The lifelong shedding and persistence of virus is of epidemiologic importance. The persistently infected BD ewe remains a source of infection for the flock both through horizontal transmission (virus shedding) and congenital transmission (a persistently infected ewe will always bear a BD lamb). Detection of persistently infected individuals within a flock is difficult: clinical signs abate with time and most frequently no antibody to BD is produced.

Key words: Border disease, bovine virus diarrhea virus, congenital hypothyroidism, congenital virus infection, hairy shaker disease, hypomyelination, immunotolerance, pestivirus, sheep, virus persistence.

Border Disease (BD) is a condition of newborn lambs caused by a congenital infection with a non-cytopathic (CPE) pestivirus of the togavirus family. BD virus cross reacts serologically with the viruses of hog cholera and bovine virus diarrhea (BVD) [7]. Serum neutralization and ELISA [4] tests using the cytopathic strains of BVD virus detect antibodies to BD virus. Antibodies produced against BVD virus can detect the presence of BD virus in tissues and cell culture. Conversely, antibodies to BD will detect the presence of BVD virus [8]. Monoclonal antibody reactivity indicates that the BD virus is closely related to non-cytopathic strains of BVD virus [5]. An 80 kDa polypeptide associated with the cytopathic strains of BVD is lacking in BD and provides additional evidence for a close relationship of non-CPE BVD and BD [1]. Several strains of BD may exist which differ only in pathogenesis and tissue tropism.

BD virus in adolescent and adult sheep is mild and frequently goes undetected, but infection of the fetus results in newborn lambs with a wide spectrum of clinical signs. Most dramatic are the tonic-clonic tremors. The tremors are frequently so severe as to impede nursing. The hairy birth coat, the other hallmark of "hairy shaker disease", is also striking and is present in the majority of cases. The most consistent finding however, is ataxia: weak lambs that are unable to stand. With care they may survive the immediate postnatal period, however, they will never acquire their normal potential in health, weight and stature. A frequently observed characteristic of newborn BD lambs is low birth weight. They are obviously smaller than normal lambs. This is especially apparent if the BD lamb has a normal littermate. The adolescent lamb exhibits a boxy stature which develops because, while longitudinal growth is normal, height is restricted due to the inhibited growth of the long bones. Facial malformations also exist. A "doming" of the cranium and short longitudinal axis of the head are observed. Eye abnormalities, such as microphthalmia and cataracts have been documented. These clinical signs can be present singly or in combination and in varying degrees of severity. The incidence of clinical BD can reach 50% in newborn lambs. Clinical signs vary. None are individually pathognomic for border disease, but when combined with a flock history of stillbirths and illthrift, the possibility of BD should be considered. Serology is of epidemiologic value to establish a flock or area prevalence but is of limited diagnostic value. Horizontally transmitted infection precedes overt signs by several months. Hence, paired samples would not differ. In the case of the persistently infected ewe giving birth to a BD lamb, neither will have detectable antibody. Isolation and identification of the virus and flock history are the only definitive means of diagnosis of border disease.

The variations observed in the expression of the BD virus relate directly to the age of the fetus at the time of infection. Prior to 16 days, the zygote is refractory to infection. After 90 days, the fetus is able to overcome the infection with no detectable evidence of infection. Infection occurring within this "window of susceptibility" (between 16 and 90 days) produces four distinct disease syndromes: (1) early embryonic death, (2) abortion and stillbirth (3) birth of lambs with malformations, and (4) birth of small, weak lambs lacking characteristic clinical signs but which are immunosuppressed. In California, breeding is generally restricted to two to three weeks. Therefore all ewes in the flock will be in the same stage of gestation at the time the infection is introduced and the pattern of disease will be consistent within the flock. If infection is introduced early in the susceptible period, the prevalent clinical sign will consist of open ewes. Infection occurring between 45 and 80 days of pregnancy results in a large proportion of affected lambs with hairy fleece and clonic-tonic tremors. A variation, often encountered, is a difference in clinical signs and degrees of disease severity among littermates. One sibling can exhibit severe clinical signs and another be normal in every respect. Differences in pathogenicity between strains of BD virus and differences in dose and route of infection also produce variations in disease expression.

In the newborn lamb, the BD virus can be recovered from all tissues. Teratologic lesions are found in the endocrine, nervous, skeletal, integumentary and immune systems. These effects are manifest in the clinical signs characteristic of the disease. Infection of the thyroid has diverse ramifications. T3 and T4 thyroid hormone levels have been reported to be significantly lower at birth [2] and at 10 weeks of age [9]. T3 and T4 deficiency adversely effects amount of 2′,3′-cyclic nucleotide-3′-phosphodiesterase (CNP), an enzyme essential for normal myelination. CNP deficiency results in hypomyelination, causing tremors. Deficient thyroid function disrupts normal development of the skeletal and integumentary systems. The results of which are retarded intrauterine growth and hyperplasia of the primary hair follicles which produces the hairy birth coat.

The most subtle and pervasive effects of infection occur in the immune system. Leukocyte numbers remain within the normal range but there is a shift in lymphocyte sub-populations [3], and a depressed lymphocyte response to phytolectins [10] indicating a functional defect. The compromised immune system results in an increased susceptibility to infection, a failure to produce specific antibody to BD virus, and an inability to clear the virus, all typical of an immunotolerant state.

The lifelong shedding and persistence of virus is of epidemiologic importance. The persistently infected BD ewe remains a source of infection for the flock both through horizontal transmission (virus shedding) and congenital transmission (a persistently infected ewe will always bear a BD lamb). Over a period of time, outbreaks of BD due to congenital transmission are distinguished by a low disease incidence in the flock and the repeated occurrence in the progeny of the same (persistently infected) ewes. On the other hand, outbreaks due to introduction of persistently infected replacement ewes into a susceptible flock (at the critical time of pregnancy) are distinguished by a high incidence of BD in the lamb crop, all showing a similar disease syndrome pattern. Control consists of identification and elimination of the persistently infected animals within the flock. Detection of these individuals can be difficult: clinical signs abate with age and there is no antibody to BD virus. Isolation of BD virus from lymphocytes provides the only definitive identification of persistently infected animals. Antigen capture ELISA [6] and polymerase chain reaction (PCR) technology offer potential as tools of epidemiologic importance.

References

1. Akkina R, Raisch KP (1990) Intracellular virus induced polypeptides of pestivirus border disease virus. Virus Res 16: 95–106
2. Anderson CA, Higgins RJ, Smith ME, Osburn BI (1987) Virus induced decrease in thyroid hormone levels with associated hypomyelination. Lab Invest 57: 168–175
3. Burrells C, Nettleton PF, Reid HW, Miller HRP, Hopkins J, McConnell I, Gorrell MD, Brandon MR (1989) Lymphocyte subpopulations in the blood of sheep persistently infected with border disease virus. Clin Exp Immunol 76: 446–451
4. Chu S, Sawyer M, Anderson C, Higgins R, Zee YC (1987) Enzyme-linked immunosorbent assay for the detection of antibodies to bovine virus diarrhea virus in sera from border disease virus–infected sheep. Can J Vet Res 51: 281–283
5. Corapi L, Donis R, Dubovi EJ (1991) Characterizing a panel on monoclonal antibodies and its use in the study of the antigenic diversity of bovine viral diarrhea virus. Am J Vet Res 51: 1388–1394
6. Fenton A, Entrican JA, Nettleton PF (1990) An ELISA for detecting pestivirus antigen in the blood of sheep persistently infected with border disease virus. J Virol Methods 27: 253–260
7. Osburn BI, Clarke GL, Stewart WC, Sawyer M (1973) Border disease-like syndrome in lambs: antibodies to hog cholera and bovine virus diarrhoea viruses. J Am Vet Med Assoc 163: 1165–1167
8. Potts BJ, Sawyer M, Shekarchi IC, Wismer T, Huddleston D (1989) Peroxidase-labeled primary antibody method for detection of pestivirus contamination in cell cultures. J Virol Methods 26: 119–124
9. Sawyer, unpublished findings
10. Sawyer M, Schore C, Menzies PI, Osburn BI (1986) Border disease in a flock of sheep: epidemiologic, laboratory, and clinical findings. J Am Vet Med Assoc 189: 61–65

Author's address: Dr. M. M. Sawyer, Department of Veterinary Pathology, University of California, Davis, CA 95616, U.S.A.

Arch Virol (1991) [Suppl 3]: 101–108

A study of some pathogenetic aspects of bovine viral diarrhea virus infection

G. Castrucci[1], **F. Frigeri**[1], **B. I. Osburn**[2], **M. Ferrari**[3], **M. M. Sawyer**[2], and **V. Aldrovandi**[4]

[1] Institute of Infectious Diseases, "Vittorio Cilli" Laboratory of Virology, School of Veterinary Medicine, University of Perugia, Italy
[2] Department of Pathology, School of Veterinary Medicine, University of California, Davis, U.S.A.
[3] C/O Zooprophylaxis Institute, Brescia, Italy, and
[4] Center for Weaning of Calves, Tripoli S. Giorgio, Mantova, Italy

Accepted March 14, 1991

Summary. The cytopathic (CP) strain TVM-2 of bovine virus diarrhea virus (BVDV) induced in calves a severe disease, whereas the calves inoculated with the non-cytopathic (NCP) New York-1 strain, remained clinically normal. When calves were immunosuppressed with dexamethasone (DMS) they underwent an overt, generally fatal disease. This result was obtained with either the CP and the NCP strain of BVDV. It was speculated that the immunosuppressive activity of BVDV could be a property peculiar to certain isolates of the virus.

Key words: Bovine viral diarrhea virus, pathogenesis, experimental infection, calves.

Introduction

It is generally known that there are 2 distinct biotypes of Bovine Viral Diarrhea Virus (BVDV): one is readily cytopathic (CP) the other is not. Several studies have suggested an explanation for the mechanism by which the two biotypes can induce fatal disease [6]. It begins with an early transplacental infection of the fetus with NCP BVDV prior to the complete maturation of the fetal immune system and the animals that eventually survive fetal infection might become persistently viremic with no neutralizing antibody response [9]. Should these animals be subsequently super-infected with a CP BVDV, they will undergo fatal disease. It has also been proven [3] that acute form of the disease is observed in persistently infected cattle with NCP BVDV when they are superinfected with homologous BVDV. In contrast, the persistently infected cattle develop a chronic form of the disease if superinfected with a heterologous BVDV. The hypothesis that immunosuppression plays a paramount role [11] in the pathogenesis of the disease, seems to be appropriate in view of the evidence that at least a proportion of persistently infected animals may have a permanently impaired immune response [8, 10, 12].

The present work was planned with the purpose of making an attempt to answer to the following questions: 1. Is the response of calves to the infection significantly different depending on the biotype of BVDV? 2. Is the response of cattle to BVDV infection influenced by an immunosuppressive state? 3. If BVDV is an immunosuppressive agent itself, could a simultaneous infection with the 2 biotypes of the virus result in the classically fatal expression of the disease?

Materials and Methods

Virus

The CP strain TVM-2 [5] and NCP strain New York-1, grown in bovine embryo kidney (BEK) cells, were used. The CP strain was at its 6th passage level and had a titer of $10^{6.50}$ median tissue culture infectious doses ($TCID_{50}$) per 1 ml. The NCP strain had an undetermined number of passages in tissue culture and its titer, as determined by immunofluorescence in BEK cells, was $10^{5.50}$ $TCID_{50}$/ml.

Experimental design: inoculation of calves

Twenty-two 30–40 days old Friesian calves, devoid of neutralizing antibody to BVDV, were subdivided into 5 groups of 6, 6, 4, 4 and 2 respectively. The calves of Group 1 and 2 were inoculated with CP or NCP strains of BVDV, respectively. In Groups 3 and 4, 2 calves in each group were inoculated as above, i.e. with CP or NCP BVDV, whereas the remaining 2 calves in each group were given dexamethasone (DMS) in addition to BVDV. Finally, calves of Group 5, received simultaneously both, the CP and the NCP BVDV. The CP and NCP BVDV were inoculated intravenously (i.v.), each calf receiving 5 ml of virus. The DMS was given i.v. and each calf was injected daily for 4 consecutive days with 0.1 mg/kg of body weight per day, with the first injection being made a few hours before virus inoculation.

Viral isolation

Attempts to isolate virus in BEK cells were carried out on pharyngeal swabbings and leukocytes obtained from each calf at pre-determined intervals after infection. Samples were considered negative for virus if cytopathic effects (CPE) failed to appear or immunofluorescence (fluorescein-isothiocyanate conjugated with USDA anti-pestivirus 445 BVD serum of porcine origin) was negative after 2 serial subpassages of the inoculum had been made. The virus recovered from the samples was identified as BVDV by serum neutralization tests.

Results

Response of calves to experimental infection with BVDV

—Group 1: CP BVDV

As shown in Table 1 all 6 animals inoculated with CP strain of BVDV had a clinical response very similar to that observed previously [1] with this viral strain. The calves developed a febrile reaction, leukopenia, respiratory symptoms and diarrhea and, in one case, mouth lesions. All animals recovered in about 2 weeks. Virus was consistently recovered from the pharyngeal swabbings of all calves from post infection day (PID) 2 through PID 11, whereas virus recovery from the buffy coat was less regular. At PID 30 neutralizing antibody were found at titers ranging 1:32–1:128.

—Group 2: NCP BVDV

The 6 calves inoculated with NCP strain New York-1 of BVDV did not show any clinical sign of the disease. The body temperatures, as well as the leukocyte counts, remained normal. However, the virus was isolated from pharyngeal swabbings and from buffy coat of all calves. The immunologic response of these calves was similar to that described for calves exposed to CP BVDV.

Response to BVDV infection of calves treated or not treated with DMS

—Group 3: CP BVDV

As depicted in Table 2, the clinical response of calves to CP strain of BVDV was generally more severe in the case of those animals which were treated with DMS. The two calves (Nos. 193, 194) which were given the virus only, had a clinical response similar to that described above for the CP BVDV infection of Group 1 calves in that they reacted with fever, nasal discharge, diarrhea and leukopenia and recovered in about two weeks. The 2 calves (Nos. 191, 192) which, beside the virus, were also given DMS for 4 consecutive days, developed symptoms later than the former animals, but they were much more severe and lasted longer; moreover, one calf (No. 192) died 3 days following the last injection of DMS, i.e. on PID 7. At necropsy the mucosa of the small intestine was slight congested and the mesenteric lymph nodes appeared enlarged. Virus was isolated from both groups of calves from both pharyngeal swabbings and from buffy coat. Virus was also isolated from the spleen and from the lung of the dead calf. Neutralizing antibody titers were found in the serums of all calves which survived, and at PID 30 they were in the same range (1:32–1:128) as those observed for the other groups of calves.

Table 1. The response of calves exposed to experimental infection with cytopathic TVM-2 strain of Bovine Viral Diarrhea Virus

Calf No.	Clinical signs				Oral lesions on PID	Virus recovery in BEK cells, on PID from	
	Fever ≥40°C onset/duration[a]	Nasal discharge onset/duration	Diarrhea onset/duration	Leukopenia onset/duration		PS	LE
141	3/4	3/7	4/1	2/6	None	2, 11	2[b]
132	2/2	2/7	4/1	2/5		2, 11	2, 6, 17
133	3/7	2/4	2/5	2/7		2, 11	2, 17
134	3/5	2/10	6/1	2/6		2, 11	11, 17
135	5/6	3/7	3/5	2/10		2, 11	4, 11, 17
136	2/7	2/6	2/6	3/10	4	2, 11	N.I.

a = Post infection day (PID)/days;
b = Samples were collected on PID 2, 4, 6, 7, 11, 17;
PS = Pharyngeal swabbings;
LE = Buffy coat;
BEK = Bovine embryo kidney;
N.I. = Virus was not isolated.

Table 2. The response of calves treated or not treated with dexamethasone (DMS) and exposed to experimental infection with cytopathic TVM-2 strain of Bovine Viral Diarrhea Virus

Calf No.	DMS treatment	Clinical signs				Virus recovery in BEK cells, on PID from	
		Fever ≥40°C onset/duration[a]	Nasal discharge onset/duration	Diarrhea onset/duration	Leukopenia onset/duration	PS	LE
191	T	5/10	6/9	5/13	10/4	2	2, 6[b]
192*	T	4/4*	6/1*	5/3*	6/2*	2, 6	2, 6[§]
193	NT	3/6	9/2	7/5	3/6	2, 6	2
194	NT	5/3	7/1	7/7	3/5	6	2

a = Post infection day (PID)/days;

b = Samples (PS pharyngeal swabs; LE buffy coat) were collected on PID 2 and 6;

BEK = bovine embryo kidney;

T = treated;

NT = not treated;

* Calf 192 died on PID 7;

§ Virus was also isolated from lung and spleen of this calf.

—Group 4: NCP BVDV

In the case of those calves exposed to NCP BVDV, the simultaneous treatment with DMS induced a severe clinical disease (Table 3). The 2 DMS treated calves had high fever, nasal discharge, leukopenia and diarrhea. One calf died on PID 11 and the other on PID 17. At necropsy congestion was observed in the upper digestive tract and superficial erosions were found in the mucosa of small intestine. Mesenteric lymph nodes were hemorrhagic and enlarged. In contrast, the 2 calves which were only injected with the NCP BVDV showed no clinical signs of disease, with the exception of a slight nasal discharge. Virus was consistently recovered from pharyngeal swabbings and buffy coat samples without any difference being observed between the two groups of calves. Virus was also isolated from spleen and lung of the calf that died on PID 11.

—Group 5: Response of calves to simultaneous infection with CP and NCP BVDV

The clinical response of the 2 calves was analogous to that observed in the calves of Group 1 which were used to test the pathogenicity of the CP BVDV. It seems, therefore, that the mixed infection did not result in any particular unexpected pathological situation.

Discussion

In this study it was demonstrated that CP TVM-2 strain of BVDV induced in calves a severe disease, whereas the calves inoculated with the NCP New York-1 strain remained clinically normal. The data are still insufficient to decide whether the key factor for a different pathogenic role of the virus can be related to its ability to induce cytopathology or not. However, it becomes increasingly obvious that among BVDV isolates there can be a potential diversity in their pathogenic pathway toward the host. This study confirms that the clinical response of cattle to BVDV infection is significantly affected by an immunosuppressive state of the host. Evidently, as already suggested [11], the immunosuppression represents a further "key factor" in the pathogenesis of bovine viral diarrhea. To conclude: The BVDV infection should be regarded as a multifactorial syndrome [7] where several factors might be involved in altering the pathogenesis, such as: 1. An immunotolerant state induced in persistently infected cattle with NCP (and/or CP also?) BVDV. These cattle when superinfected with an homologous BVDV [3] undergo an acute and, eventually, fatal expression of the disease. 2. The biotype of BVDV, seems to influence significantly the development of the disease. 3. The immunosuppressive state of the animals is responsible for a severe form of bovine viral diarrhea. In this case the biotype does not seem to play any particular role in the development of the disease

Table 3. The response of calves treated or not treated with dexamethasone (DMS) and exposed to experimental infection with non-cytopathic New York-1 strain of Bovine Viral Diarrhea Virus

| Calf No. | DMS treatment | Clinical signs | | | | Virus recovery in BEK cells*, on PID, from | |
		Fever $\geq 40\,^\circ$C onset/duration[a]	Nasal discharge onset/duration	Diarrhea onset/duration	Leukopenia onset/duration	PS	LE
149[§]	T	5/10[§]	6/8[§]	6/10[§]	6/8[§]	2, 4, 11[b]	2, 6, 8, 11
150[°]	T	6/5[°]	6/2[c]	6/5[°]	6/5[°]	2, 4, 8	2, 6, 8['']
151	NT	None	4/2	None	None	2, 11	2, 6, 8, 11
152	NT	None	10/1	None	None	4, 11	2, 6, 8, 11

a = Post infection day (PID)/days;
b = Samples (PS pharyngeal swabs; LE buffy coat) were collected on PIDs 2, 4, 6, 8, 11;
* Virus detection was made by immunofluorescence;
Died on PID 17 ([§]) or PID 11 ([°]);
''Virus was also detected in the spleen and lung of this calf;
T = treated;
NT = not treated.

which, whatever biotype is involved, is almost always fatal. 4. The immuno-suppressive activity of BVDV could be a property peculiar to certain strains of the virus. On the other hand, there is evidence that under natural conditions BVDV might be responsible of immunologic dysfunctions in cattle [11].

Acknowledgements

This work was financed by the Dept. of Public Health, Directory of Veterinary Service, the National Research Council and the Dept. of University and Research.

References

1. Avellini G, Castrucci G, Morettini B, Cilli V (1968) Bovine virus diarrhea in Italy—II. Reproduction of the disease. Arch Ges Virusforsch 24: 65–75
2. Bolin SR, McClurkin AW, Cutlip RC, Coria MF (1985) Severe clinical disease induced in cattle persistently infected with noncytopathic bovine viral diarrhea virus by super-infection with cytopathic bovine viral diarrhea virus. Am J Vet Res 46: 573–576
3. Brownlie J, Clarke MC, Howard CJ, Pocock DH (1986) The development of disease following infection of cattle with non-cytopathic and cytopathic bovine virus diarrhea virus. In: Proceedings of the 9th International Symposium W.A.V.M.I., Perugia, Italy 1986. Esculapio Editrice, Bologna (Italy), pp 35–38
4. Brownlie J, Clarke MC, Howard CJ (1984) Experimental production of fatal mucosal disease in cattle. Vet Rec 114: 535–536
5. Castrucci G, Cilli V, Gagliardi G (1968) Bovine virus diarrhea in Italy—I. Isolation and characterization of the virus. Arch Ges Virusforsch 24: 48–64
6. Corapi WV, Donis RO, Dubovi EJ (1988) Monoclonal antibody analyses of cytopathic and noncytopathic viruses from fatal bovine viral diarrhea virus infections. J Virol 62: 2823–2827
7. Harkness JW (1987) The control of bovine viral diarrhea virus infection. Ann Rech Vét 18: 167–174
8. Johnson DW, Muscoplat CC (1973) Immunologic abnormalities in calves with chronic bovine viral diarrhea. Am J Vet Res 34: 1139–1141
9. McClurkin AW, Littledike ET, Cutlip RC, Frank GH, Coria MF, Bolin SR (1984) Production of cattle immunotolerant to bovine viral diarrhea virus. Can J Comp Med 48: 156–161
10. Muscoplat CC, Johnson DW, Teuscher E (1973) Surface immunoglobulin of circulating lymphocytes in chronic bovine diarrhea: abnormalities in cell populations and cell function. Am J Vet Res 34: 1101–1103
11. Potgieter LND (1988) Immunosuppression in cattle as a result of bovine viral diarrhea virus infection. Agri-Practice—Virology/Immunology 9: 7–14
12. Roth JA, Bolin SR, Dacmar EF (1986) Lymphocyte blastogenesis and neutrophil function in cattle persistently infected with bovine viral diarrhea virus. Am J Vet Res 47: 1139–1141

Author's address: Dr. G. Castrucci, Istituto di Malattie Infettive, Facoltà di Medicina Veterinaria, Via S. Costanzo, 4, I-06100 Perugia, Italy.

Arch Virol (1991) [Suppl 3]: 109–124
© Springer-Verlag 1991

Distribution of antigen of noncytopathogenic and cytopathogenic bovine virus diarrhea virus biotypes in the intestinal tract of calves following experimental production of mucosal disease

E. M. Liebler[1], J. Waschbüsch[1], J. F. Pohlenz[1], V. Moennig[2], and B. Liess[2]

[1] Institute of Pathology, Hannover Veterinary School, Hannover
[2] Institute of Virology, Hannover Veterinary School, Hannover,
Federal Republic of Germany

Accepted March 14, 1991

Summary. Mucosal disease can be experimentally induced by inoculating calves persistently viremic with noncytopathogenic (ncp) Bovine Virus Diarrhea Virus (BVDV) with an antigenetically closely related cytopathogenic (cp) BVDV strain. Calves suffering from mucosal disease develop severe intestinal lesions causing breakdown of the gastrointestinal barrier and death. Knowledge about tissue distribution of ncp/cp biotypes of BVDV may contribute to the understanding of the pathogenesis of these lesions. Distribution of cpBVDV versus ncpBVDV was demonstrated in the intestinal tract of nine calves with experimentally induced mucosal disease and in five persistently viremic calves. Biotypes were distinguished immunohistochemically in organ tissues using monoclonal antibodies against marker epitopes on the viral surface glycoprotein gp53. In persistently viremic calves ncpBVDV was present in a few epithelial cells, mononuclear cells and intramural ganglia. A multifocal reaction was observed in vascular walls. In calves with mucosal disease a striking increase of antigen containing cells occurred. Viral antigen in these cells reacted with marker antibodies for cpBVDV. A distinct tissue distribution of biotypes was observed in intramural ganglia and duodenal glands. Severe tissue damage was correlated to the presence of cpBVDV antigen. This indicates the importance of cpBVDV for the development of lesions. Interactions of cpBVDV and immune-mediated mechanisms will need further investigation.

Key words: BVDV infection, cattle, immunohistochemistry, intestine, monoclonal antibodies, mucosal disease.

Supported by the German Research Association (DFG) project no. En65/15-3

Introduction

Bovine virus diarrhea (BVD) was first described by Olafson et al. (1946). In 1953 first cases of mucosal disease were reported [36]. Gillespie and Baker (1959) recognized that both clinically different diseases are caused by the same virus. Later the bovine virus diarrhea virus (BVDV) was classified together with hog cholera virus and border disease virus in the genus pestivirus of the togaviridae. Despite its known etiology, the pathogenesis of mucosal disease remained unclear until it was recognized that mucosal disease could only be produced in immunotolerant calves, which had been persistently infected with a noncytopathogenic (ncp) BVDV [22, 24, 27, 38]. In these calves superinfection with cytopathogenic (cp) BVDV resulted in mucosal disease in case the cpBVDV was antigenetically closely related to the persistent ncpBVDV [7, 12, 31]. Under field conditions the matching cpBVDV may originate from mutation of the persistent ncpBVDV [14, 19].

In mucosal disease severe lesions causing death of animals are found in mucus membranes of the gastrointestinal tract, especially at sites with gut-associated lymphoid tissue [35, 37]. The pathogenesis of these lesions is still inconclusive [5]. They develop within a short time after experimental superinfection [7, 12, 31]. It has been suggested that interactions between ncp and cp biotypes of BVDV and the immune system are essential for the development of lesions [5]. BVDV antigen has been demonstrated with polyclonal antisera and monoclonal antibodies in cattle with persistent viremia [4, 15, 39] or with spontaneous mucosal disease [2, 28, 39]. Distinction between the different biotypes in organ tissues using monoclonal antibodies has, however, only been attempted in the central nervous system [18]. Using monoclonal antibodies (MABs) against marker epitopes on cpBVDV we were able to demonstrate the antigen distribution of cpBVDV versus ncpBVDV in experimentally produced mucosal disease in comparison to the localization of ncpBVDV antigen in persistently viremic calves.

Material and methods

Animals

Tissues were collected from animals originating from a herd of Holstein Frisian cattle in Lower Saxony. This herd was under virological surveillance for approximately one year [31]. Fourteen clinically healthy calves of this herd were transfered at an age of two to five months to an isolation unit at the Hannover Veterinary School. All animals were persistently viremic with BVDV and showed no neutralizing antibodies against BVDV.

Experimental production of mucosal disease

The experimental procedure has been published in detail [31]. Briefly, BVDV isolated from the peripheral blood of all calves with permanent viremia was of ncp biotype and had identical reaction patterns when tested in infected monolayer enzyme immune assays with

15 MABs directed against the viral surface glycoprotein gp53 [9, 29, 30]. Therefore these ncpBVDV isolates were considered to represent a uniform virus of common source. In the following it will be refered to this virus as herd-specific virus. For superinfection by intranasal and intravenous inoculation BVDV strains which had been tested with the same MABs were selected with regard to biotype and to homology/heterology compared with the herd-specific ncpBVDV.

One calf served as control and was not inoculated. Two calves were inoculated with cpBVDV strain A1138/69 which displayed the highest degree of heterology compared to the herd-specific ncpBVDV and two calves were inoculated with a mixture of three ncpBVDV strains (Auburn, New York, 9762). These ncpBVDV strains exhibited a close homology to the herd-specific ncpBVDV. These five animals remained clinically healthy and were euthanized on days 23 to 27 post inoculation (group A, Table 1).

Six calves were inoculated with a mixture of three cpBVDV strains (Indiana, Lamspringe/735 and MD1) which displayed the closest homology to the herd-specific ncpBVDV; three calves were inoculated with cpBVDV strain Indiana only. All nine animals developed signs of mucosal disease and were euthanized in extremis on day 8 to day 14 post infection (group B, Table 1). Route of inoculation and virus dose used for superinfection are summarized in Table 1.

Tissue collection

Calves were immobilized with Xylazin (Rompun^R) and euthanized with Pentobarbital-Na (Narcoren^R). Intestinal tissues were collected from duodenum, mid jejunum containing a Peyer's patch, ileum next to the ileocecal entrance, lymphoid patches at the ileocecal entrance and in the proximal colon, mid colon and rectum. All tissues were placed in 0.15 M phosphate buffered saline at 2 °C immediately. 2×3 cm^2 large pieces of tissue were excised, placed with the mucosa on thin slices of liver for surface protection and then attached to corrugated card board [21]. Tissues were snap-frozen in chilled isopentane and stored at -70 °C. Frozen sections of 4 μm thickness were cut with a cryostat (Frigocut, Reichert & Jung), collected on chromalaun gelatine coated slides and dried for 30 minutes with a fan. Sections were used for immunohistochemical reactions immediately or briefly fixed for 30 sec in 100% acetone at room temperature and stored at -20 °C.

Immunohistochemistry

Out of the fifteen MABs against viral surface protein gp53 which had been used to characterize the BVDV strains in vitro, nine were selected as primary antibodies for immunohistochemistry [9, 29, 30, Table 2]. The pestivirus-specific MAB C16 was included as control [34]. Several of the MABs have been used for immunohistochemical reactions on alcohol- or formalin-fixed paraffin-embedded tissue [1, 18]. However, the majority of epitopes recognized by MABs are destroyed by these preparation methods. Since the whole panel of MABs had to be available for our purpose, antigen was demonstrated by indirect immunoperoxidase method in cryostat sections. With this method the pattern of MAB reactivity established in vitro could be used in tissue.

Sections were fixed in acetone for 10 minutes at room temperature, air dried and rehydrated in phosphate buffered saline pH 7.6 (PBS). Endogenous peroxidase was inhibited by incubation with 0.05% phenylhydrazine in PBS at 37 °C for 40 minutes. Non specific binding sites were blocked with inactivated rabbit serum diluted 1:10 in PBS with addition of 0.5% Tween 20 (Tween PBS). After 20 minutes serum was decanted and sections were incubated with primary antibodies in a humid chamber at room temperature for

Table 1. Experimental protocol

Group	Animal	Age (months)	BVDV used for superinfection	Route of inoculation	Dose (TCID50)	Necropsy (days p.s.[1])
A				no superinfection		
(persist. viremic, clin.	4 gr	3				
	10 gr	3	A1138/69	iv+in[2]	$10^{9.3}$	23
	11 gr	3	A1138/69	iv+in	$10^{9.3}$	23
healthy)	12 gr	3	Auburn+New York	iv+in	$10^{8.8}$	27
	14 gr	3	+9762	iv+in	$10^{8.8}$	27
B	384	3		iv+in	$10^{7.8}$	14
	385	3	Indiana	iv+in	$10^{7.8}$	10
(mucosal	387	3	+MD1	iv+in	$10^{7.8}$	11
disease)	388	3	+Lamspringe/735	iv+in	$10^{7.8}$	12
	394	2		iv+in	$10^{7.8}$	12
	399	2		iv+in	$10^{7.8}$	13
	3 gr	4	Indiana	iv+in	$10^{8.1}$	9
	15 gr	4	Indiana	iv+in	$10^{8.1}$	14
	33 gr	2	Indiana	in	$10^{7.2}$	8

[1] Days post superinfection.
[2] Intravenous and intranasal.

Table 2. Monoclonal antibodies (MABs) used to differentiate immunohistochemically between the herd-specific ncpBVDV and cp/ncp BVDV strains used for superinfections

MABs BVDV	CA1	CA3	CA25	CA36	CA80	CT2	CT3	CT6	CT9
herd-specific virus (ncp)	+	+	−	−	+	−	−	−	−
Auburn (ncp)	+	+	+	−	+	−	−	−	−
New York (ncp)	+	+	+	−	+	−	−	−	−
9762 (ncp)	+	+	−	−	+	−	+	−	−
A1138/69 (cp)	−	+	−	+	−	+	+	+	+
Indiana (cp)	+	+	−	−	+	−	+	+	−
MD1 (cp)	+	+	−	−	+	+	−	−	−
Lamspringe/735 (cp)	+	+	−	−	+	+	−	−	−

90 minutes. Primary MABs were diluted in Tween PBS. Peroxidase-conjugated rabbit anti-mouse immunoglobulin (Dakopatts P161) diluted 1:80 in normal calf serum and Tween PBS (1:1) was applied as second antibody for 60 minutes in a humid chamber at room temperature. Peroxidase activity was detected by incubation in a solution of 0.05% 3,3′-diaminobenzidine tetrahydrochloride and 0.03% hydrogen peroxide in TRIS-HCl buffer pH 7.6. Reaction was stopped after 5 minutes by rinsing slides in distilled water. Slides were counterstained in Mayer's haemalaun for 30 seconds or in an aquous 2% solution of methylene green for 10 minutes, rinsed in tap water and coverslipped.

Results

Animals of group A remained clinically healthy until necropsied on days 23 to 27 post inoculation. Macroscopic examination of the intestinal tract revealed mild hyperemia of lymphoid patches both in small and large intestine. In one calf the ileal wall was edematous, in another necrotic plugs were present in the colonic lymphoid patch and in a third petechial hemorrhages were found in rectal mucosa. Histologically moderate crypt hyperplasia, mild focal necrosis of crypt cells and follicle-associated epithelial cells and mild depletion of mucosa-associated lymphoid follicles were observed.

In all calves of group A, MABs CA1, CA3 and CA80 which recognized epitopes on herd-specific ncpBVDV stained viral antigen as diffuse, granular precipitate in the cytoplasm of epithelial cells, mononuclear cells and ganglionic neurons. An identical staining pattern was observed with the broadly reacting pestivirus-specific MAB C16. The presence of BVDV strains used for inoculation was determined by marker MABs, which recognized epitopes on the BVDV strains used for superinfection, but not on the herd-specific ncpBVDV. MABs CA25, CA36, CT3, CT6 and CT9 served as marker antibodies for cpBVDV strain A1138/69, MAB CA25 for ncpBVDV strains Auburn and New York, and MAB CT3 for ncpBVDV strain 9762 (Table 1). In intestinal tissues, viral antigen corresponding to these BVDV strains used for inoculation was not detectable.

The tissue distribution of herd-specific ncpBVDV antigen was identical in all calves of group A. In small intestinal epithelium viral antigen was present in small groups of crypt cells (Fig. 1), while in large intestinal epithelium only a few, single crypt cells were positive. In the lamina propria a small number of mononuclear cells containing viral antigen was found. Intense reactions were observed in epithelial cells of glandulae duodenales and neurons of plexus submucosus and myentericus. Lacteals and walls of blood vessels especially small arterioles in the submucosa stained in a multi-focal pattern. In lymphoid tissue of Peyer's patches (Fig. 1) and colonic and rectal lymphoid patches a few positive mononuclear cells were present. Comparing the different compartments increased staining was found in interfollicular areas.

Animals of group B developed typical clinical signs of mucosal disease. There were erosions and ulcerations in the oral cavity and profuse diarrhea. Detailed macroscopic findings have been reported [31]. Macroscopically

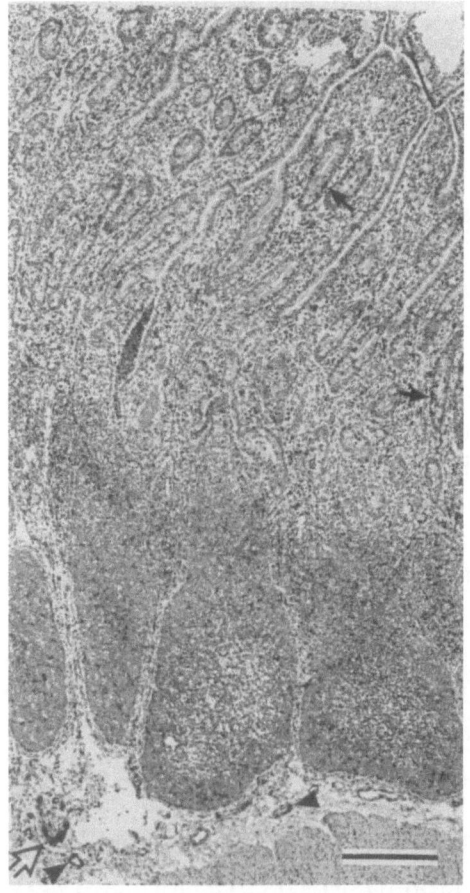

Fig 1. Distribution of herd-specific ncpBVDV antigen in the ileal Peyer's patch of a persistently viremic calf. Small groups of epithelial cells (arrows), a few mononuclear cells in the lamina propria and scattered cells in the lymphoid follicles and domes contain viral antigen. Increased numbers of positive cells are present in interfollicular areas. Note reaction of intramural ganglia (open arrow) and vascular walls (arrowheads). Calf 4 gr, ileal Peyer's patch, MAB CA1, bar = 300 μm

lesions in small intestine were generally restricted to Peyer's patches which were indented, hyperemic and frequently covered with fibrinous exsudate. In two calves erosive to diphtheroid colitis was found. Lymphoid patches at the ileocecal entrance, in colon and rectum were characterized by hyperemia, hemorrhages, necrosis and fibrinous exsudate. Lesions in rectum were milder than in colon, but focally extensive hemorrhages were more frequent.

MABs CT3 and CT6 served as marker antibodies for cpBVDV strain Indiana and MAB CT2 for cpBVDV strains Lamspringe/735 and MD1. They reacted with the cpBVDV strains used for superinfection exclusively. MABs CA1, CA3 and CA80 recognized epitopes on the herd-specific ncpBVDV as well as epitopes on the cp viruses used for superinfection (Table 1). Therefore distribution of ncpBVDV was evaluated by comparing consecutive sections stained with marker antibodies for cpBVDV or antibodies reacting with herd-specific ncp and superinfecting cpBVDV. In tissues of calves which had been inoculated with a mixture of cpBVDV strains Indiana, Lamspringe/735 and MD1, only marker antibodies directed to strain Indiana reacted in addition to antibodies recognizing the herd-specific ncpBVDV. Therefore three calves were inoculated with cpBVDV strain Indiana only. Lesions, distribution of viral antigen and intensity of immunohistochemical reactions were similar in all calves of group B and will be discussed together.

Compared to animals of group A, numbers of epithelial cells and mononuclear cells containing viral antigen were drastically increased. Distribution and number of positive cells in intestinal tissue were similar with MABs CA1, CA3 and CA80 reacting with herd-specific ncpBVDV and cpBVDV strain Indiana, and with marker antibodies MABs CT3 and CT6 directed to cpBVDV strain Indiana. This indicates that viral antigen observed in calves with mucosal disease was predominantly of cp biotype.

While lesions in duodenum were mild, in mid and lower jejunum stunted, fused villi lined with cuboidal epithelium which contained numerous intraepithelial cells were found. Necrosis of cells in the depth of hyperplastic crypts and crypt abscesses were frequent. Most crypts stained intensely for viral antigen (Fig. 2). In the lamina propria, BVDV antigen was found in mononuclear cells and in cells lining the dilated villous lacteals. Scattered mononuclear cells and mild periganglionic and perivascular lymphohistiocytic infiltrates in the submucosa reacted positive.

A distinct distribution of biotypes was observed in duodenal glands and intramural ganglia. In parallel sections of duodenal glands ncpBVDV was present in moderate numbers of epithelial cells and cpBVDV in a few epithelial cells only (Fig. 2). While ncpBVDV was found in ganglionic neurons, marker antibodies for cpBVDV strain Indiana stained satellite cells and inflammatory infiltrates (Fig. 3).

Lesions increased in severity towards Peyer's patches, where they were most pronounced (Fig. 4). Above Peyer's patches only remnants of villi and

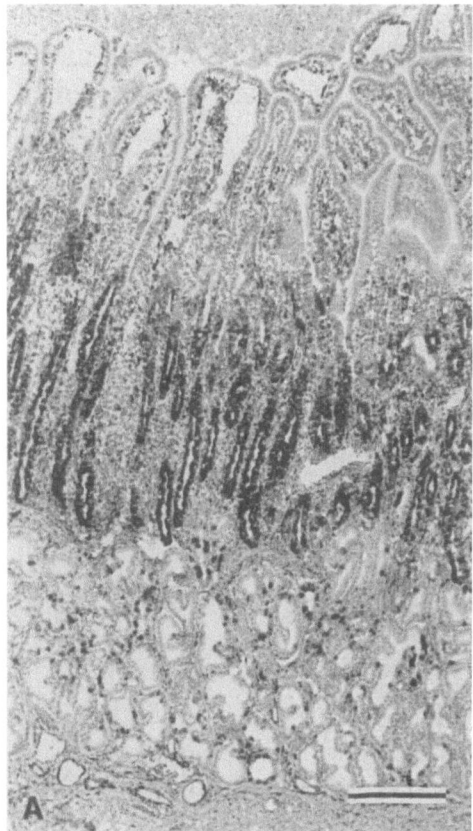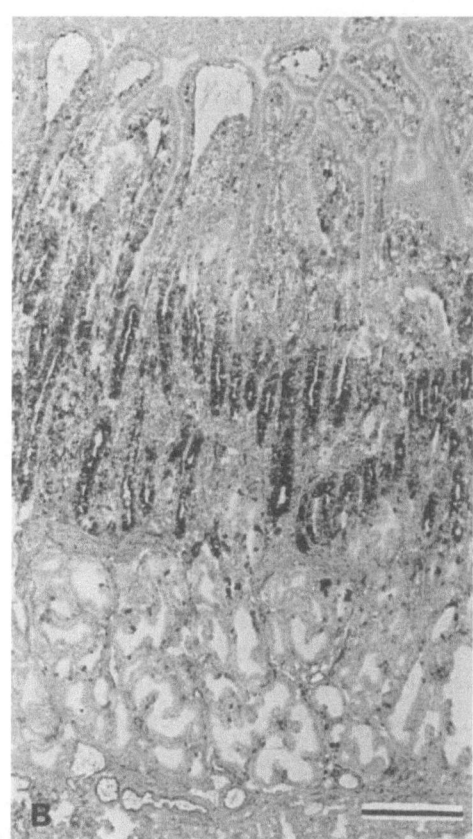

Fig. 2. Distribution of ncpBVDV versus cpBVDV in duodenal glands of a calf suffering from mucosal disease. There are increased numbers of positive cells. Most crypt cells, but only few villous epithelial cells contain viral antigen. Note reactions in lacteal walls

A Moderate numbers of duodenal gland epithelial cells are stained with MAB CA1 recognizing herd-specific ncpBVDV and cpBVDV strain Indiana
B Few duodenal gland epithelial cells are stained with MAB CT3 recognizing cpBVDV strain Indiana only

Calf 384, duodenum, consecutive sections A: MAB CA1, B: MAB CT3, bar = 300 μm

crypts were found. The epithelium was flattened and frequently blast-like cells with large, sometimes multiple nuclei were observed. BVDV antigen was present in remaining epithelial cells. Especially intense reactions occurred in the small, depleted lymphoid follicles (Fig. 5). Necrotic centers of lymphoid follicles, which were occasionally replaced by mucinous material, were surrounded by a thin layer of positively stained lymphoid cells. BVDV antigen was also present in remaining domes and follicle-associated epithelium. Sometimes only groups of follicle-associated epithelial cells or intraepithelial cells were stained. The numerous macrophages, lymphocytes and dendritic cells in the peri- and parafollicular area mostly contained viral antigen.

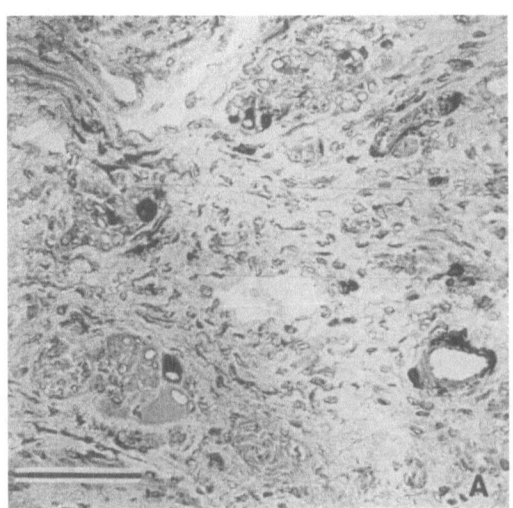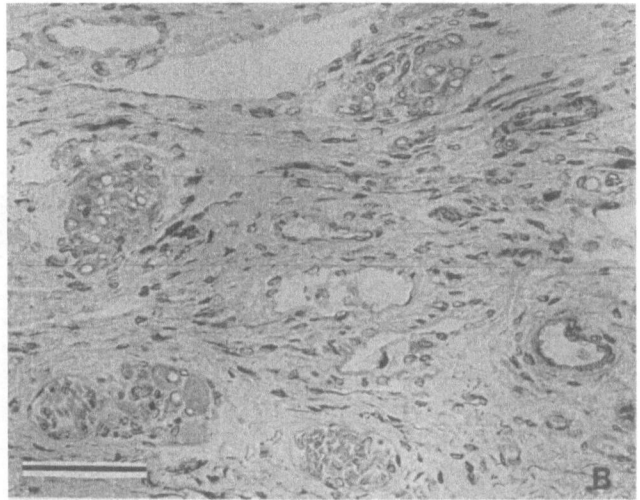

Fig. 3. Distribution of ncpBVDV and cpBVDV antigen in intramural ganglia of a calf
suffering from mucosal disease

A Ganglionic neurons and scattered mononuclear cells are stained with MAB CA1
recognizing herd-specific ncpBVDV and cpBVDV strain Indiana

B Satellite cells and scattered mononuclear cells are stained with MAB CT3 recognizing
cpBVDV strain Indiana only

Calf 15 gr, ileal Peyer's patch, consecutive sections A: MAB CA1, B: MAB CT3,
bar = 100 μm

Large intestinal lesions were more severe in calves inoculated with the
mixture of three cpBVDV strains than in calves inoculated with cpBVDV
strain Indiana only. Epithelial alterations were characterized by crypt
hyperplasia, decrease of goblet cells and multifocal crypt cell necrosis. In
areas where necrotic crypts were present almost all epithelial cells contained
viral antigen, while in other crypts only small groups or single cells were
positive (Fig. 6). Adjacent to necrotic crypts increased numbers of mast cells
and positively staining mononuclear cells were found. Findings in the
lamina propria and submucosa were comparable to small intestine. Mild to
moderate lymphohistiocytic perivascular infiltrates were more frequent than
in small intestine.

As in small intestine, most severe alterations were found in association
with lymphoid tissue of large intestine. In the area of colonic and rectal
lymphoid patches, architecture of mucosa and lymphoid tissue was completely altered and cpBVDV was demonstrated in numerous epithelial cells,
lymphocytes and macrophages. Lymphoglandular complexes were dilated
with necrotic and mucinous debris. They were surrounded by a rim of
lymphocytes, macrophages and dendritic cells containing viral antigen
(Fig. 7). Epithelial invaginations and remnants of the predominantly
necrotic follicle-associated epithelium reacted positively.

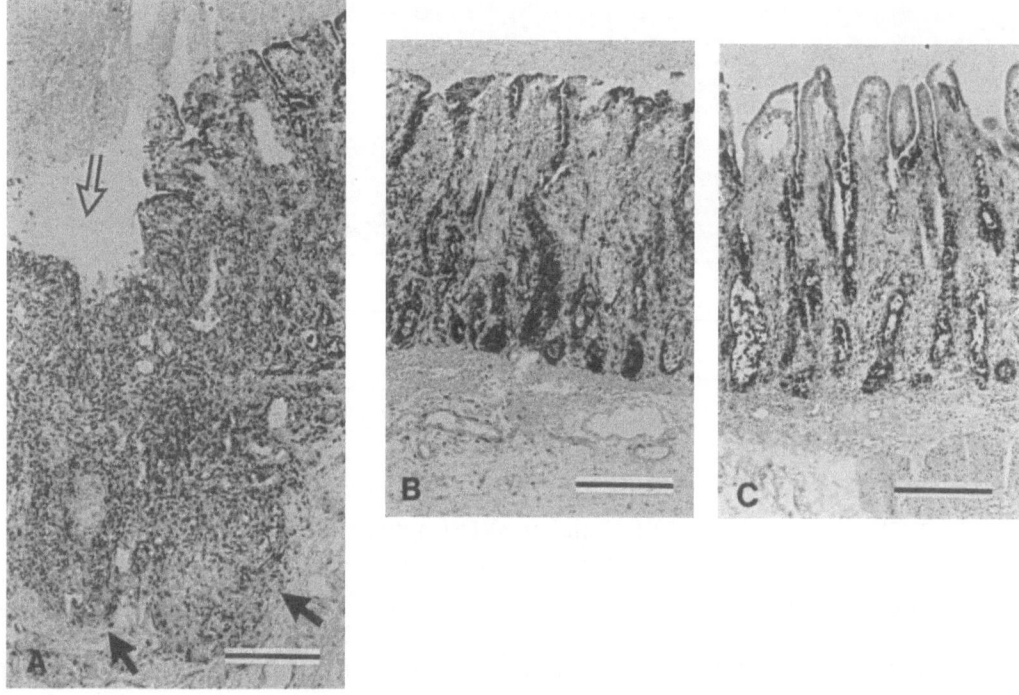

Fig. 4. Increasing severity of lesions and numbers of cells containing cpBVDV towards a Peyer's patch in a calf suffering from mucosal disease. A, B, C are from the same section

A Within Peyer's patch: complete loss of mucosal architecture (open arrow), remnants of lymphoid follicles (arrows)

B Next to Peyer's patch: stunted villi, increased numbers of positive cells

C About 2 cm distance from Peyer's patch: crypt necrosis, short villi

Calf 385, jejunal Peyer's patch, MAB CA1, bar = 300 µm

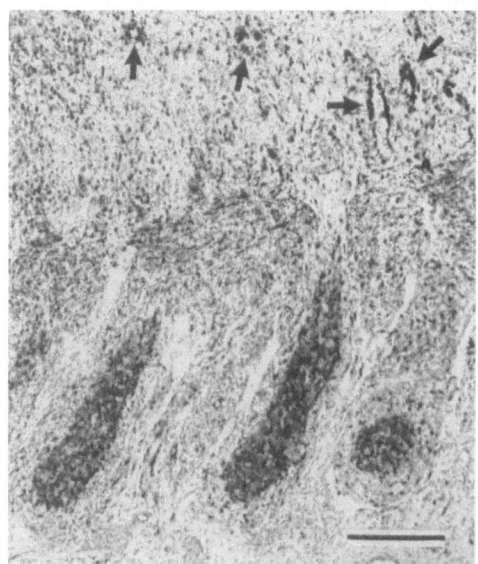

Fig. 5. Numerous cells containing cpBVDV antigen in remnants of depleted lymphoid follicles in a calf suffering from mucosal disease. Note reactions in remaining epithelial cells (arrows). Calf 15 gr, ileal Peyer's patch, MAB CT3, bar = 300 µm

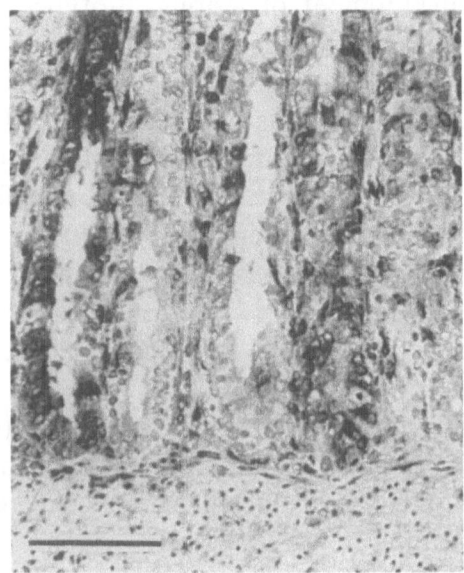

Fig. 6. Numerous epithelial cells containing cpBVDV antigen in necrotic crypts in large intestine of a calf suffering from mucosal disease. In adjacent crypts a few positive cells are present only. Calf 384, mid colon, MAB CT3, bar = 100 µm

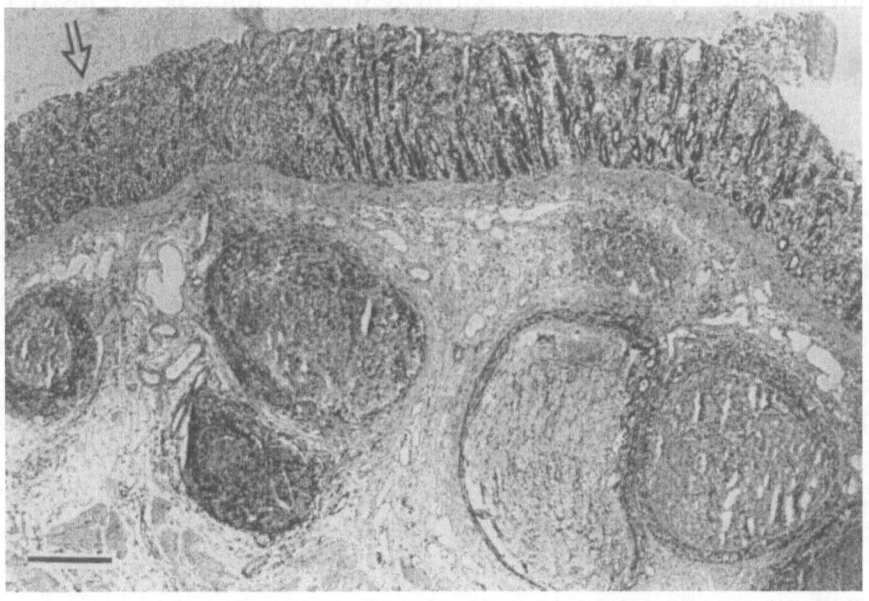

Fig. 7. Dilated and severely depleted lymphoglandular complexes in proximal colon of a calf suffering from mucosal disease. BVDV antigen is present in remaining rims of mononuclear cells surrounding lymphoglandular complexes. Note atrophic mucosa (open arrows) and staining of remaining crypt cells. Calf 384, proximal colon, MAB CA1, bar = 500 µm

Discussion

The objective of this investigation was to distinguish ncp and cp biotypes of BVDV in intestinal tissues of calves with experimentally produced mucosal disease. Knowledge about the distribution of cpBVDV in tissue may contribute to the understanding of the development of lesions. Biotypes of BVDV are defined according to their behaviour in cell cultures. The molecular basis for the distinction of biotypes is the non-structural protein p80, which is found in cp strains of BVDV only [13]. p80 is considered to be the product of a proteolytic cleavage of p125, a non-structural protein present in both ncpBVDV and cpBVDV. Antibodies which recognize p125 usually also recognize p80. Therefore this non-structural protein can currently not be used for distinction of biotypes in tissue. Distinction of biotypes in this investigation was based on marker epitopes on the viral surface glycoprotein gp53 which were present on superinfecting cpBVDV and absent on the herd-specific ncpBVDV.

In persistently viremic calves MABs which recognized the herd-specific ncpBVDV reacted only. Viral antigen was present in a few epithelial and mononuclear cells and in intramural ganglia as has been reported in previous studies [4, 39]. The multifocal staining of vascular walls which was also described by Meyling (1970) and Liess (1985) might be due to attaching or traversing mononuclear cells. Higher frequency of virus containing cells in interfollicular T cell areas than in lymphoid follicles is consistent with viral distribution in peripheral blood leucocytes. A high percentage of T lymphocytes and monocytes, but comparatively few B lymphocytes contain viral antigen in persistently viremic cattle [3, 11].

In superinfected calves, which remained healthy, marker antigens of BVDV strains used for inoculation were not detected in organ tissues. The calves inoculated with heterologous cpBVDV developed neutralizing antibodies which might have eliminated or masked the virus. This is a further confirmation for the specificity of the immunotolerance [8, 10, 25]. In persistently infected calves homologous strains of ncpBVDV might not be able to infect cells. Alternatively viral antigen may not have been detectable because of slow replication of superinfecting ncpBVDV [26, 32] or low affinity of marker antibodies [20]. In all calves superinfected with either heterologous cpBVDV or homologous ncpBVDV, distribution and number of cells containing herd-specific ncpBVDV were such as in the persistently viremic, not superinfected calf. The persistent ncpBVDV appears not to be affected by superinfection or host immune response [25].

In animals suffering from mucosal disease, high numbers of cells containing viral antigen have been reported [1, 2]. In calves inoculated with matching cpBVDV which developed mucosal disease, a striking increase of cells containing viral antigen was observed. These cells reacted with marker antibodies for cpBVDV. In consecutive sections stained with antibodies which recognized cpBVDV as well as ncpBVDV an identical staining pattern was observed in most locations. Therefore it was concluded that the large numbers of viral antigen containing cells in calves with mucosal disease are predominantly infected by cpBVDV, whereas the number of ncpBVDV-infected cells seemed to be unchanged compared to clinically healthy, persistently viremic controls. cpBVDV may be able to infect cells and replicate rapidly without protective immune reaction of the host animal.

Simultaneous infection of cells with both biotypes cannot be excluded in intestinal epithelium, lymphoid tissue and infiltrating mononuclear cells. It appears, however, unlikely, because in several locations a distinct distribution of biotypes was observed in consecutive sections. In intramural ganglionic neurons ncpBVDV was present, while cpBVDV was found only in satellite cells and infiltrating inflammatory cells. In duodenal glands moderate numbers of epithelial cells stained for ncpBVDV, but only few for cpBVDV. Comparing distribution of biotypes in Peyer's patches of calves

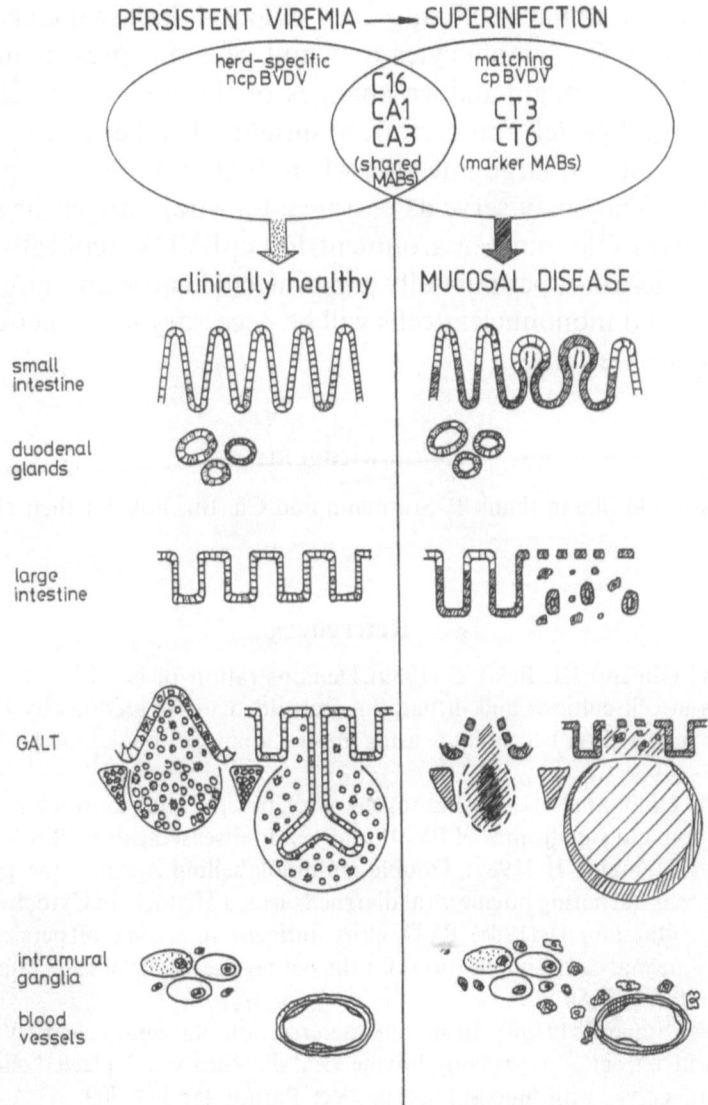

Fig. 8. Distinct distribution of ncpBVDV and cpBVDV in intestinal tissues of calves with persistent viremia and mucosal disease using MABs against marker epitopes

with persistent viremia and mucosal disease it was striking that in persistently viremic cattle most of ncp viral antigen was present in interfollicular areas, while in calves with mucosal disease cpBVDV was predominantly found in lymphoid follicles. This distinct preference of target cells might be caused by interference phenomena of cells already infected with ncpBVDV. A similar effect has been observed in cell cultures. Cell cultures infected with ncpBVDV did not exhibit cp effects after superinfection with cpBVDV [17].

In calves with mucosal disease, a correlation of severe tissue damage and presence of cpBVDV was observed (Fig. 8). This indicates the importance of cpBVDV for the development of lesions and suggests a direct destruction of cells by cpBVDV. Immune-mediated mechanisms can, however, not be excluded, yet. In animals suffering from mucosal disease increased numbers of macrophages, T_8 lymphocytes and null cells are present in the lamina propria [5]. They might induce changes of the epithelium directly or by cytokines [6]. Special consideration should also be given to lymphoid patches in small and large intestine where high amounts of cpBVDV were accumulated. They may serve as entrance for viral antigen or represent an especially favorable microenvironment for cpBVDV replication. Time sequential studies on experimentally superinfected cattle and double labelling for biotypes and mononuclear cells will be necessary to further elucidate the development of lesions.

Acknowledgements

The authors would like to thank P. Stutmann and Ch. Birkholz for their skilful technical assistance.

References

1. Belak K, Gimeno EJ, Belak S (1989) Demonstration of bovine viral diarrhea virus antigens in cell cultures and in paraffin-embedded tissue sections by the peroxidase-antiperoxidase (PAP) technique using monoclonal antibodies. Acta Vet Scand 30: 231–233

2. Bielefeldt Ohmann H (1983) Pathogenesis of bovine viral diarrhea-mucosal disease: distribution and significance of BVDV antigen in diseased calves. Res Vet Sci 34: 5–10

3. Bielefeldt Ohmann H (1987) Double-immunolabelling systems for phenotyping of immune cells harboring bovine viral diarrhea virus. J Histochem Cytochem 35: 627–633

4. Bielefeldt Ohmann H (1988) BVD virus antigens in tissues of persistently viremic, clinically normal cattle: implications for the pathogenesis of clinically fatal disease. Acta Vet Scand 29: 77–84

5. Bielefeldt Ohmann H (1988) In situ characterization of mononuclear leucocytes in skin and digestive tract of persistently bovine viral diarrhea virus-infected clinically healthy calves and calves with mucosal disease. Vet Pathol 25: 304–309

6. Bielefeldt Ohmann H, Babiuk LA (1988) Influence of interferons $\alpha_1 1$ and γ and of tumor necrosis factor on persistent infection with bovine viral diarrhea virus in vitro. J Gen Virol 69: 1399–1403

7. Bolin SR, McClurkin AW, Cutlip RC, Coria MF (1985) Severe clinical disease induced

in cattle persistently infected with noncytopathogenic bovine viral diarrhea virus by superinfection with cytopathogenic bovine viral diarrhea virus. Am J Vet Res 46: 573–576

8. Bolin SR, McClurkin AW, Cutlip RC, Coria MF (1985) Response of cattle infected with noncytopathogenic bovine viral diarrhea virus to vaccination for bovine viral diarrhea and to subsequent challenge exposure with cytopathogenic bovine viral diarrhea virus. Am J Vet Res 46: 2467–2470

9. Bolin SR, Moennig V, Kelso Gourley NE, Ridpath J (1988) Monoclonal antibodies with neutralizing activity segregate isolates of bovine viral diarrhea virus into groups. Arch Virol 99: 117–123

10. Bolin SR, Roth JA, Uhlenhopp EK, Pohlenz JF (1987) Immunologic and virologic findings in a bull chronically infected with noncytopathogenic bovine viral diarrhea virus. J Am Vet Med Assoc 190: 1015–1017

11. Bolin SR, Sacks JM, Growder SV (1987) Frequency of association of noncyto-pathogenic bovine viral diarrhea virus with mononuclear leucocytes from persistently infected cattle. Am J Vet Res 48: 1441–1445

12. Brownlie J, Clarke MC, Howard CJ (1984) Experimental production of fatal mucosal disease in cattle. Vet Rec 114: 535

13. Collett M, Moennig V, Horzinek MC (1989) Recent advances in pestivirus research. J Gen Virol 70: 253–266

14. Corapi WV, Donis RO, Dubovi EJ (1988) Monoclonal antibody analysis of cytopathic and noncytopathic viruses from fatal bovine viral diarrhea virus infections. J Virol 62: 2823–2827

15. Fernandez A, Hewicker M, Trautwein G, Pohlenz J, Liess B (1988) Viral antigen distribution in the central nervous system of cattle persistently infected with bovine viral diarrhea virus. Vet Pathol 26: 26–32

16. Gillespie JH, Baker JA (1959) Studies on virus diarrhea. Cornell Vet 49: 439–443

17. Gillespie JH, Madin SH, Darbry NB (1962) Cellular resistance in tissue culture, induced by noncytopathogenic strains, to a cytopathogenic strain of virus diarrhea virus in cattle. Proc Soc Exp Biol Med 110: 248–250

18. Hewicker M, Wöhrmann T, Fernandez A, Trautwein G, Liess B, Moennig V (1990) Immunohistochemical detection of bovine viral diarrhea virus antigen in the central nervous system of persistently infected cattle using monoclonal antibodies. Vet Microbiol 23: 203–210

19. Howard CJ, Brownlie J, Clarke C (1987) Comparison by the neutralization assay of pairs of noncytopathogenic and cytopathogenic strains of bovine virus diarrhea virus isolated from cases of mucosal disease. Vet Microbiol 13: 361–369

20. Körke J (1989) Herstellung und Charakterisierung von monoklonalen Antikörpern gegen den Stamm A1138/69 des Virus der bovinen Virusdiarrhoe. Hannover, Tierärztl. Hochsch., Diss

21. Landsverk T, Lium B, Matovelo JA, Liven E, Nordstoga K (1990) Peyer's patches in experimental Salmonella dublin infection in calves. Acta Pathol Microbiol Immunol Scand 98: 255–268

22. Liess B (1973) Pathogenese virusbedingter Darminfektionen der Tiere. Dtsch Tierärztl Wochenschr 80: 360–364

23. Liess B (1985) Bedeutung der Immuntoleranz für die Pathogenese der bovinen Virus-diarrhoe (BVD). Berl Münch Tierärztl Wochenschr 98: 420–423

24. Liess B, Frey H-R, Kittsteiner H, Baumann F, Neumann W (1974) Beobachtungen und Untersuchungen über die "Mucosal Disease" des Rindes—einer immunbiologisch erklärbaren Spätform der BVD-MD Virusinfektion mit Kriterien einer "slow virus infection"? Dtsch Tierärztl Wochenschr 81: 481–487

25. Liess B, Frey H-R, Orban S, Hafez SM (1983) Bovine Virus Diarrhoea (BVD)-"Mucosal Disease". Persistente BVD Feldvirusinfektionen bei serologisch selektierten Rindern. Dtsch Tierärztl Wochenschr 94: 261–266
26. Mahnel H, Moreau H (1984) Untersuchungen über die Vermehrung von zytopathogenem und nichtzytopathogenem BVD-Kulturvirus. I. Quantitative und fluoreszenzserologische Vergleiche. Zentralbl Veterinärmed [B] 31: 131–140
27. McClurkin AW, Bolin SR, Coria MF (1985) Isolation of cytopathogenic and non-cytopathogenic bovine viral diarrhea virus from the spleen of cattle acutely and chronically affected with bovine viral diarrhea. J Am Vet Med Assoc 186: 568–569
28. Meyling A (1970) Demonstration of BVD-virus by the fluorescent antibody technique in tissues of cattle affected with bovine viral diarrhea (mucosal disease). Acta Vet Scand 11: 59–72
29. Moennig V, Bolin SR, Coulibaly COZ, Kelso Gourley NE, Liess B, Mateo A, Peters W, Greiser-Wilke I (1987) Untersuchungen zur Antigenstruktur von Pestiviren mit Hilfe monoklonaler Antikörper. Dtsch Tierärztl Wochenschr 94: 572–576
30. Moennig V, Mateo A, Greiser-Wilke I, Bolin SR, Kelso Gourley NE, Liess B (1989) Topological and functional epitope map of the surface of bovine viral diarrhea virus. In: Abstr. 89th Annual Meeting of the Am Soc Microbiol, New Orleans, Louisiana, p 390
31. Moennig V, Frey H-R, Liebler E, Pohlenz J, Liess B (1990) Reproduction of mucosal disease with cytopathogenic bovine viral diarrhea virus selected in vitro. Vet Rec 127: 200–203
32. Nuttal PA (1980) Growth characteristics of two strains of bovine virus diarrhoea virus. Arch Virol 66: 365–369
33. Olafson P, MacCallum AD, Fox FH (1964) An apparently new transmissible disease of cattle. Cornell Vet 36: 205–213
34. Peters W, Greiser-Wilke I, Moennig V, Liess B (1986) Preliminary serological characterization of bovine viral diarrhea virus strains using monoclonal antibodies. Vet Microbiol 12: 195–200
35. Ramsey FK (1956) Pathology of a mucosal disease of cattle. Iowa State University, PhD Thesis
36. Ramsey FK, Chivers WH (1953) Mucosal disease of cattle. North Am Vet 34: 629–633
37. Schulz L-Cl (1959) Pathologisch-anatomische Befunde bei der sogenannten "Mucosal Disease" (Schleimhautkrankheit) des Rindes. Dtsch Tierärztl Wochenschr 66: 586–588
38. Steck F, Lazary S, Frey H-R, Wandeler A, Huggler Ch, Oppliger G, Baumberger H, Kaderli R, Martig J (1980) Immune reponsiveness in cattle fatally affected by bovine virus diarrhea–mucosal disease. Zentralbl Veterinärmed [B] 27: 429–445
39. Wilhelmsen CL (1989) Pathogenesis of mucosal disease and acute bovine viral diarrhea. Iowa State University, PhD Thesis

Authors' address: Dr. E. M. Liebler, Institute of Pathology, Hannover Veterinary School, Bünteweg 17, D-W-3000 Hannover, Federal Republic of Germany.

Arch Virol (1991) [Suppl 3]: 125–132

Clinical and virological observations of a mucosal disease outbreak with persistently-infected seropositive survivors

S. Edwards, L. Wood, S. Brockman, and **G. Ibata**

Central Veterinary Laboratory, Weybridge, Surrey, U.K.

Accepted March 14, 1991

Summary. A group of 14 four to nine month old calves, clinically healthy but persistently infected with bovine virus diarrhoea virus (BVDV), was obtained from a single farm, and reared as a group. Ten of them were male and were castrated soon after arrival. Signs of mucosal disease (MD) developed within a month and eight of the males had died or been killed on humane grounds by 2 months after purchase. The other two males and one of the females developed more chronic but progressive signs of MD and were killed during the next four months. The remaining three females showed only transient signs of MD followed by clinical recovery. They subsequently remained healthy up to slaughter at 2, 2.5 and 5 years respectively. These three survivors were persistently infected with BVDV, and shed virus in their mucous secretions, although two of them were also seropositive to the virus with fluctuating neutralizing antibody titres (at times as high as 1/960) to a range of BVDV strains including their own persisting virus.

Key words: Bovine virus diarrhoea, mucosal disease, BVD, pestivirus, cattle.

Introduction

Mucosal disease (MD) is a progressive, usually fatal, disease of cattle which occurs only in animals which are persistently infected (PI) with, and immuno-tolerant to, bovine virus diarrhoea virus (BVDV) as a consequence of in utero infection. MD appears to be associated with an interaction between cytopathogenic and non-cytopathogenic biotypes of the virus which are antigenically closely related [5]. This paper describes a spontaneous outbreak of MD among a cohort of PI cattle which had been homebred on a single farm, and were likely therefore to have been infected in utero with the same strain of virus. The range of clinical and virological responses seen in the individual cattle highlights the complexity of this disease, and warns against an oversimplistic approach to its diagnosis or control.

Materials and methods

A group of 20 clinically healthy four to nine month old calves was obtained from a single farm and reared in group housing. Fourteen of the calves were BVDV-PI, 10 males and 4 females. The males were castrated one week after arrival. They were clinically monitored regularly over the next six months, and sporadically thereafter. The only experimental interventions were periodic blood sampling from the coccygeal vein using evacuated blood tubes, swabbing of nasopharyngeal, oral and rectal mucosae, and the intravenous administration of frusemide to facilitate collection of urine samples. Apart from the first case, which died, the cattle were killed on humane grounds when they became severely sick with MD. Tissue samples taken at post mortem examination were stored at $-70\,^{\circ}C$ pending virological testing.

Virus isolation from blood serum samples was done in microtitre plate cultures of bovine turbinate cells, followed by immunoperoxidase labelling to detect growth of noncytopathogenic BVDV [12]. In the case of swabs and tissues, sample suspensions were inoculated into secondary bovine kidney cells in Leighton tubes, with fluorescent antibody staining for BVDV antigen after 5 days incubation. For serology the sera were heated at $56\,^{\circ}C$ for 30 min to inactivate the intrinsic virus. Antibodies were detected by a microtitre virus neutralization test using 100 $TCID_{50}$ BVDV per well, 2 hours neutralization at $37\,^{\circ}C$, and bovine turbinate cells. The BVDV strains used were NADL, Oregon C24V, an isolate (A619) from an unrelated PI animal, and several isolates from within the present outbreak. Monoclonal antibody reaction patterns were determined using the Weybridge pestivirus panel in the peroxidase-linked assay as described previously [10].

Results

Slight to moderate clinical signs leading to suspicion of mucosal disease appeared among the male PI calves 2–3 weeks after arrival at the institute. These included inappetance, intermittent diarrhoea, and small ulcers around the gums. By the fourth week it was evident that a MD episode was in progress; the calves could not be used for their original purpose which was related to other experimental studies, so intensive monitoring of the clinical progression of the MD was commenced. The signs were typical of those previously described for MD [1], including nasal and ocular discharges, salivation, diarrhoea, inappetance, apathy, loss of body condition with a dry, scurfy skin, and ulceration of the gums, cheeks, tongue and hard palate. The heart and respiratory rates remained normal except for a terminal tachypnoea in one animal, and pyrexia was not a feature of the outbreak. No clinical signs were recorded in the six virus-negative calves. The temporal progression of the outbreak is indicated in Fig. 1. Eight of the males died or were killed in extremis between weeks 6 and 10 of the study. The remaining two males and one of the females developed signs of chronic MD leading to slaughter in weeks 18, 19 and 28. The other three females showed mild and intermittent signs (oral ulceration, nasal and ocular discharges) but eventu-

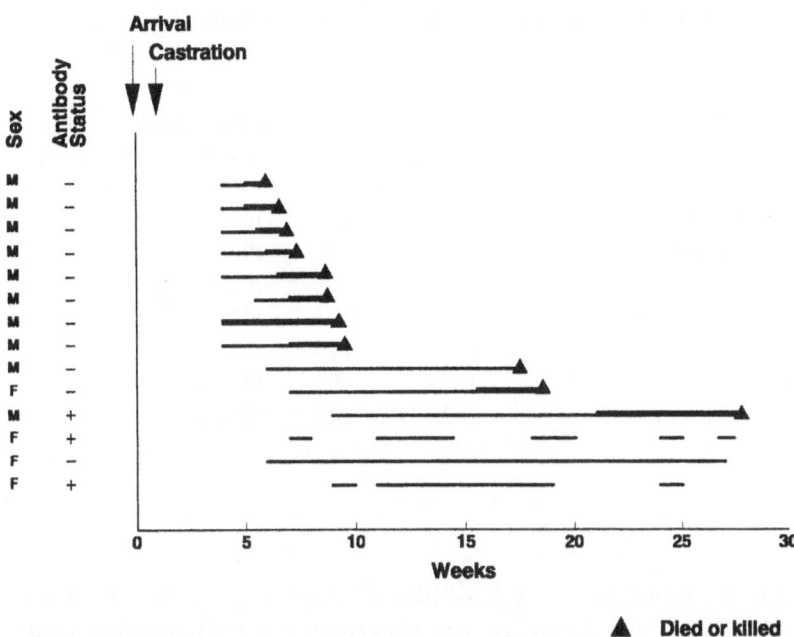

Fig. 1. Clinical progression of a mucosal disease outbreak in a group of 14 BVDV-persistently infected calves from a single farm

Each horizontal bar indicates the presence of clinical signs in one animal, with the thickness of the bar giving an indication of severity

ally recovered and remained healthy up to slaughter at 2, 2.5 and 5 years respectively.

At necropsy, the animals with MD showed extensive ulceration of the alimentary tract, in particular linear erosions in the oesophagus, and erosions on the ruminal pillars, the tips of the omasal leaves, and at the pylorus and the ileo-caeco-colic junction. Abomasitis, with or without ulceration, was a constant finding, as was mucosal damage in the small and/or large intestines although there was considerable variation in the distribution of the lesions between individual calves.

As expected, virus was isolated from a wide range of tissues of the PI cattle at necropsy, including cytopathogenic viruses from the mucosal disease cases. Cytopathogenic virus was also isolated from the mesenteric lymph nodes of one of the survivors at slaughter 2 years after the start of the study. In the live animals BVDV was readily isolated from buffy coat, blood clot, serum and nasopharyngeal swabs, including samples from those animals shown later to have antibodies to the virus. A lower, but still significant, isolation rate was recorded from samples of urine, oral swabs and rectal swabs (Table 1).

For the monoclonal antibody reaction patterns, comparisons were made between the earliest and the last isolate from each animal in the study, and between the isolates from different tissues taken at necropsy from each

S. Edwards et al.

Table 1. BVDV isolation rates from samples taken during the course of
a mucosal disease outbreak in persistently infected cattle

Sample	Total tested	No. (and percentage) positive for BVDV isolation	
Buffy coat	209	196	(94%)
Blood clot	226	223	(99%)
Serum	220	216	(98%)
Naso-pharyngeal swab	185	174	(94%)
Urine	173	116	(67%)
Oral swab	183	57	(31%)
Rectal swab	163	45	(28%)

animal. All the isolates reacted identically, corresponding to "pattern 1" as defined previously [10], except for the reaction with monoclonal WH216 which labelled only plaques within completely infected cell sheets, from a proportion of tissues from MD affected cases only. This monoclonal is believed to react selectively with cytopathogenic isolates of BVDV [10]. The labelling with WH216 did not precisely correlate with observed cyto-pathogenicity, although exhaustive attempts were not made to separate mixtures of the two biotypes of virus. The monoclonal reaction patterns also remained constant after four passages in different bovine cell types (turbinate, kidney, and MDBK).

Eleven of the PI calves remained seronegative throughout the study, to all strains of BVDV tested. This included all those which developed MD up to the 19th week of the study and one of the survivors which was killed at 2.5 years. One male was seronegative to NADL and A619, but showed a low neutralizing antibody titre to Oregon C24V and to four isolates from animals in the study group including the homologous isolate from the animal itself (Fig. 2). These titres varied between 1/20 and 1/240 with a median of 1/80. From day 160 this animal became seronegative to these viruses, coinciding with the worsening of clinical signs which led to its eventual slaughter suffering from, chronic MD.

The remaining two females were those which showed mild fluctuating signs followed by clinical recovery. They were seropositive throughout the study and remained so up to slaughter at 2 and 5 years respectively. Their sera neutralized NADL, Oregon C24V and A619 viruses with titres in the range 1/20 to 1/160, but the highest titres were found to the four isolates from animals in the study group, which included their individual homol-ogous viruses. In one case the homologous neutralizing titre fluctuated between 1/20 and 1/120, and in the other between 1/10 and 1/960 (Fig. 2). The pattern of neutralizing response to the serum titrations in the test plates

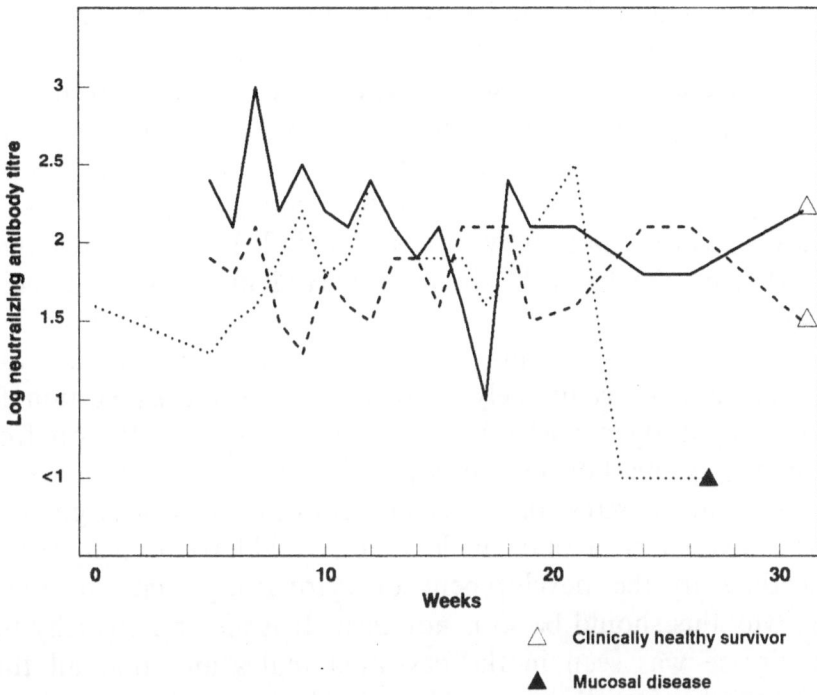

Fig. 2. BVDV neutralizing antibody titres in three seropositive BVDV-persistently infected calves

Each line represents one animal. The titres shown are those directed against the homologous persisting virus for each individual calf. Similar patterns were found to other strains of BVDV as described in the text

was more erratic than generally observed with non-PI seropositive cattle, suggesting that the neutralization was incomplete. Although they were definitely persistently infected, the isolation of virus from blood samples was less consistent than with the seronegative calves.

Discussion

Because it occurred on institute premises, this spontaneous outbreak of MD among a group of PI calves gave the opportunity for more detailed monitoring than is usually possible with disease episodes on farms. The PI animals fell into 3 groups according to their clinical responses, namely acute MD cases, chronic MD and mild disease with clinical recovery. Eight calves showed the classical signs and lesions of acute MD [1], with a time course of 1–2 weeks from the appearance of severe clinical signs, although early signs were detectable on clinical examination up to 3 weeks before this (Fig. 1). It is interesting to note the absence of pyrexia in these calves. The role of "homologous" cytopathogenic BVDV in the pathogenesis of MD is now well established [4, 5, 6, 7] although the reasons for the emergence of such a virus among a group of cattle remain obscure. These calves were homebred on a single farm and, although it cannot be proved conclusively, it is likely that they were born to dams who were exposed to the same strain of BVDV during early pregnancy. This is supported by the monoclonal typing of isolates from the calves, and by the neutralization test results, which indicated a high degree of antigenic homology among the persisting viruses in the 14 calves, as also reported by Corapi et al. [8]. These tests also indicated that, apart from the biotypic change involved in cytopathogenicity, the viruses were antigenically stable within the group for the whole of the study period. The calves were maintained in isolation so an external source of cytopathogenic virus is unlikely. The most plausible suggestion for the origin of cytopathogenic BVDV in MD cases is by mutation from the persistent non-cytopathogenic virus [8, 11]. It may be speculated in the present case that the stress of movement from the farm of origin, followed a week later by castration of the male calves, could have been a precipitating factor leading to the development of cytopathogenicity in the virus, although why this should be so is not clear. It is also noteworthy that the severest disease was seen in the castrated males and that all three PI survivors were females. Roeder & Drew [13] suggested a possible role for hormonal changes at puberty in the pathogenesis of MD. It has been suggested that both physiological events and the immune system may be implicated in MD pathogenesis [2]. The possible influence of such factors on molecular changes in the virus in PI cattle further investigation.

Three of the calves survived for 17 to 27 weeks before succumbing to chronic MD [1]. The onset of signs in this group was later and more insidious than in the acute cases, and the course of disease more prolonged. All the calves in the study would have had equal exposure to any circulating cytopathogenic virus so the different response may be attributed to host factors, possibly related to subtleties in the degree of BVDV-immunotolerance exhibited. The longest survivor of the three was seropositive, even to its own persisting virus, and only developed frank MD at the time when neutralizing antibodies were no longer detectable in its serum (Figs. 1 and 2). An alternative hypothesis for the pathogenesis of chronic MD, the super-infection of PI cattle with "partially homologous" cytopathogenic virus [5], would not seem tenable in the present case.

The final group of 3 heifers included 2 seropositives and may be presumed to have been only partially immunotolerant to the virus, with persistent virus and a defective immune response co-existing in some sort of equilibrium, which appeared to become more stable after about 6 months. In one case it appears that this mechanism extended to the cytopathogenic variant of the virus, which was isolated from a healthy survivor at slaughter 2 years from the start of the study. Although it is known that PI cattle can mount an immune response to "heterologous" extraneous BVDV [3, 7] it seems an unlikely explanation for the seropositivity in this case, in view of the neutralizing activity against the homologous persisting virus and the absence of neutralizing antibody in the majority of PI calves in the group. The degree of immune responsiveness may be related to the stage of gestation at which the fetuses were originally exposed to become persistently infected. Unfortunately no detailed information was available on this aspect of the animals' history. Studies of the cell-mediated immune effector systems in similar animals would clearly be of value.

This case has important implications for the diagnosis and control of BVDV. Great caution should be expressed about schemes for the identification of PI cattle which rely on preliminary serological screening, with virus screening only of seronegatives. A small proportion of PI cattle are not fully immunotolerant to the virus, and may have high titres of neutralizing antibody [9]. Such animals may be doubly difficult to detect since they are the very ones for which virus isolation tests, particularly those relying on serum samples, may be the least reliable.

References

1. Baker JC (1990) Clinical aspects of bovine virus diarrhoea virus infection. Rev Sci Tech Off Int Epiz 9: 25–41
2. Bielefeldt Ohmann B (1988) BVD virus antigens in tissues of persistently viraemic, clinically normal cattle: implications for the pathogenesis of clinically fatal disease. Acta Vet Scand 29: 77–84
3. Bolin SR (1988) Viral and viral protein specificity of antibodies induced in cows persistently infected with noncytopathic bovine viral diarrhea virus after vaccination with cytopathic bovine viral diarrhea virus. Am J Vet Res 49: 1040–1044
4. Bolin SR, McClurkin AW, Cutlip RC, Coria MF (1985) Severe clinical disease induced in cattle persistently infected with noncytopathic bovine viral diarrhea virus by superinfection with cytopathic bovine viral diarrhea virus. Am J Vet Res 46: 573–576
5. Brownlie J (1990) The pathogenesis of bovine virus diarrhoea virus infections. Rev Sci Tech Off Int Epiz 9: 43–59
6. Brownlie J, Clarke MC, Howard CJ (1984) Experimental production of fatal mucosal disease in cattle. Vet Rec 114: 535–536
7. Brownlie J, Clarke MC, Howard CJ (1987) Clinical and experimental mucosal disease—defining a hypothesis for pathogenesis. In: Harkness JW (ed) Pestivirus infections of ruminants. CEC Agriculture series, EUR 10238, Brussels, pp 147–157
8. Corapi WV, Donis RO, Dubovi EJ (1988) Monoclonal antibody analyses of cytopathic and noncytopathic viruses from fatal bovine viral diarrhea virus infections. J Virol 62: 2823–2827
9. Duffell SJ, Harkness JW (1985) Bovine virus diarrhoea–mucosal disease infection in cattle. Vet Rec 117: 240–245
10. Edwards S, Sands JJ, Harkness JW (1988) The application of monoclonal antibody panels to characterize pestivirus isolates from ruminants in Great Britain. Arch Virol 102: 197–206
11. Howard CJ, Brownlie J, Clarke MC (1987) Comparison by the neutralisation assay of pairs of non-cytopathogenic and cytopathogenic strains of bovine virus diarrhoea virus isolated from cases of mucosal disease. Vet Microbiol 13: 361–369

12. Meyling A (1984) Detection of BVD virus in viraemic cattle by an indirect immuno-peroxidase technique. In: McNulty MS, McFerran JB (eds) Recent advances in virus diagnosis. Martinus Nijhoff Publishers, Boston The Hague Dordrecht Lancaster, pp 37–46
13. Roeder PL, Drew TW (1984) Mucosal disease of cattle: a late sequel to fetal infection. Vet Rec 114: 309–313

Authors' address: Dr S. Edwards, Central Veterinary Laboratory, Weybridge, Surrey, KT15 3NB. U.K.

Arch Virol (1991) [Suppl 3]: 133–142

Insertion of cellular sequences in the genome of bovine viral diarrhea virus

G. Meyers[1], **T. Rümenapf**[1], **N. Tautz**[1], **E. J. Dubovi**[2], and **H.-J. Thiel**[1,a]

[1] Federal Research Centre for Virus Diseases of Animals, Tübingen
[2] New York State College of Veterinary Medicine, Cornell University, Ithaca, New York, U.S.A.

Accepted March 14, 1991

Summary. The genomic sequences of four pestiviruses, two BVDV strains (Osloss and NADL, both of which are cytopathogenic) and two HCV strains, were analyzed. Comparative studies revealed the presence of small insertions of cellular sequences in the genomes of both BVDV strains; the insertions are located in a region coding for a nonstructural protein. Such insertions are not present in the HCV sequences. The insertion identified in BVDV Osloss encodes a complete ubiquitin-like element. The sequence inserted in the BVDV NADL genome shows no homology to a ubiquitin gene but is almost identical with another bovine mRNA sequence.

Molecular characterization of a BVDV "pair", isolated from an animal with mucosal disease, led to the detection of a ubiquitin-like sequence in the genome of the cytopathogenic strain but not of the noncytopathogenic strain. It is proposed that recombination between viral and cellular RNA leads to formation of cpBVDV genomes. This hypothesis has direct implications for the pathogenesis of mucosal disease.

Key words: Bovine viral diarrhea virus, RNA recombination, mucosal disease, pestivirus, ubiquitin.

Introduction

Fatal mucosal disease (MD) of cattle is unique with respect to its dependence on the coexistence of antigenically closely related noncytopathogenic bovine

---

[a] To whom requests for reprints should be addressed.

viral diarrhea virus (ncpBVDV) and cytopathogenic BVDV (cpBVDV) in one animal [16, 3]. It is generally assumed that cpBVDV arises from the noncytopathogenic virus in the infected animal by some kind of mutation [3].

Comparison of the genomic sequences of the two cpBVDV isolates Osloss [20] and NADL [1] with the ones of HCV [10, 15] led to the identification of insertions in the BVDV genomes [2, 10]. Such insertions are not present in the HCV sequence. The amino acid sequences encoded by the insertions are highly homologous to cellular sequences [11, 12, 13]. This led us to suggest a novel model for pathogenesis of MD [11]. Accordingly, a noncpBVDV mutates in persistently infected animals to a cpBVDV by taking up cellular sequences during a recombination event.

Detection of ubiquitin-coding sequence

The genome of the cpBVDV Osloss strain was the first BVDV strain to be completely sequenced [20]. The insertion identified in the Osloss genome comprises 228 nucleotides. A data bank search revealed homology of 97% between the amino acid sequence deduced from the Osloss insertion and animal ubiquitin (Fig. 1). In fact, the 228 nucleotides identified in the RNA genome of the BVDV Osloss strain code for a complete ubiquitin protein of 76 amino acids [11, 13]. In comparison with the ubiquitin sequence conserved in all animals examined so far [18], two amino acid exchanges were detected in the Osloss ubiquitin (Fig. 1).

Fig. 1. Amino acid sequence comparison of parts of the polyproteins encoded by the genomes of BVDV strains Osloss [20] and NADL [1] and the animal ubiquitin [18]. The amino acids differing between animal ubiquitin and the protein encoded by the Osloss insertion are boxed

To examine whether the ubiquitin-coding sequence is specific for the Osloss strain, a BglII fragment derived from a porcine polyubiquitin cDNA clone (pCL208 [5]) was hybridized to total RNA of Madin-Darby bovine kidney (MDBK) cells infected with five different strains of cpBVDV. The ubiquitin probe recognized only genomic RNA from the Osloss strain (Fig. 2A, lane 1), whereas viral RNA of all strains was clearly detectable with a pestivirus-specific probe [10, 13] (Fig. 2B). Hybridization of the ubiquitin probe to poly(A)$^+$ RNA of noninfected MDBK cells indicated that the three bands visible in all lanes of Fig. 2A represent ubiquitin mRNAs (data not shown). Ubiquitin mRNA species varying in number and size have been described for other mammalian species [5, 18, 23].

Insertion in the BVDV strain NADL

The NADL insertion of 270 nucleotides is located in the same genomic region as the Osloss insertion (see Conclusions). However, the two inserted sequences exhibit no nucleotide or deduced amino acid sequence homology

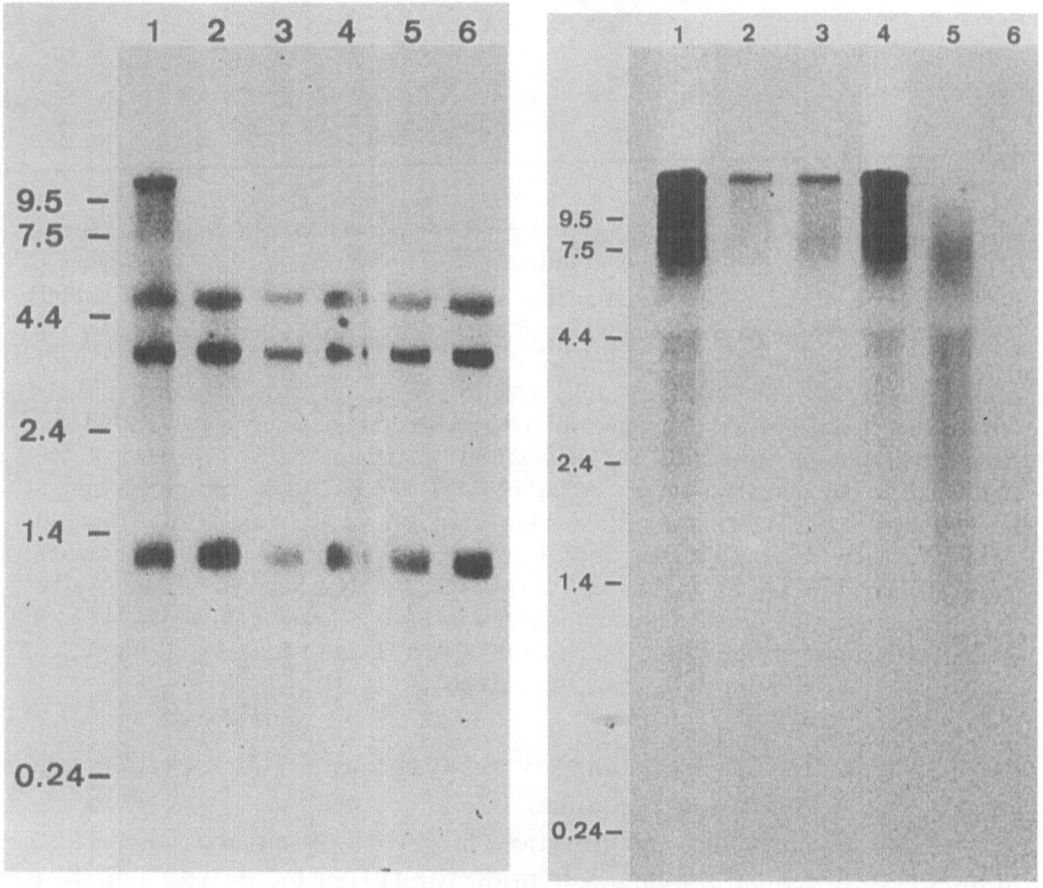

Fig. 2 A, B

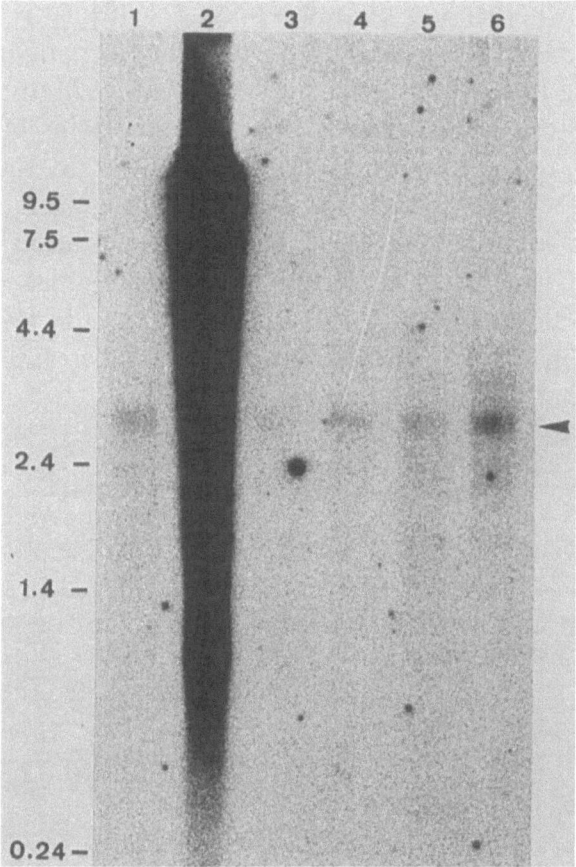

Fig. 2. Northern blot analyses of total RNA from noninfected MDBK cells (lanes 6) and MDBK cells infected with BVDV strains Osloss (lanes 1), NADL (lanes 2), Oregon (lanes 3), Danmark (lanes 4), and Singer (lanes 5). The blots were hybridized with a 0.2 kb BglII fragment isolated from the porcine polyubiquitin cDNA clone pCL208 [5] (**A**), a 2.3 kb SalI fragment derived from the hog cholera virus cDNA clone 4.5 [10] (**B**), or two oligo-nucleotides complementary to nucleotides 4994 to 5093 and 5094 to 5182 of the BVDV NADL genome [1] (**C**). A 5 µg (**A** and **B**) or 10 µg (**C**) amount of glyoxylated RNA was separated in phosphate-buffered 1% agarose gels containing 5.5% formaldehyde and transferred to Duralon membranes (Stratagene) [19]. Hybridization with probes labeled with ^{32}P by nick lanslation (nick translation kit; Amersham Corp.) (**A** and **B**) or polynucleo-tide kinase (New England BioLabs, Inc.) (**C**) was performed in 0.5 M sodium phosphate (pH 6.8) − 1 mM EDTA–7% SDS at 68 °C (**A**) or 54 °C (**B** and **C**). Posthybridization washes were carried out with 40 mM sodium phosphate (pH 6.8) − 1 mM EDTA–5% SDS and 40 mM sodium phosphate (pH 6.8) − 1 mM EDTA–1% SDS two times for 30 min each at hybridization temperature

to each other. In addition similarity between the NADL insertion and known sequences could not be found.

To find out whether the insertion in BVDV strain NADL is also homologous to cellular sequences, hybridization experiments with a mixture of two oligonucleotides complementary to this insertion were carried out. A cellular RNA of about 2.9 kb could be detected on the resulting Northern (RNA) blots (arrow in Fig. 2C). After longer exposure times, two additional bands were observed (data not shown). All three RNA species bind to oligo(dT)-cellulose (see below) and thus probably represent mRNAs. The hybridization experiments also showed that the NADL insertion is strain specific, at least with regard to the five strains tested here (Fig. 2C).

For further investigation, a cDNA library was constructed from poly(A)$^+$ RNA of noninfected MDBK cells and screened with the NADL insertion-specific oligonucleotide mixture; one positive clone (pcINS) containing an insert of about 1 kb hybridized to genomic RNA of BVDV NADL and to three species of poly(A)$^+$ RNA of noninfected MDBK cells (Fig. 3). Nucleotide sequencing revealed that the respective cDNA fragment contains a sequence highly homologous to the insertion identified in the genome of BVDV strain NADL (Fig. 4). The complete sequence spanning the insertion is conserved except for two nucleotide exchanges, whereas the flanking regions show almost no homology (Fig. 4).

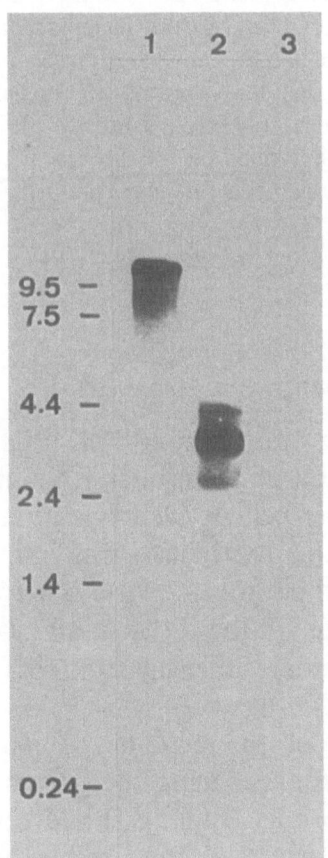

Fig. 3. Northern hybridization analysis of 0.5 μg of total RNA of MDBK cells infected with BVDV NADL (lane 1) and 10 μg of poly(A)$^+$ (lane 2) or 20 μg of poly(A)$^-$ (lane 3) RNA of noninfected MDBK cells, using the insert of pcINS as a probe. RNA electrophoresis, transfer and hybridization were done as described for Fig. 2A

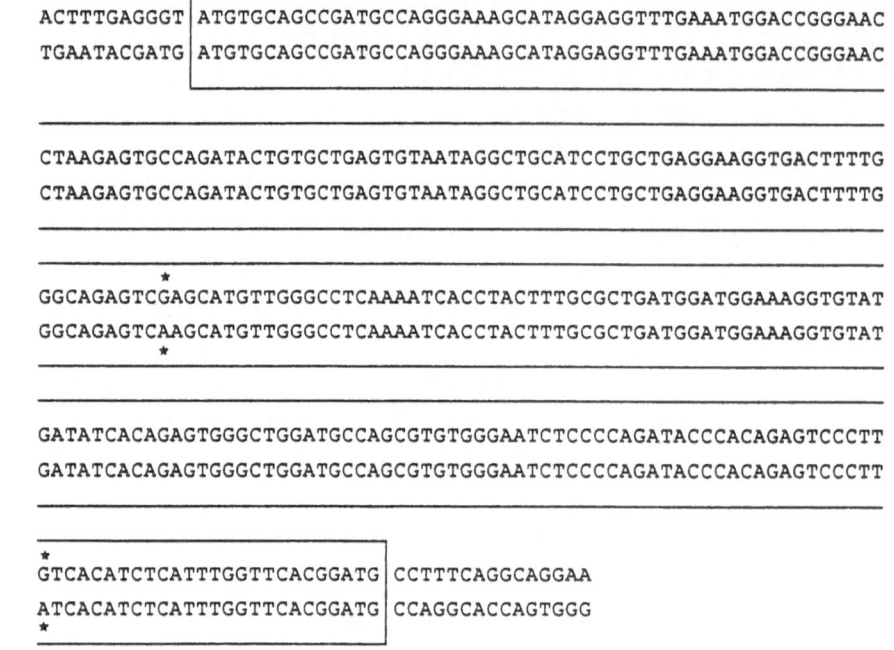

```
BVDV:   ACTTTGAGGGT | ATGTGCAGCCGATGCCAGGGAAAGCATAGGAGGTTTGAAATGGACCGGGAAC
cINS:   TGAATACGATG | ATGTGCAGCCGATGCCAGGGAAAGCATAGGAGGTTTGAAATGGACCGGGAAC


        CTAAGAGTGCCAGATACTGTGCTGAGTGTAATAGGCTGCATCCTGCTGAGGAAGGTGACTTTTG
        CTAAGAGTGCCAGATACTGTGCTGAGTGTAATAGGCTGCATCCTGCTGAGGAAGGTGACTTTTG

                       *
        GGCAGAGTCGAGCATGTTGGGCCTCAAAATCACCTACTTTGCGCTGATGGATGGAAAGGTGTAT
        GGCAGAGTCAAGCATGTTGGGCCTCAAAATCACCTACTTTGCGCTGATGGATGGAAAGGTGTAT
                       *

        GATATCACAGAGTGGGCTGGATGCCAGCGTGTGGGAATCTCCCCAGATACCCACAGAGTCCCTT
        GATATCACAGAGTGGGCTGGATGCCAGCGTGTGGGAATCTCCCCAGATACCCACAGAGTCCCTT

        *
        GTCACATCTCATTTGGTTCACGGATG | CCTTTCAGGCAGGAA
        ATCACATCTCATTTGGTTCACGGATG | CCAGGCACCAGTGGG
        *
```

Fig. 4. Comparison of the nucleotide sequences of the region of the BVDV strain NADL genome [1] containing the insertion (upper line) and of part of a cDNA clone isolated from an MDBK poly(A)$^+$ cDNA library (cINS, lower line). The region identified as an insertion in the viral genome and the corresponding part of the cellular sequence are boxed. Nucleotide exchanges in this region are indicated by asterisks. cDNA synthesis [19], cloning in lambda ZAPII bacteriophages (Stratagene), screening of the library with oligonucleotides [10] (see legend to Fig. 2), and nucleotide sequencing [10] were performed as described before

Investigation of a BVDV "pair"

According to our working hypothesis the genome of a given cpBVDV has developed from that of a ncpBVDV by integration of additional RNA in a recombination process. As a consequence, host cell derived insertions should be specific for cpBVDV. Verification of this hypothesis was hampered by the fact that sequence data from noncytopathogenic viruses were not available. At this point, molecular characterization of a "pair" of ncpBVDV and cpBVDV isolated from one animal suffering from MD represented the most obvious approach.

As a first step towards molecular analysis of the genomes of the cytopathogenic BVDV strain CP1 and its noncytopathogenic counterpart NCP1 [3] hybridization experiments were performed (Fig. 5). While the genome of BVDV strain NCP1 had the expected size of about 12.5 kb the genomic RNA of BVDV CP1 was much larger with an estimated size of 14 to 15 kb (Fig. 5).

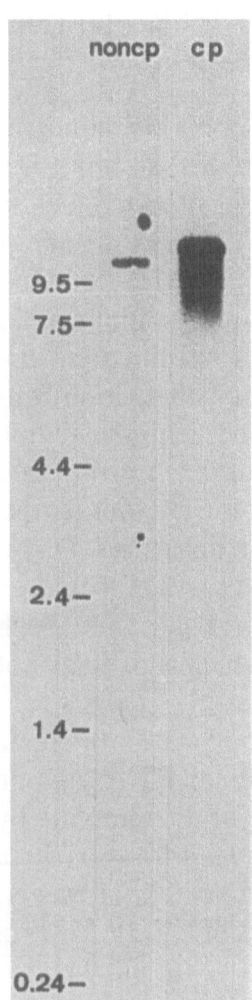

Fig. 5. Northern blot analysis of total RNA from MDBK cells infected with noncpBVDV (left lane) and cpBVDV (right lane) isolated from one animal suffering from mucosal disease. The blot was hybridized with a 2.3 kb SalI fragment derived from the HCV cDNA clone 4.5 [10]. The lanes are mounted from autoradiographs with different exposure times of the same blot. Sizes of an RNA ladder are indicated in kilobases at the left

Northern hybridization experiments with the porcine ubiquitin clone had shown that among five common cpBVDV laboratory strains the ubiquitin-coding insertion was specific for the Osloss strain (see above). An analogous experiment was conducted with RNA of cells infected with CP1 and NCP1. Surprisingly, a positive signal was obtained with genomic CP1 RNA after stringent hybridization with the ubiquitin probe (data not shown). The genomic RNA of BVDV NCP1 was not recognized by this DNA fragment; therefore at least part of the additional RNA in the CP1 genome represents a ubiquitin-like sequence.

Conclusions

The data outlined above demonstrate that three different cpBVDV strains (Osloss, NADL, CP1) have integrated cellular sequences into their genomes.

In two of these strains ubiquitin-coding sequences are present (see "note added in proof" for CP1). In contrast, the genomes of a ncpBVDV strain NCP1 (see "note added in proof") as well as generally noncytopathogenic HCV strains [10, 15] do not contain such insertions.

One remarkable difference between cpBVDV and ncpBVDV concerns the processing of a nonstructural 125 kDa protein (p125) of unknown function to a product of about 80 kDa. This cleavage can be detected only in tissue culture cells infected with cpBVDV, whereas the 125 kDa protein is not processed after infection with ncpBVDV [4, 17]. The 80 kDa products resulting from the cpBVDV specific processing of this protein exhibit similar migration rates upon SDS-PAGE for all BVDV strains examined so far, indicating similar positions at which cleavage occurs [4, 17]. The insertions of cpBVDV strains are located in close proximity in the genomic region coding for p125 [11, 12, 13]. Our calculations revealed that the insertions are in proximity of the putative processing site(s) of p125. It is therefore tempting to speculate that cleavage of p125 is influenced by the insertions. Direct evidence for the molecular basis of cytopathogenicity of cpBVDV could be obtained by analyses involving mutagenesis of the viral genome. Such experiments require, however, infectious BVDV cDNA which is thus far not available.

The most reasonable explanation for development of cytopathogenic viruses is a recombination process between viral and cellular RNA. As one mechanism for recombination of RNA viruses, switching of the template by the viral RNA polymerase has been proposed [7, 8, 9]. Most of these recombinations could be explained by a single template switch. However, the integration of host cellular sequences into the genomes of BVDV NADL, Osloss and CP1 by a "copy choice" mechanism would represent heterologous reactions requiring two template switches between viral and cellular RNA. As the cellular reaction partner is of positive polarity such a process should occur during viral negative strand synthesis to result in the observed integration of cellular sequences in coding orientation. As already indicated in the Introduction the hypothesis has direct implications for the pathogenesis of mucosal disease [11].

To our knowledge BVDV is the first classical positive stranded RNA virus for which integration of host cellular protein coding sequences has been demonstrated. Acquisition of new properties from the host cell by recombination has been proposed as an important force in the evolution of RNA viruses [21, 22]. Identification of the BVDV recombination was facilitated by the biological selection system "mucosal disease" but analogous processes may also occur in other RNA viruses [14]. Recombination between influenza virus RNA and ribosomal RNA led to a virus with altered biological properties [6]. Without any doubt future investigations of BVDV "pairs" will have a major impact on studies concerning RNA recombination and pathogenesis of virus induced diseases.

Note added in proof

Molecular characterisation of the BVDV pair mentioned in this article was recently published. Meyers G, Tautz N, Dubovi EJ, Thiel H-J (1991) Viral cytopathogenicity correlated with integration of ubiquitin-coding sequences. Virology 180: 602–616

References

1. Collett MS, Larson R, Gold C, Strinck D, Anderson DK, Purchio AF (1988) Molecular cloning and nucleotide sequence of the pestivirus bovine viral diarrhea virus. Virology 165: 191–199
2. Collett MS, Moennig V, Horzinek MC (1989) Recent advances in pestivirus research. J Gen Virol 70: 253–266
3. Corapi WV, Donis RO, Dubovi EJ (1988) Monoclonal antibody analysis of cytopathic and noncytopathic viruses from fatal bovine viral diarrhea virus infections. J Virol 62: 2823–2827
4. Donis RO, Dubovi EJ (1987) Molecular specificity of the antibody responses of cattle naturally and experimentally infected with cytopathic and noncytopathic bovine diarrhea virus biotypes. Am J Vet Res 48: 1549–1554
5. Einspanier R, Sharma HS, Scheit KH (1987) An mRNA encoding polyubiquitin in porcine corpus luteum: identification by cDNA cloning and sequencing. DNA 6: 395–400
6. Khatchikian D, Orlich M, Rott R (1989) Increased viral pathogenicity after insertion of a 28S ribosomal RNA sequence into the haemagglutinin gene of an influenza virus. Nature 340: 156–157
7. King AMQ, Ortlepp SA, Newman JWI, McCahon (1987) Genetic recombination in RNA viruses. In: Rowlands DJ, Mayo MA, Mahy BWJ (eds) The molecular biology of the positive stranded RNA viruses. Academic Press, London, pp 129 152
8. Kirkegaard K, Baltimore D (1986) The mechanism of RNA recombination in poliovirus. Cell 47: 433–443
9. Lazzarini RA, Keene JD, Schubert M (1981) The origins of defective interfering particles of the negative-strand RNA viruses. Cell 26: 145–154
10. Meyers G, Rümenapf T, Thiel H-J (1989) Molecular cloning and nucleotide sequence of the genome of hog cholera virus. Virology 171: 555–567
11. Meyers G, Rümenapf T, Thiel H-J (1989) Ubiquitin in a togavirus. Nature 341: 491
12. Meyers G, Rümenapf T, Thiel H-J (1990) Insertion of host cell derived sequences identified in the RNA genome of a Togavirus. In: Lerner RA, Ginsberg H, Chanock RM, Brown F (eds) Vaccines 90, modern approaches to new vaccines including prevention of Aids. Cold Spring Harbor Laboratory, pp 47–51
13. Meyers G, Rümenapf T, Thiel H-J (1990) Insertion of ubiquitin coding sequence identified in the RNA genome of a Togavirus. In: Brinton MA, Heinz FX (eds) New aspects of positive strand RNA viruses. American Society for Microbiology, Washington DC, pp 25–30
14. Miller RH, Purcell RH (1990) Hepatitis C virus shares amino acid sequence similarity with pestiviruses and flaviviruses as well as members of two plant virus supergroups. Proc Natl Acad Sci USA 87: 2057–2061
15. Moormann RJM, Warmerdam PAM, Van der Meer B, Schaaper WMM, Wensvoort G, Hulst MM (1990) Molecular cloning and nucleotide sequence of hog cholera virus strain brescia and mapping of the genomic region encoding envelope protein E1. Virology 177: 184–198

16. Pocock DH, Howard CJ, Clarke MC, Brownlie J (1987) Variation in the intracellular polypeptide profiles from different isolates of bovine viral diarrhea virus. Arch Virol 94: 43–53
17. Purchio AF, Larson R, Collett MS (1984) Characterization of bovine viral diarrhea viral proteins. J Virol 50: 666–669
18. Rechsteiner M (1987) Ubiquitin-mediated pathways for intracellular proteolysis. Ann Rev Cell Biol 3: 1–30
19. Rümenapf T, Meyers G, Stark R, Thiel H-J (1989) Hog cholera virus — characterization of specific antiserum and identification of cDNA clones. Virology 171: 18–27
20. Renard A, Dino D, Martial J (1987) Vaccines and diagnostics derived from bovine diarrhea virus. European Patent Application number 86870095.6 Publication number 0208672 14 January 1987
21. Steinhauer DA, Holland JJ (1987) Rapid evolution of RNA viruses. Ann Rev Microbiol 41: 409–433
22. Strauss JH, Strauss EG (1988) Evolution of RNA viruses. Ann Rev Microbiol 42: 657–683
23. Wiborg O, Pedersen MS, Wind A, Berglund LE, Marcker KA, Vuusi J (1985) The human ubiquitin multigene family: some genes contain multiple directly repeated ubiquitin coding sequences. EMBO J 4: 755–759

Authors' address: Dr. H.-J. Thiel, Federal Research Centre for Virus Diseases of Animals, P.O. Box 1149, D-W-7400 Tübingen.

Arch Virol (1991) [Suppl 3]: 143–148
© Springer-Verlag 1991

Congenital curly haircoat as a symptom of persistent infection with bovine virus diarrhoea virus in calves

B. Larsson[1], **S.-O. Jacobsson**[1], **B. Bengtsson**[1], and **S. Alenius**[2]

[1] Department of Cattle and Sheep Diseases, Swedish University of Agricultural Sciences, Uppsala, Sweden, and
[2] National Veterinary Institute, Uppsala, Sweden

Accepted March 14, 1991

Summary. Ten calves were born small and with a curly haircoat in a dairy herd which comprised approximately 185 milking animals. These calves commonly developed diarrhoea and/or signs of respiratory disease at the age of 2 to 4 weeks. Two of the calves died and 5 were chronically ill and poor doers and were therefore euthanized. This susceptibility to disease of the curly haired calves was quite different from what was observed among other calves in the herd.

Sera from seven of the curly haired calves were examined and were all found to be free from detectable antibodies to bovine virus diarrhoea virus (BVDV) and to harbour a non-cytopathic strain of BVDV. One of the calves was retested after 7 weeks and was still seronegative and viraemic. Of 49 non-curly haired calves examined in the herd 44 were BVDV seropositive. The other 5 were seronegative to BVDV but attempts to isolate BVDV from their sera failed.

Key words: Curly haircoat, bovine virus diarrhoea virus.

Introduction

Infection with bovine virus diarrhoea virus (BVDV) causes a wide spectrum of clinical syndromes in cattle throughout the world. A postnatal infection with this pestivirus is commonly mild or even subclinical but severe disease has been observed [1]. BVDV is also pathogenic for the bovine foetus. The outcome of a transplacental infection includes foetal death, teratogenic effects and immunotolerance to BVDV [10]. The tolerance is a sequel to a foetal exposure to a non-cytopathic strain of BVDV in early pregnancy and such calves are born persistently infected with the virus [8]. Due to the tolerance, persistently infected animals have no, or only a low level of, antibodies to BVDV. However, if they have been exposed to a strain somewhat different to the persistent infecting strain, the animals may produce antibodies to the recent strain [3, 11].

Persistently infected cattle may perform normally and reach breeding age, and in case of pregnancy, give birth to calves also persistently infected. Thus, families of persistently infected cattle may develop in a herd. However, such cattle are often recognized as being small at birth, having a poor growth rate and an unthrifty appearance. Persistently infected cattle also appear to have an increased susceptibility to other infections [2] and are the population at risk of developing mucosal disease [4]. In this paper the serial birth of calves with a congenital curly haircoat, which occurred in a Swedish dairy herd is described and this clinical condition is related to a persistent infection with BVDV.

Materials and methods

Herd history

The herd comprised approximately 185 milking cows and heifers mostly of Swedish Red and White breed (SRB). Ten calves of this breed were born with a curly haircoat in 1989, one of them in February and 9 in August and September. The calves were of both sexes and were born to 9 heifers and one biparous cow. Artificial insemination was used and sperm from different donors had been used to serve the dams of the curly haired calves. The curly haired calves were born small and some of them had a narrow head and small ears with hypotrichosis. Typically, when the curly haired calves were 2 to 4 weeks old they developed diarrhoea and/or signs of respiratory disease and there was, in some calves, a regional loss of hair on the posterior legs in association with the diarrhoea. In the end of September 1989, two of the curly haired calves (nos. 151 and 156) were hospitalized at the clinic of Department of Cattle and Sheep Diseases, Swedish University of Agricultural Sciences, Uppsala.

Herd investigation

In early November 1989, blood samples were collected from the 5 curly haired calves still alive in the herd and from 9 of the 10 dams giving birth to the curly haired calves. The 10th dam had earlier been culled as a poor milk producer. Blood samples were also collected from 49 calves with a normal haircoat born between May and September. Milk samples were obtained from 93 cows. In late December 1989, a second blood sample were obtained from the only curly haired calf (no. 168) still available.

Virological techniques

An enzyme-linked immunosorbent assay (ELISA) (SVANOVA Biotech AB, Uppsala, Sweden) was used to detect antibodies to BVDV both in serum [6] and milk [9]. Presence of BVDV in sera was tested by inoculating sera on cultures of embryonic bovine turbinate cells. After incubation for four days the cultures were examined for cytopathic effect and the presence of BVDV was determined by an indirect immunofluorescence test, using a monoclonal antibody to BVDV.

Results

Examination of the hospitalized calves

The hospitalized calves nos. 151 and 156 were both male, aged 7 and 5 weeks respectively, and in poor bodily condition. Both calves had a curly haircoat (Fig. 1) and they had regional areas of alopecia on their hind legs. Body temperature varied between 39.2 and 39.3 °C and 38.8 and 39.6 °C, respectively, during a 4-day period. Both calves coughed spontaneously and on ascultation of the lungs, crackles were heard particularly over the cranioventral part. The faeces were yellowish and pasty. A few small erosions were seen in the mucosa of the hard palate in calf no. 151. Routine haematological examination of blood from both calves showed a low level of protein in

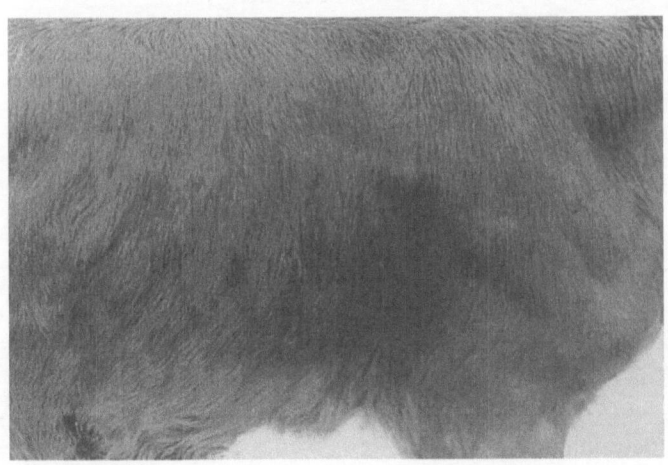

Fig. 1. Close-up of congenital curly haircoat (**a**) and normal haircoat (**b**) of Swedish Red and White breed

serum (55 and 48 g/l), leukocytosis (14.4 and 22.3×10^9 leukocytes/l), absolute neutrophilia (9.2 and 18.7×10^9 neutrophils/l) which included an increased number of band neutrophils (3.1 and 11.1×10^9/l).

Sera from both calves were free from detectable antibodies to BVDV (titre < 10). A non-cytopathic strain of BVDV was isolated from serum of both calves.

Herd investigation

Table 1 shows the identity and fate of the 10 curly haired calves. Two of the calves died following a period of diarrhoea and respiratory disease, two calves were hospitalized and subsequently euthanized because of their severe disease (see above), and three of the calves were chronically ill and poor doers and were euthanized. Three calves were considered to be in a fairly good condition and two of them were sold and the third was still alive within the herd in March 1990.

Five of the curly haired calves were available for sampling in early November and all were found to be seronegative to BVDV (titre < 10) and to harbour a non-cytopathic strain of BVDV. One of these calves (no. 168) was bled again in late December and was still antibody negative and virus positive. All examined dams of the curly haired calves were seropositive to BVDV. Seventyfive out of 93 cows were found to be antibody positive in milk and 44 out of the 49 non-curly haired calves were found to be antibody positive to BVDV in serum. The five seronegative calves were virus negative.

Discussion

The outcome of infection with BVDV of pregnant cattle depends largely on the stage of gestation at which the infection occurs and the biotype of the

Table 1. Status to BVDV and fate of calves born with a curly haircoat

Calf (No.)	Date of birth	Sex	Antibody titre to BVDV	BVDV in serum	Fate
692	890219	F	nd	nd	Died in spring 1989
151	890805	M	< 10	yes	Euthanized 890928
154	890807	M	nd	nd	Sold 891013
156	890816	M	< 10	yes	Euthanized 890928
158	890829	M	nd	nd	Sold 891013
733	890907	F	< 10	yes	Euthanized 891124
737	890916	F	< 10	yes	Euthanized 891124
740	890918	F	< 10	yes	Euthanized 891124
168	890920	M	< 10	yes	Alive in March 1990
175	890930	M	< 10	yes	Died 891117

F, female; M, male; nd, not determined

infecting virus. A broad spectrum of abnormalities is reported ranging from death in utero to a lifelong subclinical infection. Between these two extremes lies a variety of teratogenic effects such as cerebellar hypoplasia, ocular defects, thymus hypoplasia, hydrancephaly, and hypotrichosis [10]. In this paper we add another clinical symptom as a sequel to an intra-uterine infection with BVDV, i.e., a congenital curly haircoat of calves of a breed which normally are born with straight hairs. Calves born with a curly haircoat were generally small at birth, were persistently infected with a non-cytopathic strain of BVDV and were at risk of developing clinical disease. This susceptibility to disease of the curly haired calves was quite different from what was observed among the normal non-curly haired calves in the herd. The low rate of survival of the curly haired calves was probably attributed to a BVDV-induced suppression of the defense mechanisms of these calves to other infections. The haematological data and clinical findings of the hospitalized calves supported the diagnosis of a suppurative pneumonia. This diagnosis was also confirmed at necropsy. There are only a few reports in the literature concerning malformations of hair and skin as a consequence of a prenatal infection with BVDV in cattle. Intra-uterine inoculation of foetuses younger than 150 days has been reported to cause partial destruction of the germative cells of epidermis [5]. Partial alopecia and irregular length of hair of a calf was observed after an experimental infection of its dam at 93 days of pregnancy [7]. This calf had precolostral neutralizing antibodies to BVDV and, thus, was not likely to be persistently BVDV infected. These sporadic reports of hair defects due to BVDV infection is in contrast to the strong relation between curly haircoat and persistent infection observed in the herd of this study. All calves found to be infected with BVDV had a curly haircoat. However, in our experience, a curly haircoat is not a common sign of a persistent BVDV infection in SRB calves. Therefore, the pathogenicity of BVDV isolated from curly haired calves may differ from that of other field strains of the virus. Hair abnormalities is often observed in lambs after an intra-uterine infection with border disease virus, a virus which is closely related to, if not the same as, BVDV. The main lesion in skin of lambs with border disease is an enlargement of the primary hair follicles and a concurrent reduction in the number of secondary follicles. The resulting hairiness, which is due to the presence of medullated primary fibres, only becomes manifest in normally smooth coated breeds [12]. No closer histological examination of the skin of the curly haired calves was performed to determine the degree of similarities, if there is any, to border disease.

Since the relation between a congenital curly haircoat and a persistent infection became established, we have received blood samples from curly haired calves in 3 other herds. The calves, not only of SRB but also of Swedish Friesian breed, have been seronegative to BVDV and virus positive, indicating a persistent infection.

Acknowledgement

We wish to thank Mrs. Hjort for skilful laboratory assistance. This work was supported by grants from the Farmers' Council for Information and Development.

References

1. Baker JC (1987) Bovine viral diarrhea virus. A review. J Am Vet Med Assoc 190: 1449–1458
2. Barber DML, Nettleton PF, Herring JA (1985) Disease in a dairy herd associated with the introduction and spread of bovine virus diarrhoea virus. Vet Rec 117: 59–464
3. Bolin SR, McClurkin AW, Cutlip RC, Coria MF (1985) Response of cattle persistently infected with noncytopathic bovine viral diarrhea virus to vaccination for bovine viral diarrhea and to subsequent challenge exposure with cytopathic bovine viral diarrhea virus. Am J Vet Res 46: 2467–2470
4. Brownlie J, Clarke MC, Howard CJ (1984) Experimental production of fatal mucosal disease in cattle. Vet Rec 114: 535–536
5. Casaro APE, Kendrick JW, Kennedy PC (1971) Response of the bovine fetus to bovine viral diarrhea–mucosal disease virus. Am J Vet Res 32: 1543–1562
6. Juntti N, Larsson B, Fossum C (1987) The use of monoclonal antibodies in enzyme-linked immunosorbent assays for detection of antibodies to bovine viral diarrhoea virus. J Vet Med B 34: 356–363
7. Kendrick JW (1971) Bovine viral diarrhea–mucosal disease virus infection in pregnant cows. Am J Vet Res 32: 533–544
8. McClurkin AW, Littledike ET, Cutlip RC, Frank GH, Coria MF, Bolin SR (1984) Production of cattle immunotolerant to bovine viral diarrhea virus. Can J Comp Med 48: 156–161
9. Niskanen R, Alenius S, Larsson B, Juntti N (1989) Evaluation of an enzyme-linked immunosorbent assay for detection of antibodies to bovine virus diarrhoea virus in milk. J Vet Med B 36: 113–118
10. Oirschot JT van (1983) Congenital infections with nonarbo togaviruses. Vet Microbiol 8: 321–361
11. Steck F, Lazary S, Fey H, Wandeler A, Huggler C, Oppliger G, Baumberger H, Kaderli R, Martig J (1980) Immune responsiveness in cattle fataly affected with bovine virus diarrhea–mucosal disease. Zentralbl Veterinarmed [B] 27: 429–445
12. Terpstra C (1985) Border disease: a congenital infection of small ruminants. In: Pandey R (ed) Prog Vet Microbiol Immun, Vol 1. Infection and immunity in farm animals. Karger, Basel, pp 175–198

Authors' address: Dr. B. Larsson, Department of Cattle and Sheep Diseases, Swedish University of Agricultural Sciences, Box 7019, S-75007 Uppsala, Sweden.

Arch Virol (1991) [Suppl 3]: 149–156

Identification and production of pestivirus proteins for diagnostic and vaccination purposes

C. Lecomte[1], **D. Vandenbergh**[1], **N. Vanderheijden**[1], **L. De Moerlooze**[1], **J. J. Pin**[2],
G. Chappuis[2], **Ph. Desmettre**[2], and **A. Renard**[1]

[1] Eurogentec Campus du Sart Tilman Liège, Belgium, and
[2] Rhône-Mérieux Laboratoires IFFA Lyon, France

Accepted March 14, 1991

Summary. Using a panel of monoclonal antibodies (MAbs) previously characterized by seroneutralization, immunofluorescence and radio-immunoprecipitation, we have identified Pestivirus proteins useful for diagnostic purposes from the cytopathic Osloss isolate of bovine viral diarrhea virus (BVDV). Proteins that should be useful for vaccination have also been analysed. Cell-free translation of RNA from glycoprotein-coding cDNA fragments produced, when synthesized in the presence of canine pancreatic microsomes, two glycosylated proteins that were independently recognized and immunoprecipitated by two distinct classes of neutralizing MAbs. A similar in vitro procedure was carried out on nonstructural protein-coding sequences and allowed to identify a viral translation product that specifically reacted with MAbs directed against the 80 kDA protein of a number of Pestivirus strains. Its positioning within the polyprotein encoded by the viral genome was refined by epitope scanning using synthetic hexameric peptides. This viral antigen was further expressed in E. coli, produced as inclusion bodies and used successfully as an ELISA antigen in both competitive and indirect assays for the detection of BVD antibodies in cattle sera.

Key words: Cattle, pestivirus, bovine viral diarrhea, ELISA, diagnosis, glycoprotein, antigenic determinants, epitope scanning.

Introduction

Bovine viral diarrhoea virus (BVDV), hog cholera virus (HCV) and border disease virus (BDV) are positive-stranded RNA viruses currently classified in the Pestivirus genus of the Togaviridae family [8, 17]. The viral RNA genome of four Pestivirus strains has been recently cloned and sequenced, two BVDV strains, Osloss [15] and NADL [1], and two HCV strains, Alfort [12] and Brescia [13]. From the analysis of the sequence and genomic organization of the Pestivirus genus, it seems obvious that it should no longer be grouped within the Togaviridae family but rather be related to the Flaviviridae [14, 3].

The nucleotide sequence of BVD virus is a single large open reading frame (ORF) with a protein coding capacity of 450 kilodaltons. Two ORFs were first reported in the viral sequence of BVDV Osloss strain [15], but the presence of a single major ORF has been recently confirmed by direct sequencing after RNA amplification using the PCR technology (data not shown). Nucleotide sequence comparison between the Osloss and NADL strains of BVDV shows that they are closely related [4]. A preliminary map of the genomic organization of BVDV NADL strain has been presented [2], but to date, there are no sequence data to allow the precise positioning of the BVDV gene products within the polyprotein encoded by the large ORF.

Here we report on the use of an in vitro procedure allowing to identify BVDV Osloss proteins which may be useful for diagnostic and vaccination purposes. Epitope scanning using synthetic peptides was further carried out to achieve a precise positioning of the diagnostic viral antigen. This viral protein was expressed in microorganisms, and the purified recombinant protein was used in competitive and indirect ELISA for the detection of BVD specific antibodies in a number of bovine sera.

Material and methods

Preparation and characterization of monoclonal antibodies

The preparation of monoclonal antibodies (MAbs) has been previously described [11] as well as their characterization by immunofluorescence, by virus neutralization and by radioimmunoprecipitation (RIP) of BVD-infected cell extracts.

Construction of pSP174 and pSPgp plasmids

The pSP174 recombinant plasmid was prepared by subcloning the 3,500-bp 174 cDNA clone of the cytopathic BVDV Osloss strain [14] (Fig. 1) into the pSP65 transcription vector (Promega) between the EcoRI and BamHI sites. The pSPgp/1 and pSPgp/2 plasmids were constructed by respectively inserting between the EcoRI and BamHI sites of pSP65 a 1,400-bp or a 2,200-bp DNA fragments constructed from the 63 and 36 cDNA clones of the BVDV Osloss strain genome [14] (Fig. 1).

In vitro transcription and translation

Synthesis of RNA transcripts was carried out in a typical transcription reaction [7] using 2 µg of restricted pSP174, pSPgp/1 or pSPgp/2 as DNA templates and 15 U of SP6 RNA

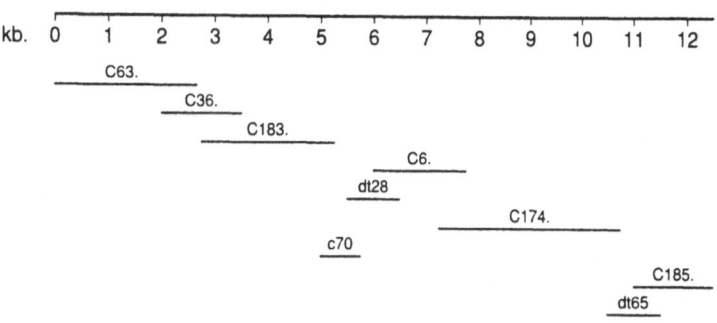

Fig. 1. Mapping of the cDNA clones covering the entire genome of the cytopathic Osloss strain of BVDV [14]

polymerase (Boehringer). Cell-free translation of SP6-derived transcripts was performed in a nuclease-treated rabbit reticulocyte lysate system (Promega) containing [^{35}S]-methionine at 1,200 µCi/ml. Some reactions were supplemented with one equivalent of canine pancreatic microsomes (Promega). After 1 h incubation at 30 °C, translation products were analysed on 10% Laemmli gels [10].

Further radioimmunoprecipitation of the in vitro translated products was carried out as described [6] using the corresponding monoclonal antibodies.

Epitope scanning

Determination of antibody specificity was performed on hexameric peptides synthesized using the epitope scanning kit commercialized by CRB (Cambridge Research Biochemicals).

Preparation of ELISA antigens and ELISA procedure

The cytopathic Osloss BVDV strain was grown in foetal ovine kidney cells (OCK) and total RNA was used as a template for PCR (Polymerase Chain Reaction) amplification as described elsewhere [5], using two amplimers located in the nonstructural-coding region of the viral genome. The amplified fragment was further cloned in an inducible bacterial expression system under control of a T7 promoter [16]. Inclusion bodies were purified as described [9], then denatured in 7 M urea, 0.25 M TrisHCl pH 8 and renatured by dialysis against 0.25 M TrisHCl pH 8.

The cellular antigen was prepared from OCK cells infected with the Singer strain of BVDV; infected cells were harvested before cytopathic effects became apparent [11].

Assays were performed as previously described [11] on 350 bovine sera from the "Centre de sélection de Ciney" and from a number of farms in Belgium.

Results

Analysis of the Osloss glycoprotein region

pSPgp/1 and pSPgp/2 were used as DNA templates for in vitro transcription assays. pSPgp/1 contains the Osloss-coding sequence which corresponds to the NADL gp62 region [2]. The pSPgp/2 plasmid contains, in addition to this Osloss-gp62 sequence, 800 nucleotides encoding a part of

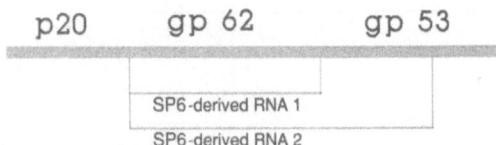

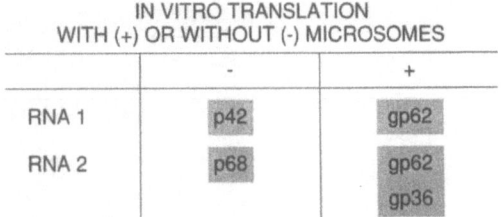

Fig. 2. In vitro expression of the Osloss glycoprotein region

the second glycoprotein (NADL gp53 [2]). SP6-derived RNA transcripts were synthesized from both linearized plasmids, and further in vitro translated in a rabbit reticulocyte system (Fig. 2). Cell-free translation of RNA transcripts 1 and 2 respectively produced proteins with an apparent molecular size of 42 and 68 kilodaltons (kDa). When the RNA transcripts were translated in the presence of canine pancreatic microsomes, the RNA 1 translation product was a glycosylated protein with an estimated molecular weight of 62 kDa, whereas the RNA 2 gave rise to two glycosylated polypeptides (estimated sizes: 62 and 36 kDa; Fig. 2).

Radioimmunoprecipitation of these respective translation products using neutralizing monoclonal antibodies showed that we could distinguish between two groups of MAbs directed against two separate epitopes, one being located in the first glycoprotein and the other in the second one. However, when used in RIP assays on BVDV infected cell extracts, both groups of neutralizing MAbs showed identical patterns of immunoprecipitated material: two major glycosylated proteins that appeared at approximately 46 and 44 kDa after endoglycosidase F treatment [11].

Analysis of the Osloss nonstructural region

A similar in vitro procedure was carried out to position within the Osloss genome the viral determinant specifically recognized and immunoprecipitated by non-neutralizing MAbs directed against the respective 80 kDa and 120 kDa nonstructural proteins of the cytopathic or noncytopathic BVDV biotypes (anti-80 kDa MAbs) [11]. The pSP174 recombinant plasmid was successively digested with a number of restriction endonucleases from which site locations are shown in Fig. 3. SP6-derived RNA transcripts were synthesized from all these restricted plasmids and in vitro translated as described above. The translation products were then immunoprecipitated using anti-80 kDa MAbs. These experiments allowed to deter-

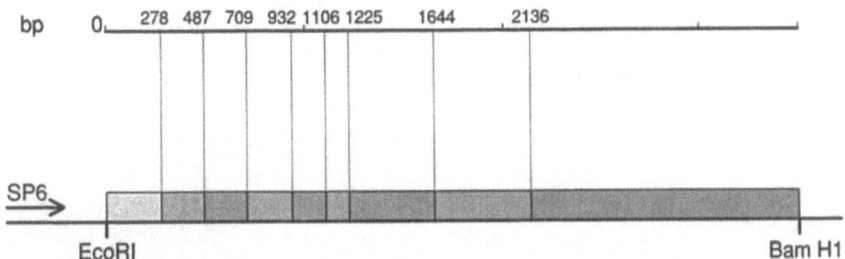

Fig. 3. In vitro analysis of the Osloss nonstructural region. Numbers indicate the size of the different fragments obtained by digestion of the 174 cDNA clone with a number of restriction endonucleases

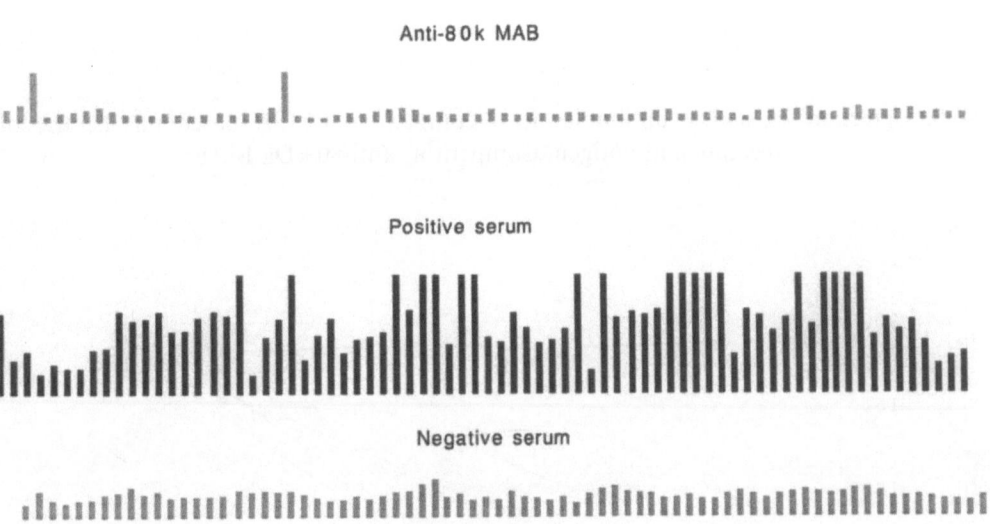

Fig. 4. Epitope scanning using hexameric synthetic peptides, of a 70 amino acids region identified in the nonstructural part of the polyprotein encoded by the Osloss viral genome

mine the presence of an immunoreactive region located in a 70 amino acids-region corresponding to the second fragment illustrated in Fig. 3. This region was further scanned using 70 hexameric peptides that were synthesized in microtiter plates and directly tested in indirect ELISA using anti-80 kDa MAbs and bovine positive and negative sera (Fig. 4).

Production and test of the ELISA antigen

A 2,200-bp Osloss genomic fragment including the sequence encoding the nonstructural immunoreactive region was expressed in Escherichia coli under control of an inducible T7 promoter [16] and purified as described for the recombinant ELISA antigen that we had previously produced [11].

The purified recombinant protein was assayed in both indirect and competitive ELISA for the detection of BVDV antibodies in cattle sera, and

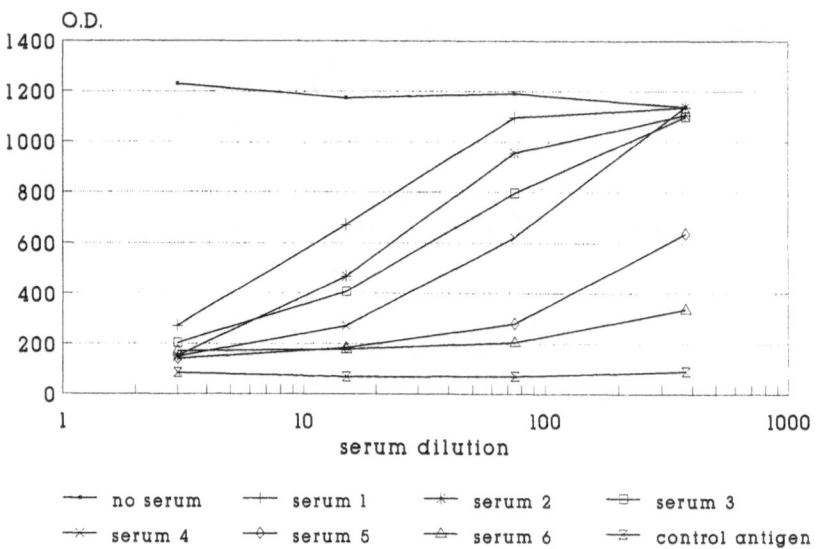

Fig. 5. Detection of BVDV antibodies in a competitive ELISA based on the use of a recombinant antigen. Competitor: anti-80 kDa MAb

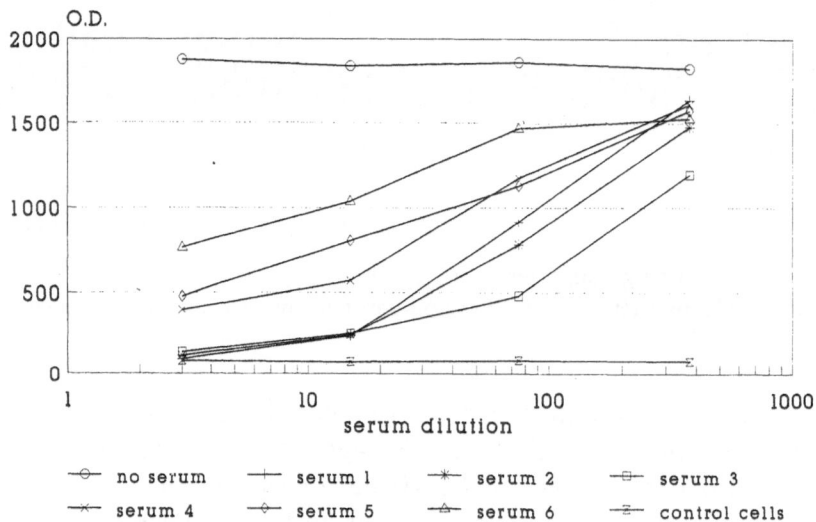

Fig. 6. Detection of BVDV antibodies in a competitive ELISA using antigen extracted from Singer infected cells

the results were compared with those obtained using an antigen extracted from BVDV-infected cells (Fig. 5 and 6). A blind trial was performed on a sample of 350 bovine sera to compare the results obtained in the competitive ELISA with those obtained in a virus neutralization assay carried out against the Singer BVDV isolate. This comparison is illustrated in Fig. 7.

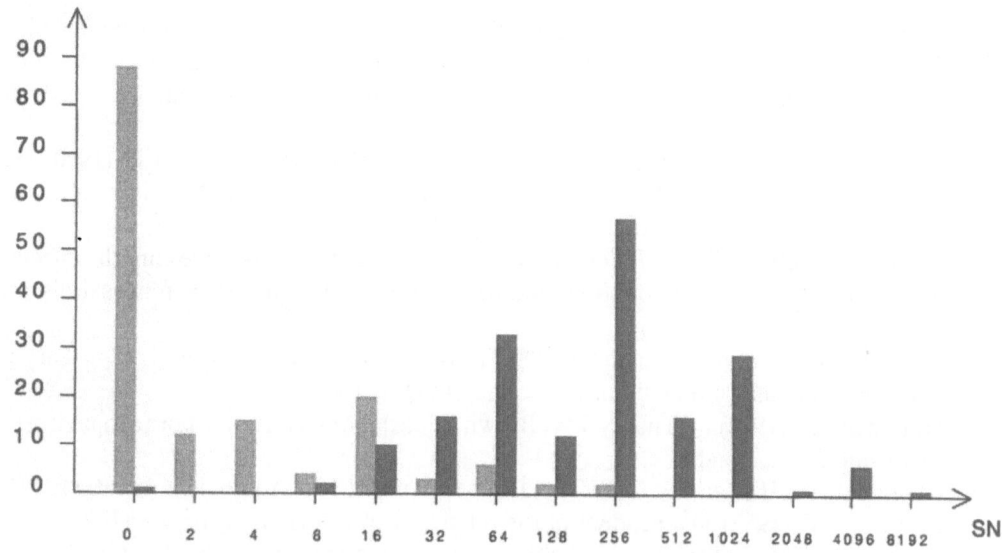

Fig. 7. ELISA detection of BVDV antibodies in a sample of 350 bovine sera and comparison with their seroneutralization titers. Dark columns represent positive ELISA values

Discussion

Using previously cloned cDNA sequences of the cytopathic Osloss strain of BVDV [14, 15] as well as a panel of monoclonal antibodies previously developed and characterized [11], we have identified, using an in vitro transcription–translation system, viral determinants that should be useful for vaccination and diagnostic purposes.

This in vitro procedure allowed us to identify structural determinants which are now being expressed in usual in vivo systems. Moreover, the nonstructural BVDV protein containing anti-80 kDa epitopes was positioned within the polyprotein encoded by the Osloss genome and further expressed in an inducible E. coli system. This protein was efficiently used as an ELISA antigen for the detection of BVDV specific antibodies in cattle sera (Fig. 7). The specificity of the resulting ELISA exactly compares with a similar assay that uses an antigen extracted from BVDV-infected cells. The non-specific response previously obtained with a recombinant antigen produced as a fusion protein with the β-galactosidase [11] was not observed using this new recombinant protein.

References

1. Collett MS, Larson R, Gold C, Strick D, Anderson DK, Purchio AF (1988) Molecular cloning and nucleotide sequence of the Pestivirus bovine viral diarrhea virus. Virology 165: 191–199
2. Collett MS, Larson R, Belzer SK, Retzel E (1988) Proteins encoded by bovine viral diarrhea virus: the genomic organization of a Pestivirus. Virology 165: 200–208

3. Collett MS, Anderson DK, Retzel E (1988) Comparisons of the Pestivirus bovine viral diarrhoea virus with members of the Flaviviridae. J Gen Virol 69: 2637–2643

4. Collett MS, Moennig V, Horzinek MC (1989) Recent advances in Pestivirus research. J Gen Virol 70: 253–66

5. De Moerlooze L, Desport M, Renard A, Lecomte C, Brownlie J, Martial JA (1990) The coding region of the 54 kDa protein of several Pestiviruses lacks host insertions but reveals a "zinc finger-like" domain. Virology 177: 812–815

6. Donis RO, Dubovi EJ (1987) Differences in virus-induced polypeptides in cells infected by cytopathic and non cytopathic biotypes of bovine virus diarrhea–mucosal disease virus. Virology 158: 168–173

7. Hobman TC, Shukin R, Gillam S (1988) Translocation of rubella virus glycoprotein E1 into the endoplasmic reticulum. J Virol 62: 4259–4264

8. Horzinek MC (1981) In: Tinsley TW, Brown F (eds) Non-arthropod-borne togaviruses. Academic Press, London

9. Klempnauer KH, Ramsay G, Bishop JM, Moscovici MG, Moscovici C, McGrath JP, Levinson AD (1983) The product of the retroviral transforming gene v-myb is a truncated version of the protein encoded by the cellular oncogen c-myb. Cell 33: 345–355

10. Laemmli UK (1970) Cleavage of structural proteins during the assembly of the head of bacteriophage T4. Nature 227: 680–685

11. Lecomte C, Pin JJ, De Moerlooze L, Vandenbergh D, Lambert AF, Pastoret PP, Chappuis G (1990) ELISA detection of bovine viral diarrhoea virus specific antibodies using recombinant antigen and monoclonal antibodies. Vet Microbiol 23: 193–201

12. Meyers G, Rümenapf T, Thiel H-J (1989) Molecular cloning and nucleotide sequence of the genome of hog cholera virus. Virology 171: 555–567

13. Moorman RJM, Warmerdam PAM, Vandermeer B, Schaaper WMM, Wensvoort G, Hulst MM (1990) Molecular cloning and nucleotide sequence of hog cholera virus strain Brescia and mapping of the genomic region encoding envelope protein E1. Virology 177: 184–198

14. Renard A, Brown-Shimmer S, Schmetz D, Guiot C, Dagenais L, Pastoret PP, Dina D, Martial JA (1985) Molecular cloning, sequencing and expression of BVDV RNA. In: "Pestivirus infections of ruminants", a seminar in the CEC programme of coordination of research on animal husbandry held in Brussels, 10–11 September 1985. Edited by Harkness JW in 1987

15. Renard A, Dina D, Martial JA (1987) Vaccines and diagnostics derived from bovine viral diarrhea virus. European patent application number 86870095.6. Publication number 0208672, 14 January 1987

16. Studier FW, Moffat BA (1986) Use of bacteriophage T7 RNA polymerase to direct selective high-level expression of cloned genes. J Mol Biol 189: 113–130

17. Westaway EG, Brinton MA, Gaidamovitch SY, Horzinek MC, Igarashi A, Kääriäinen L, Lvov DK, Porterfield JS, Russell PK, Trent DW (1985) Togaviridae. Intervirology 24: 125–139

Authors' address: C. Lecomte, Eurogentec S.A., Campus du Sart Tilman, Allée du Six Août, B6, B-4000 Liège, Belgium.

Arch Virol (1991) [Suppl 3]: 157–164

Surveillance of cattle herds for bovine virus diarrhoea virus (BVDV)-infection using data on reproduction and calf mortality

H. Houe[1, 2] and A. Meyling[1]

[1] National Veterinary Laboratory, Copenhagen, Denmark
[2] Royal Veterinary & Agricultural University, Copenhagen, Denmark

Accepted March 14, 1991

Summary. The effect of bovine virus diarrhoea virus (BVDV)-infection on pregnancy rate, on stillbirths and mortality of neonatal calves and the size of newborn calves was evaluated in 8 herds in which persistently infected (PI)-animals had been identified. Data from 9 herds without PI-animals were used as controls.

At the time of conception of the oldest PI-animal a significant drop in pregnancy rate to about half the herd average was found. About 6 months later a 3-fold rise in calf mortality was seen. This pattern was found in 4 of the herds. In the remaining 4 herds the pattern was less clear, probably reflecting different immune states of the herds.

Rough estimation of the size of newborn calves showed that PI-animals were significantly smaller than normal animals.

Monitoring of herds for the above-mentioned parameters may be a means of pointing out herds with PI-animals. Most of the data necessary for such surveillance schemes are already available and may readily be used.

Key words: Bovine virus diarrhoea virus, BVDV, pregnancy rate, calf mortality, cattle.

Introduction

The clinical signs following introduction of bovine virus diarrhoea virus (BVDV)-infection into a herd vary considerably. The reasons for this variability among different outbreaks may be due to the different proportions of seronegative animals in the critical period of pregnancy [7]. Some of the variation may be ascribed to variability in virulence among strains of BVDV [4].

The clinical signs during the course of a BVDV-infection in a susceptible herd can be divided into the following entities: (1) symptoms among cattle undergoing acute infection, (2) altered reproductive performance, i.e. repeat breeding and abortions, (3) birth of congenitally malformed, weak and undersized calves, (4) unthriftiness and (5) mucosal disease (MD) either acute or chronic. Herd outbreaks are highly variable mixtures of these different clinical manifestations, which can occur during a period of 2–3 years or longer.

Often the first clinical signs that cause the farmer to call veterinary assistance is the development of mucosal disease or other clinical diseases in persistently infected (PI)-animals. By this time much damage due to BVDV-infection has already occurred but has not necessarily been associated with BVDV [1].

Often PI-animals run a chronic course of disease rather than acute MD [5] and often these cases are not diagnosed.

Registration of disease and reproduction data is performed in many Danish dairy herds. These data are available through the data base of the agricultural organizations and makes continuous surveillance of the herd health possible. It is a purpose of this study to see if these data might be used for presumptive diagnosis of BVDV in dairy herds. For this purpose data comprising pregnancy rate, stillbirths, neonatal mortality and size of new-born calves in herds with known BVD-status were analysed.

Material and methods

Selection of herds for the study

Approximately all animals in 17 herds were examined for BVD by virus isolation and neutralization test [6]. None of the herds had shown typical clinical evidence of BVDV-infection. PI-animals were found in 8 herds. All the PI-animals in this study had been conceived in the herds to which they belonged. In the remaining 9 herds no PI-animals were found.

Beyond the ordinary registration procedures for Danish cattle herds the herds were subjected to regular pregnancy control that was made by rectal examination 6 weeks after insemination.

Data collection

The data records on inseminations, pregnancy control, stillbirths, mortality of newborn calves and size of newborn calves were gathered by the National Institute of Animal Science.

The dates of insemination, pregnancy control, birth and death were recorded.

The size of newborn calves was roughly estimated by the farmer as one of four scores. Small calves were designated 1, calves below normal 2, calves above normal 3 and big calves 4.

All the data were attainable in a SAS data set.

Calculations

Pregnancy rate

The pregnancy rate was calculated as the percentage of inseminations that was followed by pregnancy. Inseminations followed by positive pregnancy diagnosis 6 weeks later were counted as successful, whereas inseminations followed by return to heat or negative pregnancy diagnosis 6 weeks after insemination were counted as unsuccessful. Inseminations occurring less than 10 days apart were counted as one.

If the outcome of an insemination was not known because of slaughter of an animal before pregnancy control the observation was omitted from the calculations.

In order to reduce unwanted fluctuations a so-called moving average was used when calculating the pregnancy rate, i.e. the average for each month was calculated for periods comprising the month before and the month after the month in question.

Stillbirths and mortality among newborn calves

The number of stillborn calves added to the number of calves that had died within the first month was calculated as a percentage of all calves born. This percentage was also calculated for each month as a moving average over 3 months.

Size

The sizes of the PI-animals in these herds were compared to the sizes of all calves delivered in all the herds during a 3-year-period (1986–1988), i.e. the period in which PI-animals were born. The sizes of PI-animals were tested against the sizes of all calves by means of the Mantel-Haenszel technique in order to correct for confounding due to parity. This is necessary because the parity of the cow might have an effect on the size of the calf and the majority of PI-calves are born by first and second parity cows.

Study period

For most herds data from a 3-year-period were used i.e. 1986–1988. In two herds it was necessary to include 1985 in the calculation of pregnancy rate as the oldest PI-animals in these herds had been conceived during that year.

Results

The latest time for the introduction of the infection into the herd must be shortly after the conception of the oldest PI-animal. This time is marked with an arrow in the graphs showing the pregnancy rate (Fig. 1). In the graphs showing stillbirths and neonatal calf mortality the birth date of the oldest PI-animal found in the herd is marked by an arrow (Fig. 2).

In four herds a significant drop in pregnancy rate from about 60% to about 20% was seen (Fig. 1, only 2 herds shown).

Six months later the calf mortality (stillbirths and deaths within the first month of life) in the same 4 herds were about three times the average for the herd (Fig. 2, only 2 herds shown).

In herds without PI-animals a more constant pregnancy rate was seen (Fig. 3, only 1 herd shown), whereas the neonatal mortality might sometimes

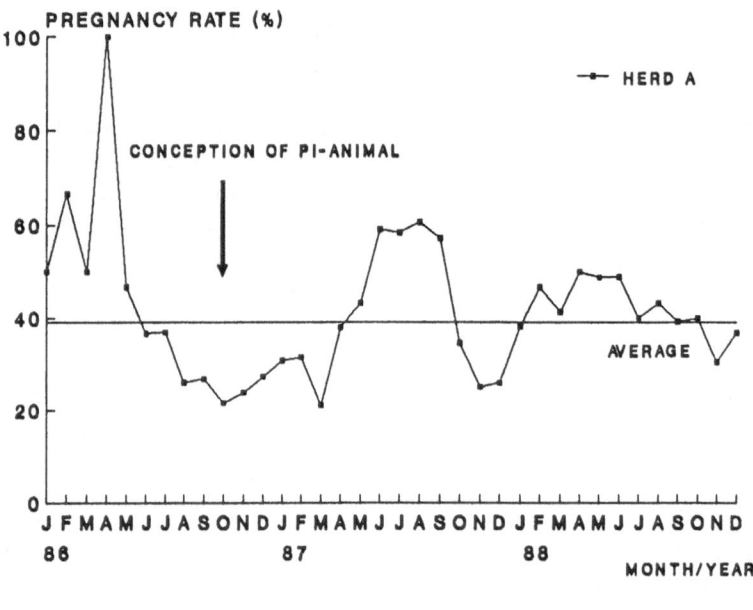

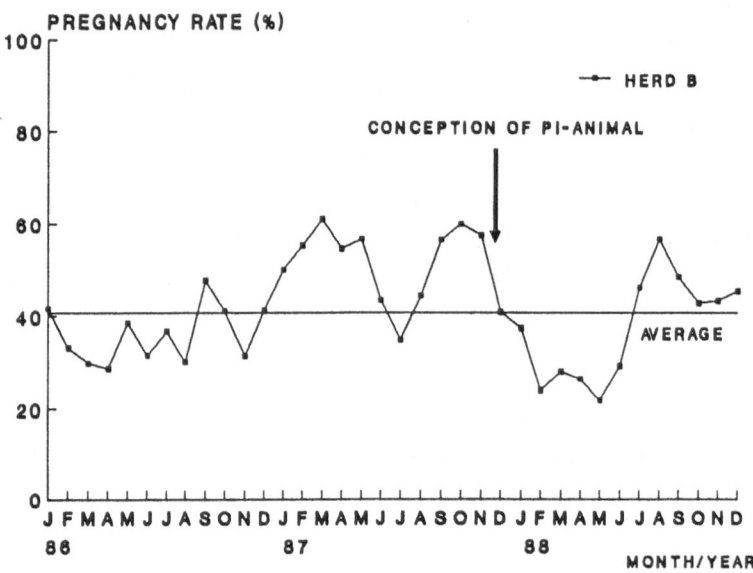

Fig. 1. Pregnancy rate calculated as the percentage of successful inseminations. Herds with PI-animals

also reach high levels in herds without PI-animals (Fig. 3, only 1 herd shown).

Of 29 PI-animals found in the 8 herds the size scores were known of 28. The total number of calvings in all herds during a 3-year-period was 2952. Among the PI-animals 24 were estimated as size score 1 or 2 whereas only 4 PI-animals were estimated as size score 3 or 4. The PI-animals were significantly smaller than normal calves ($P < 0.001$). But because the numbers of PI-animals in many herds were low, and the random variation is big,

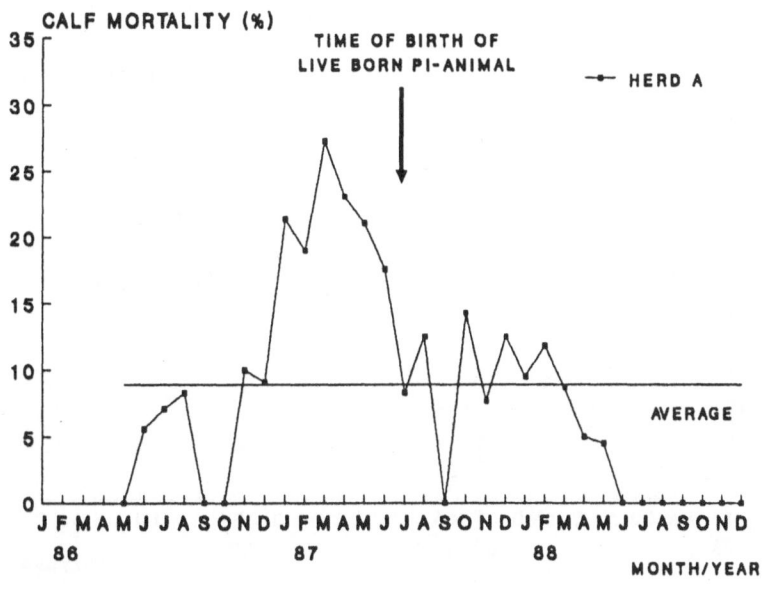

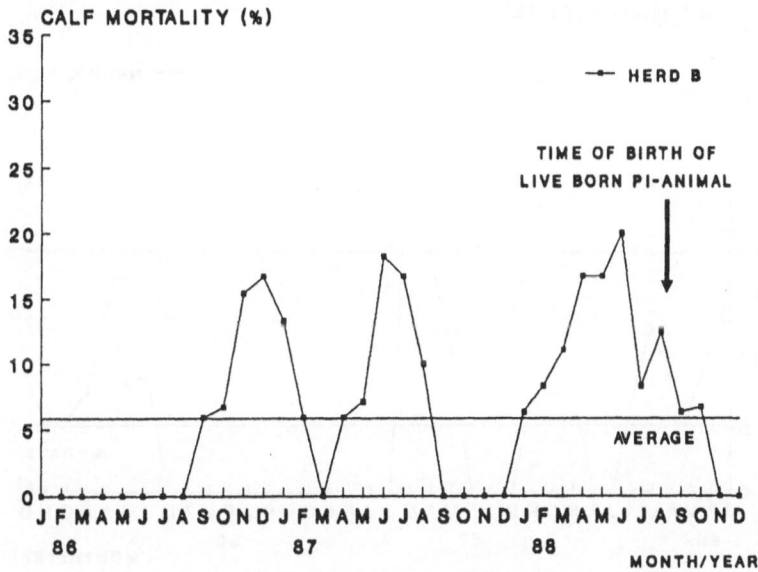

Fig. 2. Stillbirths and neonatal mortality calculated as a percentage of all parturitions. Herds with PI-animals

this finding will seldom seem significant in the individual herds. The per cent distribution of the size of PI-animals and normal animals is shown in Fig. 4.

Discussion

Only few investigations have demonstrated the effect of acute BVDV-infection on pregnancy rate.

Intrauterine infusion of BVD-virus suspension after insemination have been shown to reduce fertilization [3].

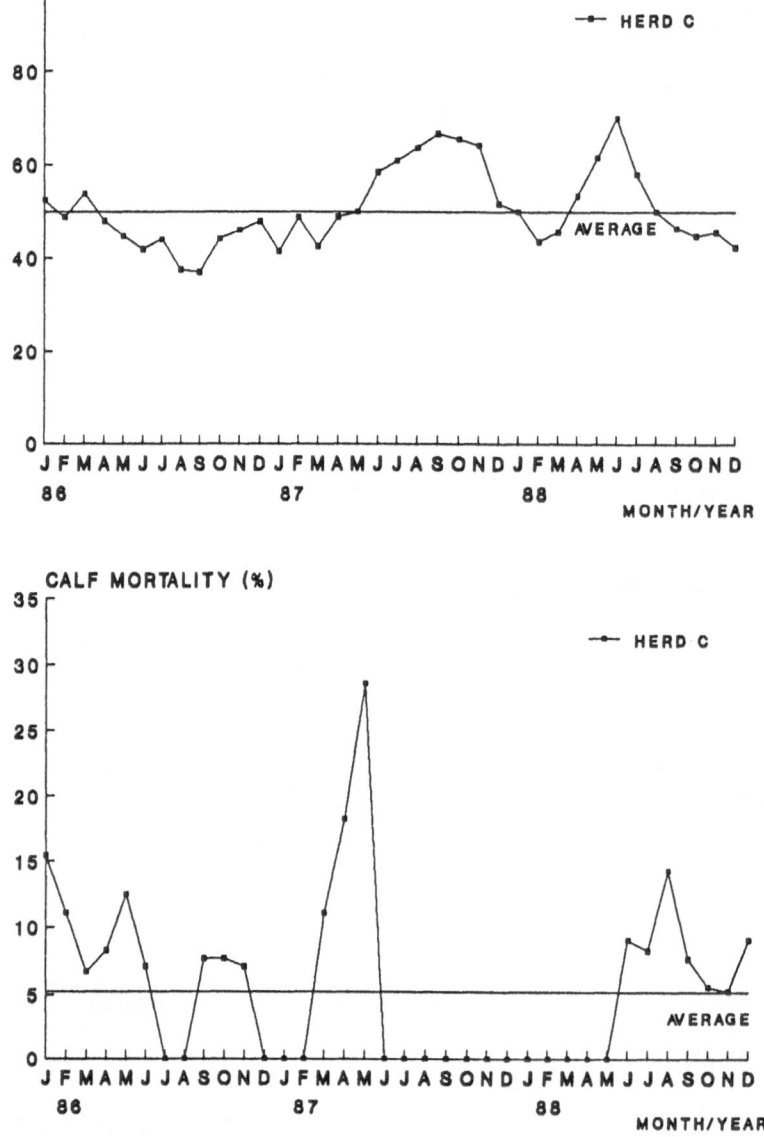

Fig. 3. Pregnancy rate calculated as the percentage of successful inseminations and still-births and neonatal mortality calculated as a percentage of all parturitions. Herds without PI-animals

In another experimental study inoculation of the virus within two hours after breeding reduced pregnancy rate when virus were inoculated by the intrauterine route whereas inoculation by the nasal and oral route did not affect pregnancy rate [12].

Twelve heifers inseminated with semen from a persistently infected bull all became pregnant despite the fact that all the heifers seroconverted within two weeks after insemination [8].

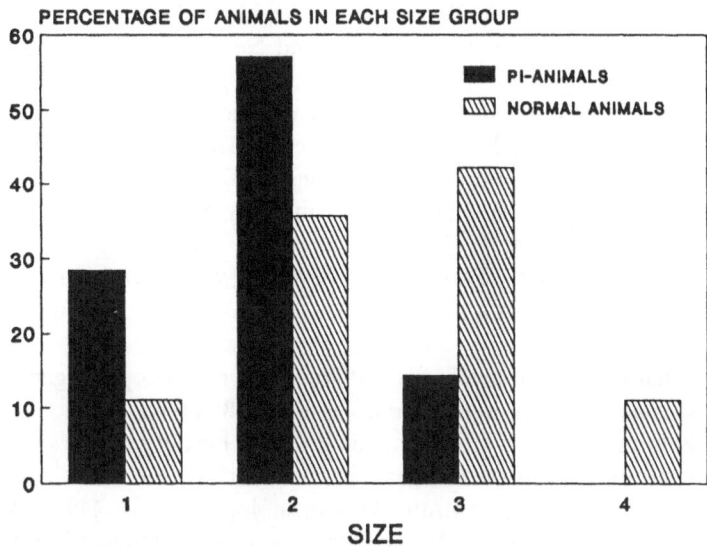

Fig. 4. Relative distribution of size of newborn calves. Size was estimated by the farmer as score 1–4. PI-animals were significantly smaller than normal animals (P < 0.001)

When breeding seronegative cows just before seroconversion these showed a first service conception rate of 22.2% compared to 78.6% among cows that were seropositive when bred [10].

So reports on the effect of BVDV-infection in early pregnancy seem somewhat contradictory.

The results in the present study indicate that the introduction of BVDV into a herd at least sometimes may give rise to repeat breeding.

The effect of BVDV-infection on stillbirths and neonatal mortality is well documented [2, 9, 11]. Often these calves are infected during mid gestation and are therefore not persistently infected [9]. They are born shortly before birth of PI-animals (Fig. 2).

The data obtained in this work showed that monitoring of herds for pregnancy rate, neonatal mortality and size may be a means of pointing out herds with PI-animals.

References

1. Brownlie J (1985) Clinical aspects of the bovine virus diarrhoea/mucosal disease complex. In Practice 7: 195–202
2. Done JT, Terlecki S, Richardson C, Harkness JW, Sands JJ, Patterson DSP, Sweasey D, Shaw IG, Winkler CE, Duffel SJ (1980) Bovine virus diarrhoea–mucosal disease virus: pathogenicity for the fetal calf following maternal infection. Vet Rec 106: 473–479
3. Grahn TC, Fahning ML, Zemjanis R (1984) Nature of early reproductive failure caused by bovine viral diarrhoea virus. J Am Vet Med Assoc 185: 429–432
4. Harkness JW, Roeder PL (1988) The comparative biology of classical swine fever. In: Liess B (ed) Classical swine fever and related viral infections. Martinus Nijhoff Publishers, Boston, Dordrecht, Lancaster, pp 233–288

5. Littlejohns IR, Horner GW (1990) Incidence, epidemiology and control of bovine pestivirus infections and disease in Australia and New Zealand. Rev Sci Tech Off Int Epiz 9: 195–205
6. Meyling A (1984) Detection of BVD virus in viremic cattle by an indirect immuno-peroxidase technique. In: McNulty MS, MacFerran JB (eds) Recent advances in virus diagnosis. Martinus Nijhoff Publishers, Boston, pp 37–46
7. Meyling A, Houe H, Jensen AM (1990) Epidemiology of bovine virus diarrhoea virus. Rev Sci Tech Off Int Epiz 9: 75–93
8. Meyling A, Jensen AM (1988) Transmission of bovine virus diarrhoea virus (BVDV) by artificial insemination (AI) with semen from a persistently-infected bull. Vet Microbiol 17: 97–105
9. Roeder PL, Jeffrey M, Cranwell MP (1986) Pestivirus fetopathogenicity in cattle: changing sequelae with fetal maturation. Vet Rec 118: 44–48
10. Virakul P, Fahning ML, Joo HS, Zemjanis R (1988) Fertility of cows challenged with a cytopathic strain of bovine viral diarrhea virus during an outbreak of spontaneous infection with a non cytopathic strain. Theriogenology 29: 441–449
11. Werdin RE, Ames TR, Goyal SM, DeVries GP (1989) Diagnostic investigation of bovine viral diarrhea infection in a Minnesota dairy herd. J Vet Diagn Invest 1: 57–61
12. Whitmore HL, Zemjanis R, Olson J (1981) Effect of bovine viral diarrhoea virus on conception in cattle. J Am Vet Med Assoc 178: 1065–1067

Authors' address: Dr. H. Houe, National Veterinary Laboratory, Copenhagen, Denmark.

Arch Virol (1991) [Suppl 3]: 165–167
© Springer-Verlag 1991

Flow cytometric detection of bovine viral diarrhoea virus

P. Qvist[1], H. Houe[1], B. Aasted[2], A. Meyling[1], L. Rønsholt[3] and B. Bloch[2]

[1] Animal Biotechnology Research Center, National Veterinary Laboratory, Copenhagen V.
[2] The Royal Veterinary and Agricultural University, Frederiksberg C. and
[3] State Veterinary Institute for Virus Research, Lindholm, Kalvehave, Denmark

Accepted March 14, 1991

Summary. Flow cytometry and virus isolation in cell culture was compared for ability to detect bovine viral diarrhoea virus in blood samples from persistently infected cattle.

Bovine viral diarrhoea virus (BVDV) in blood samples is usually demonstrated by isolation of the virus in cell culture and subsequent identification by immunostaining. As this technique has the time consuming disadvantage, that the virus has to be multiplied in living cells, it was investigated if flow cytometry could be used for the detection of BVDV directly in blood cells. Flow cytometry has previously been used for detection of viral infections in cell cultures (Jacobberger et al., 1986). The technique is based largely on the same principles as immunofluorescence, but flow cytometry has several unique qualities, e.g. quantitative capacity, lower detection limit and improved speed. The analysis is based on measurements of fluorescence associated with single cells passing through a flow chamber. The fluorochromes are excited by a laser beam and the emitted light is quantified in photomultipliers, the electric signal digitized and then stored in computer memory for analysis. Typically, several thousands cells are measured per second.

The flow cytometric assay was performed in microdilution plates on lysates of stabilized blood samples (Qvist et al., 1991). The leukocytes were fixed in solution, membranes solubilized and BVDV antigens detected using biotinylated immunoglobulin from porcine antiserum to BVDV followed by incubation with fluorescein isothiocyanate-conjugated avidin (Qvist et al., 1990). The ability of the flow cytometric analysis and virus isolation (Meyling, 1984) to detect BVDV in 143 bovine blood samples was compared. If BVDV was demonstrated by either virus isolation or flow cytometry, blood samples were recollected three weeks later for isolation of BVDV. Persistently infected (PI) animals were diagnosed after two successive isolations of virus. Serum antibodies to BVDV were determined by the neutralizing peroxidase-linked antibody assay (Holm Jensen, 1980).

Table 1. Detection of BVDV in 143 blood samples by virus isolation in cell culture and flow cytometry

Flow cytometry	Virus isolation	
	positive results	negative results
Positive results	33 (33)[a]	8 (4)
Negative results	1 (0)	101[b] (—)

[a] The number of blood samples originating from cattle later shown to be persistently infected with BVDV is given in brackets.

[b] Blood samples from these animals were not recollected and analyzed.

Virus was initially isolated from 34 samples (Table 1). Recollection of blood from these 34 bovines demonstrated that 33 were PI. One animal was found virus negative and antibody positive. Eight blood samples were negative by virus isolation but positive by flow cytometry (Table 1). Blood samples from four of these animals were virus negative, by the virus isolation method, and antibody positive in repeated tests after 3 weeks. Two were retested by flow cytometry and found negative (data not shown). Of the remaining four samples two were antibody positive and originated from cattle younger than one week. All four samples were found to originate from PI cattle as virus was isolated in blood samples obtained 3 and 6 weeks later.

Initially, blood samples from four PI cattle were negative by virus isolation and positive by flow cytometry. Antibodies to BVDV were demonstrated in two samples suggesting that neutralization of virus was responsible for the negative findings. One additional sample originated from a 4 days old calf also likely to have colostral antibodies in the blood. This, however, could not be demonstrated by serum neutralization. The reason for the remaining blood sample to be negative in cell culture was unknown. Four other samples were negative by virus isolation but positive by flow cytometry. The presence of serum antibodies to BVDV in these four animals could indicate previous acute infection. However, the time of infection was unknown. It is therefore possible that virus were present in the blood samples, but remained undetected by virus isolation because of BVDV antibodies. Virus isolation in cell culture, as performed here, tends to miss a number of blood samples from PI cattle, probably because of the presence of colostral antibodies to BVDV. On the contrary, all samples from persistently infected animals were positive by flow cytometry. Therefore, flow cytometric detection of BVDV in blood cells could be a useful alternative for identification of PI cattle.

Key words: Bovine virus diarrhoea virus, BVDV, pregnancy rate, calf mortality, cattle.

References

1. Holm Jensen M (1980) Detection of antibodies against hog cholera virus and bovine viral diarrhea virus in porcine serum. Acta Vet Scand 22: 85–98
2. Jacobberger JW, Fogleman D, Lehman JM (1986) Analysis of intracellular antigens by flow cytometry. Cytometry 7: 356–364
3. Meyling A (1984) Detection of BVD virus in viremic cattle by an indirect immunoperoxidase technique. Curr Top Vet Med Anim Sci 29: 37–46
4. Qvist P, Aasted B, Bloch B, Meyling A, Rønsholt L, Houe H (1990) Flow cytometric detection of bovine viral diarrhoea virus in peripheral blood leukocytes of persistently infected cattle. Can J Vet Res 54: 469–472
5. Qvist P, Houe H, Aasted B, Meyling A (1991) Comparison of flow cytometry and virus isolation in cell culture for identification of cattle persistently infected with bovine viral diarrhoea virus. J Clin Microbiol 29: 660–661.

Authors' address: Dr. P. Qvist, Animal Biotechnology Research Center, National Veterinary Laboratory, P.O. Box 373, DK-1503 Copenhagen V., Denmark.

Arch Virol (1991) [Suppl 3]: 169–174
© Springer-Verlag 1991

Identification of cattle infected with bovine virus diarrhoea virus using a monoclonal antibody capture ELISA

A. Fenton[1], P. F. Nettleton[1], G. Entrican[1], J. A. Herring[1], C. Malloy[1], A. Greig[2], and J. C. Low[2]

[1] Moredun Research Institute, Edinburgh, Scotland and
[2] Scottish Agricultural Colleges, Veterinary Investigation Centre, Bush Estate, Penicuik, Midlothian, Scotland

Accepted March 14, 1991

Summary. A monoclonal antibody capture enzyme linked immunosorbent assay (ELISA) has been developed to detect pestivirus-specific antigen in the leucocytes of cattle infected with bovine virus diarrhoea virus (BVDV). A blind trial was conducted to compare the specificity of the ELISA with conventional tissue culture virus isolation on 215 blood samples submitted for BVDV diagnosis from cattle throughout Scotland. One hundred and sixty seven samples were negative by both ELISA and virus isolation and 47 samples were positive by both tests. One blood was negative by ELISA and positive by virus isolation.

Key words: BVD virus, ELISA, monoclonal antibodies.

Introduction

Bovine virus diarrhoea virus (BVDV) is an important pathogen of cattle which is serologically related to hog cholera virus (HCV) and border disease virus (BDV). All three viruses have been grouped in the genus Pestivirus [5, 17]. The widespread prevalence of BVDV in cattle throughout the world is due to its ability to cross the placenta, invade the fetus and establish a persistent infection which continues into post-natal life. These persistently infected (PI) animals excrete virus continuously and remain potent sources of infectious virus usually for the rest of their lives. Such PI cattle may themselves succumb to mucosal disease (MD) at any time [4, 11].

The prevalence of PI cattle is surprisingly high with estimates varying from 0.4% to 4.5% in surveys conducted in different countries [3, 9, 12, 13,

14, 16]. To control the spread of virus and to assist with the laboratory confirmation of clinical MD the detection of virus is essential. Virus is detected readily in serum or associated with leukocytes by isolation of virus in susceptible monolayers of cultured cells. Cells co-cultivated with test leukocytes or grown in test serum are stained by immuno-fluorescence (IF) or immunoperoxidase (IPX) methods to detect the presence of the replicating non-cytopathic pestivirus; four to six days are commonly required to give a result [12, 13, 15].

Recently we have described an ELISA for the rapid detection of pestivirus antigen in the blood of sheep persistently infected with BDV [10]. In this paper a similar ELISA is described and evaluated against standard virus isolation techniques for the detection of BVDV in the blood of cattle suspected of having a BVDV viraemia.

Materials and methods

Bovine blood samples

Veterinary practitioners seeking laboratory confirmation of a clinical diagnosis of MD or wishing to screen healthy cattle for persistent BVDV infection were requested to submit duplicate 7 to 10 ml blood samples, one with no anticoagulant and one with heparin as anticoagulant. Between August 1989 and March 1990 215 duplicate samples complying with this request were received.

All samples were coded and tested blind by independent operators. Blood without anticoagulant was centrifuged at 600 g for 10 minutes, serum removed for virus isolation and serology, and virus transport medium added to the remaining clot which was stored at +4 °C. The heparinised blood was processed as described previously to extract BVDV antigen from the leucocytes [10].

ELISA

The method was a modification of that reported previously [10]. All wells of a microtitre plate (M129B, Dynatech) were coated with 100 µl of a dilution in carbonate buffer, pH 9.6 of mouse monoclonal antibody (mab) VPM22 ascitic fluid, which had been raised against lysates of cells infected with Moredun cytopathic BDV [8]. The plates were incubated at 37 °C for 2 h. Between all the following stages the plates were washed three times with PBS containing 0.05% Tween-20 (Sigma) (PBST). Plates were blocked for 1 h at 37 °C with 100 µl of PBST with 2% horse serum (PBSTH) before 100 µl of test antigen was added to each of 4 wells and incubated at 4 °C overnight. After washing duplicate pairs of wells received either 100 µl of an optimum dilution of a polyclonal bovine antiserum containing high levels of neutralising antibody to BVDV or a similar dilution of a calf serum without neutralising antibody to BVDV. The anti-BVDV serum (designated 1965J) was obtained from a calf hyperimmunised with purified BVDV (NADL strain). After incubation for 2 hours at 37 °C and washing, 100 µl of rabbit anti-bovine IgG horse-radish peroxidase (ICN immunobiologicals) diluted in PBSTH was added to all wells and incubated at 37 °C for 1 1/2 h. The substrate was freshly prepared orthophenyl diamine (0.8 mg/ml) in citrate phosphate buffer, pH 5.0, containing 0.8 µl/ml of 30% H_2O_2, 100 µl being added to each well. The reaction was allowed to proceed for 10 min in the dark before being stopped by the addition of 50 μl of 2.5 M H_2SO_4. The optical density (OD) was measured on a Flow Titertek Multiskan with a 492 nm filter.

Results for the ELISA are expressed as corrected ODs, i.e. mean OD with positive serum minus mean OD with negative serum. A corrected OD of greater than 0.1 was taken as positive.

Virus isolation

Sera were screened for the presence of BVDV using an indirect IPX technique based on that described by Meyling [13]. Fifteen microlitres of test serum were added to each of four wells before the addition of 100 μl of a suspension containing 1×10^5/ml of a semicontinuous cell line of embryonic bovine trachea cells. Four days later cells were fixed in 95% acetone at $-20\,°C$. Duplicate wells were incubated with either the hyperimmune anti-BVD serum (1965J) or the negative calf serum as used in the ELISA followed, after thorough washing, by the same anti-bovine IgG conjugate as used in the ELISA. The substrate was 3-amino-9-ethyl-carbazole in 0.05 M acetate buffer, pH 5.0. Cells were examined using a low power inverted microscope. The presence of characteristic cytoplasmic staining only in the two wells exposed to the specific anti-BVDV denoted a positive result. Blood clots from animals whose sera gave inconclusive results in the IPX or which were negative in the IPX test but had reciprocal serum neutralising antibody levels less than 22 were ground using sterile sand and centrifuged at 2000 g for 10 minutes. Supernatant fluid was filtered through 0.45 nm filters and 0.2 ml was added to each of two duplicate tubes of washed monolayer cultures of secondary bovine embryonic kidney (BEK) cells. After 10 days the cultures were frozen and thawed and passed into tubes containing BEK cells grown on coverslips. Seventy-two hours later the coverslips were fixed in acetone and stained by an indirect immunofluorescence test using serum 1965J as the first stage antiserum. Samples producing specific fluorescence in this test were considered positive for BVDV.

Serology

Heat-inactivated serum samples were tested in a micro-neutralisation test using 100 $TCID_{50}$ per well of the NADL strain of BVDV. Results were expressed as the reciprocal of the serum dilution corresponding to the 50% end-point of neutralisation.

Results

Duplicate blood samples were received from 113 cattle with symptoms suggestive of MD on 93 different farms. The remaining 102 duplicate blood samples were from apparently healthy animals on 11 different farms which had experienced BVDV related disease.

The ELISA detected specific pestivirus antigen in 47 of the heparinised blood samples. Thirty came from MD animals on 25 farms and 17 from apparently healthy cattle on 4 farms. The mean ($\pm$ SD) ELISA values of the apparently healthy cattle was higher and less variable than that from the MD animals, 1.53 ($\pm$0.49) versus 1.26 ($\pm$0.7) but the difference was not significant (t = 1.58; p = 0.12).

Infectious BVDV was isolated from the blood of the same 47 cattle that were ELISA positive and 1 other (Table 1). The IPX test gave a positive

Table 1. Bovine virus diarrhoea virus isolation and pestivirus antigen detection in blood from 215 cattle

Number of cattle	Virus isolation	Antigen detection	ELISA values	
			Median	Range
167	0	0	0.03	0–0.09
47	47	47	1.41	0.28–2.71
1	1	0		0.06

result on 43 of these sera, whereas the other 5 cattle, all clinical cases of MD, were shown to be viraemic by culturing their blood clots. Of these 5 cases, serum from 3 were toxic in the IPX test giving no result and 2 gave negative results. The one animal which was ELISA negative, infectious virus positive was one of the 2 which was negative by IPX test but positive following culture of the clot.

Serological findings on the virus positive blood showed that 45 had SN antibody titres < 4 while 3 sera, all from ELISA positive IPX test positive animals had titres of 6, 11 and 16 respectively.

Discussion

A mab-capture ELISA method for detecting a pestivirus-specific antigen in the blood of cattle with BVDV viraemia has been developed. The mab used in the ELISA has been shown by radiolabelling and immunoblotting to react with the two related p80K and p120K–130K non-structural proteins found in pestivirus infected cells [8]. Available evidence suggests that p80 is highly conserved and represents the immunodominant 'soluble antigen' common to all pestiviruses [6, 7]. Our results with the mab forming the basis of the ELISA would support this view since it has been shown to react with all pestiviruses against which it has been tested; 18 HCV, 16 BVDV and 12 BDV isolates (8, S. Edwards personal communication).

The results reported in the present study provide further evidence of the pan-pestivirus nature of the mab since it detected antigen in the blood of 47 cattle from 29 farms throughout Scotland. Failure to detect antigen in the blood of one viraemic animal may have been due either to non-reactivity of the soluble antigen of this isolate with the Mab or due to a poor antigen preparation from this animal. Non-reactivity with the mab was not the case since antigen prepared from the BEK cells in which the virus was cultured reacted strongly in the ELISA. A poor antigen preparation from this animal is a more likely explanation since the leukocyte pellets from some animals dying of MD were often smaller than expected and this was reflected in the variability of the ELISA results in this group compared to the apparently healthy animals. Since severe leukopaenia is a common finding in cattle suffering from MD (1) it was encouraging that there was sufficient antigen in all but one of the virus positive bloods to be detected by the ELISA. The major source of the antigen detected by the ELISA has yet to be determined, although B and T lymphocytes, monocytes and null cells have all been shown to contain virus in cattle persistently infected with BVDV (2).

Future work will be directed at identifying the major source of the pestivirus antigen detected by the ELISA and determining the limits of sensitivity of the assay.

Acknowledgements

We are grateful to Dr. Bernadette Dutia for characterisation of mab VPM 22. We thank all the veterinary practitioners and staff of the Scottish Agricultural Colleges Veterinary Investigation Service who submitted the samples that made this investigation possible. Kathryn Todd provided excellent technical assistance and Dr. Steven Edwards kindly tested the reactivity of the mab against a range of pestiviruses.

References

1. Baker JC (1990) Clinical aspects of bovine virus diarrhoea virus infection. Rev Sci Tech Off Int Epiz 9: 25–41
2. Bielefeldt Ohmann H, Ronsholt H, Bloch B (1987) Demonstration of bovine viral diarrhoea virus in peripheral blood mononuclear cells of persistently infected, clinically normal cattle. J Gen Virol 68: 1971–1982
3. Bolin SR, McClurkin AW, Coria MF (1985) Frequency of persistent bovine viral diarrhea virus infection in selected cattle herds. Am J Vet Res 46: 2385–2387
4. Brownlie J (1990) The pathogenesis of bovine virus diarrhoea virus infection. Rev Sci Tech Off Int Epiz 9: 43–59
5. Collett MS, Anderson DK, Retzel E (1988) Comparisons of the pestivirus bovine viral diarrhoea virus with members of the Flaviviridae. J Gen Virol 69: 2637–2643
6. Collett MS, Moennig V, Horzinek MC (1989) Recent advances in pestivirus research. J Gen Virol 70: 253–266
7. Donis RO, Dubovi EJ (1987b) Characterisation of bovine viral diarrhoea–mucosal disease virus-specific proteins in bovine cells. J Gen Virol 68: 1597–1605
8. Dutia BM, Entrican G, Nettleton PF (1989) Cytopathic and non-cytopathic biotypes of border disease virus induce polypeptides of different molecular weight and common antigenic determinants. J Gen Virol 71: 1227–1232
9. Edwards S, Drew TW, Bushnell SE (1987) Prevalence of bovine virus diarrhoea viraemia. Vet Rec 120: 71
10. Fenton A, Entrican G, Herring JA, Nettleton PF (1990) An ELISA for detecting pestivirus antigen in the blood of sheep persistently infected with border disease virus. J Virol Meth 27: 253–260
11. Harkness JW, Roeder P (1988) The comparative biology of classical swine fever. In: Liess B (ed) Classical swine fever and related viral infections, Nijhoff Publishing, Boston, pp 233–288
12. Howard CJ, Brownlie J, Thomas LH (1986) Prevalence of bovine virus diarrhoea virus viraemia in cattle in the UK. Vet Rec 119: 628–629

13. Meyling A (1984) Detection of BVD virus in viraemic cattle by an indirect immuno-peroxidase technique. In: McNulty MS, McFerran JB (eds) Recent advances in virus diagnosis. Published by Martinus Nijhoff for the CEC, pp 37–46
14. Nettleton PF (1986) The epidemiology of bovine virus diarrhoea virus. In: Thrusfield MV (ed) Proceedings of the Society of Veterinary Epidemiology and Preventive Medicine, April 1986, pp 42–53
15. Roeder PL, Drew TW (1984) Mucosal disease of cattle: a late sequel to fetal infection. Vet Rec 114: 309–313
16. Shimizu M, Satou K (1987) Frequency of persistent infection of cattle with bovine viral diarrhea–mucosal disease virus in epidemic areas. Jpn J Vet Sci 49: 1045–1051
17. Westaway EG, Brinton MA, Gaidamovich SYA, Horzinek MC, Igarashi A, Kaariainen L, Lvov DK, Porterfield JS, Russell PK, Trent DW (1985) Togaviridae. Intervirology 24: 125–139

Authors' address: A. Fenton, Moredun Research Institute, 408 Gilmerton Road, Edinburgh, EH17 7JH, Scotland.

Arch Virol (1991) [Suppl 3]: 175–180

Detection of border disease virus in sheep efferent lymphocytes by immunocytochemical and in situ hybridisation techniques

G. Entrican[1], A. Flack[2], J. Hopkins[3], M. MacLean[1] and P. F. Nettleton[1]

[1]Moredun Research Institute, 408 Gilmerton Road, Edinburgh EH17 7JH, U.K.
[2]Department of Veterinary Pathology, University College, Dublin, Eire
[3]Department of Veterinary Pathology, University of Edinburgh, Summerhall, U.K.

Accepted March 20, 1991

Summary. The prefemoral efferent lymphatics of four sheep persistently infected with a non cytopathic (NCP) isolate of border disease virus (BDV) were cannulated. Recovered lymphocytes were examined for the presence of virus by an immunocytochemical technique employing a pool of monoclonal antibodies which recognise the 120K non-structural polypeptide of NCP BDV. The results revealed that 9.5% of the lymphocytes carried virus antigen. Lymphocytes from two of the sheep were studied by in situ hybridisation using a viral antisense RNA probe complementary to the region of the BDV genome coding for the 120K polypeptide. This showed that 70–80% of the cells were infected, confirming the greater sensitivity of the in situ hybridisation technique.

Key words: Pestivirus, lymphatic cannulation, monoclonal antibodies, in situ hybridisation.

Introduction

Border disease virus (BDV) is an ovine pestivirus which is serologically related to bovine virus diarrhoea virus (BVDV) and hog cholera virus (HCV) [14]. Infection of a ewe with the noncytopathic (NCP) biotype of BDV during early pregnancy can result in the birth of lambs persistently infected (PI) with virus. These lambs continuously excrete virus and appear to be immunotolerant to BDV, thereby acting as a source of infection for other sheep [13].

PI animals may be immunosuppressed, since they are reported to succumb to intercurrent infections and their lymphocytes exhibit depressed mitogenic responses in vitro [11]. The mechanisms underlying these apparent abnormalities in immune responsiveness remain to be elucidated. However, recirculating efferent lymphocytes recovered from PI sheep have been shown to be infected with BDV by virus-specific monoclonal antibodies (Mabs) using an immunocytochemical technique (Entrican et al., submitted for publication). In this report we compare the techniques of immunocytochemistry and in situ hybridisation for identification of virus-infected lymphocytes and demonstrate that in situ hybridisation is the more sensitive of these two techniques.

Materials and methods

Animals

Two-year old PI sheep were the progeny of Dorset dams experimentally infected in early pregnancy with the Oban strain of BDV isolated from a natural outbreak of BD in the West of Scotland [3]. These sheep were viraemic and free of neutralising antibody both at birth and before cannulation. The efferent duct draining the prefemoral lymph node was cannulated as previously described [10]. Lymph was collected quantitatively into sterile, siliconised plastic bottles containing 100 IU heparin. Sheep were housed in metabolism crates and given hay and water ad libitum.

Immunocytochemistry

Efferent lymph (EL) cells cytocentrifuged onto glass slides ($60 \times g$, 5 minutes) were air dried, fixed in ice-cold acetone for 10 minutes and stored at $-20\,°C$. Cells were stained as described previously (Entrican et al., submitted for publication). Briefly, fixed cells were incubated for 2 hours with a pool of monoclonal antibodies (Mabs) which react with BDV. After washing, the cells were reacted with a sheep anti-mouse horse radish peroxidase conjugate for 1 hour. Following further washes, diaminobenzidine (Sigma) substrate was added for 10 minutes. Slides were then washed in running tap water and the cells were counterstained with haematoxylin (Gurr), dehydrated and mounted.

The number of virus-infected cells was determined by examining 500 cells per slide and counting the number of cells with positive staining.

In situ hybridisation

The pGem-4 vector, containing a 2035 base pair insert of the nucleotide sequence of the BVDV Nadl isolate [5] was a gift from Dr. M. Collett (Molecular Vaccines Inc., Gaithersburg, MD, USA). This was transcribed to make a complementary RNA sequence (riboprobe) to the p80 region of BVDV Nadl RNA as previously described [9].

For in situ hybridisation studies the efferent lymph cells were cytocentrifuged onto glass slides, which were pre-treated as follows: alcohol-cleaned slides were immersed in 3-aminopropyltriethoxysilane (Sigma, Poole, England) for 5 seconds, in methanol for 5 seconds and then incubated in 0.1% diethyl pyrocarbonate (BDH, Poole, England) at $37\,°C$ overnight. The slides were finally air dried before use. In situ hybridisation using the riboprobe was performed on the cells as previously described [9]. Briefly, 25µl of hybridisation mixture, containing approximately 100 pg of ^{35}S-UTP labelled RNA/µl, was applied to each slide (10^4 cpm/slide). Hybridisation was carried out at $25\,°C$ for 48 hrs. The slides were washed for 18 hours at room temperature in 2 × standard saline citrate (SSC) pH 7.4, 1 mM EDTA pH 8.0, 1 mM dithiothreitol and 0.1% Triton X-100. Slides were then washed in a small volume of 50% formamide, 0.3 M NaCl, 5 mM Tris pH 8.0 and 0.5 mM EDTA for 30 minutes at room temperature followed by four washes in 0.1 × SSC at 40°C for 30 minutes each. Autoradiography was carried out by coating the slides with K5 nuclear track emulsion (Ilford). The slides were then exposed for 5 days, developed for 5 minutes with Kodak 19 developer at 16°C, rinsed in water, fixed for 5 minutes in 30% (w/v) sodium thiosulphate, stained with haematoxylin, dehydrated and mounted.

Results

Immunoperoxidase staining with BDV-specific Mabs revealed that the efferent lymphocytes carried virus antigen (Plate 1). The percentage of positive-staining cells varied between sheep and between samples recovered from individual sheep on different days (Table 1). The mean of the 16 observations in the 4 sheep examined was 9.5%. The staining was specific for virus, as demonstrated by its absence in lymphocytes from PI sheep reacted with a control Mab (Plate 2). Efferent lymphocytes from a control uninfected sheep which were reacted with the BDV Mabs were also negative (not shown).

Efferent lymphocytes from 2 of the 4 sheep (numbers 3 and 4, day 2, as detailed in Table 1) were analysed for the presence of BDV RNA by in situ hybridisation. Both sheep had less than 10% virus-infected cells in efferent lymph as determined by immunocytochemistry (Table 1), but 70–80% virus-infected cells as determined by in situ hybridisation. An example is shown in Plate 3. The specificity of the hybridisation was confirmed by reacting the probe with efferent lymphocytes from a control sheep. There was no labelling of those control cells (Plate 4).

Discussion

To understand how BDV may affect the immune system of PI sheep, it is important to determine the extent of infection of lymphocytes with virus in vivo. We have used the cannulated efferent lymph model as a source of lymphocytes, since this compartment has been shown to contain almost pure populations of lymphocytes in sheep [12]. Phenotypic data has previously confirmed that the efferent lymph cells from PI sheep which we have used in this study are lymphocytes (Entrican et al, submitted for publication).

BDV-infected lymphocytes were identified by two techniques: immunocytochemistry and in situ hybridisation. The Mabs used for the immunocytochemistry studies have previously been characterised and shown to react with a 120K polypeptide in NCP BDV-infected cells [8]. This polypeptide appears to be non-structural and highly conserved among

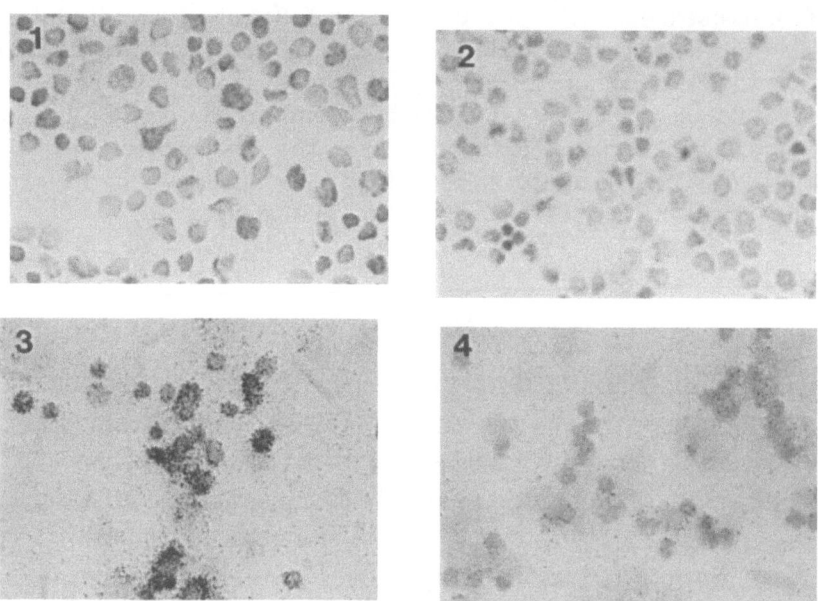

Plates. The presence of virus in efferent lymphocytes from persistently infected sheep, specifically demonstrated by immunocytochemistry and in situ hybridisation. *Plate* 1. Lymphocytes reacted with anti-BDV Mabs, showing characteristic peroxidase staining of virus-infected cells. *Plate* 2. Lymphocytes reacted with a control Mab showing absence of staining pattern. *Plate* 3. Lymphocytes reacted with the radiolabelled riboprobe. Virus-infected cells are identified by the granular reaction. *Plate* 4. Lymphocytes from a control sheep reacted with the radiolabelled probe, where no specific hybridisation has occurred. Magnification × 250

Table 1. Presence of BDV antigen in efferent lymph cells of persistently infected sheep measured by immunocytochemistry

| Sample (days post cannulation) | Percentage of virus-infected cells | | | |
	Sheep 1	Sheep 2	Sheep 3	Sheep 4
2	12.0	22.2	6.8	4.8
3	6.4	14.6	7.8	5.0
4	10.0	10.2	ND	ND
5	6.2	6.6	ND	ND
6	9.0	10.0	ND	ND
7	7.0	9.8	ND	ND

The first sample was taken 48 hr after cannulation. Subsequent samples were analysed on consecutive days. The mean value of virus-infected cells is 9.5%. ND = not determined.

pestiviruses [7]. Therefore, these Mabs do not recognise virus particles but identify cells which carry viral antigen as a result of virus replication.

The molecular probe used for in situ hybridisation studies hybridises to a sequence of the pestivirus genome predicted to code for the 80K non-structural polypeptide in BVD Nadl [6]. The 80K polypeptide was shown to be a cleavage product of a higher molecular weight 125K non-structural polypeptide in BVDV. The probe cross-hybridises with NCP BDV, presumably at the equivalent 120K non-structural polypeptide coding region [9]. There is therefore a potential correlation between the molecular probe and the Mabs, since the Mabs may recognize the protein product of the region of the genomic RNA to which the probe hybridises. Although we have so far analysed cells from only two sheep by in situ hybridisation, our preliminary results indicate that in situ hybridisation is very much more sensitive than immunocytochemistry for identifying virus-infected cells (Plates 1 and 3).

We have identified BDV-infected lymphocytes in PI sheep, but cannot as yet attribute infection to any particular lymphocyte subsets. However, progress has been made on the identification of cells from PI cattle which harbour BVDV. In experiments in cattle using a polyclonal antiserum to BVDV in conjunction with Mabs to bovine cell subsets, BVDV was detected in peripheral blood mononuclear cells from PI cattle. BVDV was found to be present in a proportion of all the mononuclear cell subsets examined, namely monocytes, T cells, B cells and null cells, with T cells being the most heavily infected population [1, 2]. It is not yet known if BDV infects subsets of peripheral blood mononuclear cells to the same extent, although PI sheep are known to have abnormal lymphocyte subpopulations in peripheral blood, resulting in a B cell hyperplasia [4].

The development of the highly sensitive in situ hybridisation technique for detecting pestivirus-infected cells is particularly relevant to studies on virus/cell subset associations, where cells previously thought to be uninfected may in fact be infected. We intend to use the techniques described in this report in conjunction with phenotype markers to address the question of cellular distribution of BDV in PI sheep.

Acknowledgements

The authors thank Dr. Marc Collett for his gift of the pGem-4 vector, Iris Campbell for the anti-rotavirus Mab and A. M. Dawson for preparing the HRP-conjugate. We thank Isobel Brown for typing the manuscript. This work was supported in part by an Agricultural and Food Research Council Link Research Grant (LRG 31) and the Wellcome Trust.

References

1. Bielefeldt Ohmann H (1987) Double-immunolabelling systems for phenotyping of immune cells harboring bovine viral diarrhea virus. J Histochem Cytochem 35: 627–633

2. Bielefeldt Ohmann H, Ronsholt L, Bloch B (1987) Demonstration of bovine viral diarrhoea virus in persistently infected, clinically normal cattle. J Gen Virol 68: 1971–1982
3. Bonniwell MA, Nettleton PF, Gardiner AC, Barlow RM, Gilmour JS (1987) Border disease without nervous signs or fleece changes. Vet Rec 120: 246–249
4. Burrells C, Nettleton PF, Reid HW, Miller HRP, Hopkins J, McConnell I, Gorrell MD, Brandon MR (1989) Lymphocyte subpopulations in the blood of sheep persistently infected with border disease virus. Clin Exp Immunol 76: 446–451
5. Collett MS, Larson R, Gold C, Strick D, Anderson DK, Purchio AF (1988) Molecular cloning and nucleotide sequence of the pestivirus bovine viral diarrhea virus. Virology 165: 191–199
6. Collett MS, Larson R, Belzer SK, Retzel E (1988) Proteins encoded by bovine viral diarrhea virus: the genomic organisation of a pestivirus. Virology 165: 200–208
7. Collett MS, Moennig V, Horzinek MC (1989) Recent advances in pestivirus research. J Gen Virol 70: 253–266
8. Dutia BM, Entrican G, Nettleton PF (1990) Cytopathic and noncytopathic biotypes of Border disease virus induced polypeptides of different molecular weight with common antigenic determinants. J Gen Virol 71: 1227–1232
9. Flack AM, Smyth JMB, Sheahan BJ, Atkins GJ, Demonstration of bovine viral diarrhea and border disease virus RNA in formalin fixed, paraffin embedded tissues using an RNA probe. Submitted for publication.
10. Hall JG (1967) A method for collecting lymph from the prefemoral lymph node of unanaesthetized sheep. Q J Exp Physiol 211: 200–204
11. Harkness JW, Roeder PL (1988) The comparative biology of classical swine fever. In: Liess B (ed) Classical swine fever and related viral infections, 1st ed. Martinus Nijhoff, Boston, pp 233–288
12. Mackay CR (1988) Sheep leukocyte molecules: a review of their distribution, structure and possible function. Vet Immunol Immunopathol 17: 1–20
13. Terpstra C (1981) Border disease: virus persistence, antibody response and transmission studies. Res Vet Sci 30: 183–191
14. Westaway EG, Brinton MA, Gaidamovich S Ya, Horzinek MC, Igarashi A, Kaariainen L, Lvov DK, Porterfield JS, Russell PK, Trent DW (1985) Togaviridae. Intervirol 24: 125–139

Authors' address: G. Entrican, Moredun Research Institute, 408 Gilmerton Road, Edinburgh EH17 7JH, U.K.

Arch Virol (1991) [Suppl 3]: 181–190
© Springer-Verlag 1991

Bovine viral diarrhea virus infection: rapid diagnosis by the polymerase chain reaction

S. Belák and **A. Ballagi-Pordány**

Department of Virology, The National Veterinary Institute, Biomedical Center,
Uppsala, Sweden

Accepted March 20, 1991

Summary. The polymerase chain reaction (PCR) was applied to detect bovine viral diarrhea virus (BVDV) by amplification of its nucleic acid sequences in cell cultures, in serum samples of persistently infected cattle, and in organ specimens of acutely diseased calves. The primers and the probes were selected from the gp48 region of the cytopathic NADL strain. The products of single PCR or double PCR were identified by electrophoresis as well as by hybridization with biotinylated probes. The results thus obtained correlated with those of conventional diagnostic procedures, i.e., virus isolation and serology. The detection assay of the BVDV genome by the PCR amplification proved to be both specific and sensitive.

Key words: Bovine viral diarrhea virus (BVDV), diagnosis, polymerase chain reaction (PCR), nucleic acid hybridization, oligonucleotide probes.

Introduction

Bovine viral diarrhea virus (BVDV) is one of the most important pathogens of cattle, causing economic losses of considerable importance throughout the world.

The genome of BVDV is infectious, positive-strand RNA [10], estimated at 2.9 to 4.4×10^6 Da in size [7, 8, 16, 17]. BVDV is currently classified as a member of the family Togaviridae, genus Pestivirus [20]. Propagation of BVDV in cell cultures allows the differentiation of cytopathic and non-cytopathic biotypes (cp-BVDV and noncp-BVDV). Both biotypes are pathogenic for cattle [4].

There are two disease entities caused by BVDV and occurring predominantly in adult calves: the acute bovine virus diarrhea with high morbidity and low mortality, and the acute or chronic form of mucosal disease with low morbidity and high mortality.

Both entities show suggestive clinical symptoms and lesions. The virus is spread from animals that carry a persistent, latent infection but are usually seronegative [13]. These animals are the offspring of cows in which, after a primary infection, a transplacental spread occurred during pregnancy. Depending upon the stage of pregnancy, the transplacental infection may result in abortion or stillbirth, in malformation or immunotolerance. Immunotolerant calves are the virus carriers and excretors but may, after a recurrent infection, develop mucosal disease [6].

BVDV also is a frequent contaminant in fetal calf serum, a commonly used component in cell culture systems. This infection may lead to the BVDV contamination of biological products, e.g., vaccines [1].

Due to the obvious economic impact of BVDV infection, attempts are made world-wide to introduce effective control measures. The aim of these is to break the cycle of transmission by identifying and eliminating the sources of infection.

The conventional diagnosis of BVDV infection has been based on the direct demonstration of the virus in clinical specimens or on indirect detection by assessment of specific antibody response. Virus isolation as well as the techniques of immunohistochemistry are the most common methods of direct detection, whereas the indirect detection is assessed by various immunoassays.

The present methods are either insensitive or unsuitable for large-scale screening. Sensitive and novel approaches are needed to trace the spread and circulation of BVDV as well as to study the pathogenesis of the disease [8].

In recent years efforts have been made to develop direct methods based on the demonstration of the BVDV RNA by nucleic acid hybridization [5, 15, 17].

In this study, we adapted the method of polymerase chain reaction (PCR) for the detection of the BVDV genome in specimens. This already well-known method makes it possible to amplify selected gene sequences to amounts that allow their detection and identification.

The experiences we gained on some viral diseases show that the PCR is several orders of magnitude more sensitive than the direct hybridization assays [2, 3].

Materials and methods

Cells and Viruses

In order to estimate the specificity of the PCR method, we tested various strains of both biotypes of BVDV (kindly provided by Professors B. Liess and V. Moennig, Hannover, FRG), as well as six local isolates. The viruses were grown on bovine turbinate (BT) cell

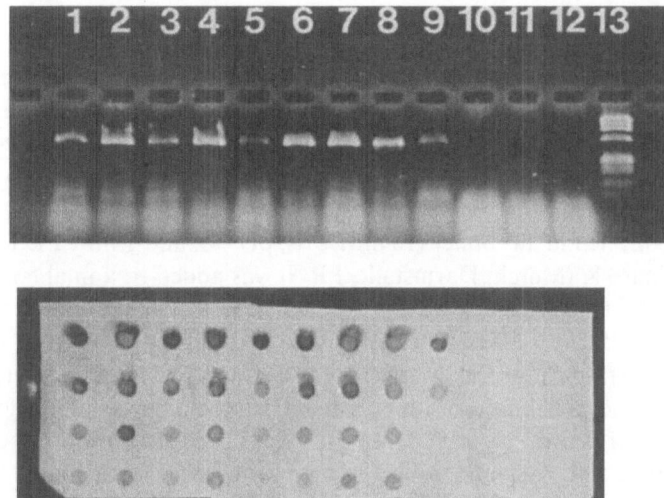

Fig. 1. Single PCR amplification of BVDV cDNA from cell cultures. Primers OBVD-5 and OBVD-6 were applied for amplification from BVDV-infected cell cultures as well as from cell cultures infected with heterologous viruses as listed below. **A** Electrophoresis of the PCR products of cells infected with BVDV strains NADL (1), Ug59 (2), Singer (3), Oregon (4), New York (5), Osloss (6), A-1138/69 (7), 0715 (8), and Swedish isolate S2 (9). PCR products from cells infected with parainfluenza-3 virus (10), bovine parvovirus (11), and bovine coronavirus (12). DNA-size marker ΦX-174-RF/HaeIII digest is seen in position 13. **B** Dot-blot hybridization of three-fold dilutions of the PCR products with the OBVD-4a probe

cultures. Bovine kidney- and BT-cells inoculated with bovine parvovirus, bovine coronavirus, and parainfluenza-3-virus strains were used as controls of specificity (Fig. 1). The cells were harvested when the cytopathic effect (CPE) appeared. The noncp-BVDV strains were harvested after 6 days of incubation. The cells were dispersed by trypsinization and washed two times in PBS. The cell number was standardized to 5×10^4 cells/ml in PBS.

Clinical samples

In order to estimate the diagnostic applicability of the PCR, samples of acutely diseased cattle as well as of persistently infected cattle were examined. The test materials consisted of samples of lungs, spleen, and placenta from the acute cases. The organs were homogenized and 10% v/w suspensions were prepared in PBS. From the persistently infected cattle serum samples were tested. For comparison, each organ- or serum-sample was examined in parallel by virus isolation and by PCR.

Conventional diagnostic procedures

Virus isolation from the specimens was attempted on BT cells by two consecutive passages, 6 days each. The results were assessed by indirect immunofluorescence, using monoclonal antibodies prepared by Juntti et al. [12], and by immunoperoxidase method using polyclonal antibodies.

Preparation of specimens for the PCR

In order to destroy cell membranes the cell- and the organ-samples (concentrations see above) were three times frozen at $-20\,^\circ$C. After the last thawing, 1 ml amounts of the samples were centrifuged in an Eppendorf centrifuge at 500 g for 5 minutes. The supernatants were recentrifuged at 5,000 g for 15 minutes and then, in the same way as the serum samples, they were pelleted at 50,000 rpm in a Kontron TST-55.5 rotor for 30 min at 15 °C. The pellets were collected in TE buffer (10 mM Tris, pH 7.5, and 1 mM EDTA) containing 1.5% SDS. Proteinase K (Merck, Darmstadt, FRG) was added to a final concentration of 100 to 200 µg/ml and the specimens were incubated at 55°C for 30 minutes. Subsequently, they were heated to 95°C for 10 minutes and 10 to 20 µg fresh yeast RNA (USB Comp., Cleavland, OH, USA) was added as carrier RNA. The samples were extracted twice with phenol/chloroform and precipitated with ethanol. The precipitates were pelleted at 14,000 rpm in an Eppendorf centrifuge for 20 min at 4 °C. The pellets were dissolved in 10 µl $0.5 \times$ TE buffer. In order to perform the reverse transcriptase reaction, the following materials were added to the 10 µl RNA: 2 µl $10 \times$ PCR buffer (0.1 M Tris-HCl, pH 8.3; 0.5 M KCl, 25 mM $MgCl_2$, 1 mg/ml BSA), 1 µl of each of the four dNTPs (10 mM stocks, Boehringer, Mannheim, FRG), 2 µl of reverse transcriptase primer (10 µM stock), 1 µl RNAsin (Pharmacia, Uppsala, Sweden), and 1 µl reverse transcriptase (200 U/µl, Moloney MuLV, Bethesda Research Laboratories, Gaithersburg, MD, USA). The specimens were allowed to stand at room temperature for 10 min; they were incubated at 37 °C for 90 min, subsequently heated to 98 °C for 10 minutes, and the cDNA products were then chilled on ice.

Primers and probes

Four primers were selected, complementary to the published sequences of the gp48 region of the cytopathic NADL strain [7].

For the identification of the PCR products, two probes were designed. The probes corresponded to two regions between the two internal primers.

The primer sequences, locations and functions as well as the oligonucleotide probe sequences and positions on the BVDV genome are shown in Table 1.

The oligonucleotides were synthesized and then purified by reverse phase chromatography at the Research Genetics, Huntsville, AL, USA. The oligonucleotide probes were biotinylated according to Guitteny et al. [11]. In order to add a tail of biotin-16-dUTP (Boehringer, Mannheim, FRG) at the 3' end, the TdT enzyme (Pharmacia, Uppsala, Sweden) was used. In the hybridization assays the oligonucleotide probes were used in a final concentration of 200 ng/ml.

The conditions of amplification

The PCR was performed by using 20 µl cDNA, 8 µl $10 \times$ PCR buffer, 3 µl of each 10 pmole/µl upstream and downstream BVDV primers, 0.5 µl (2.5 U) Taq polymerase (AmpliTaq, Perkin-Elmer Cetus, Norwalk, CT, USA), and distilled water to 100 µl. PCR was allowed to run in 32 cycles in a DNA Thermal Cycler (Perkin-Elmer Cetus). Each cycle included three segments: denaturation at 94 °C for 45 seconds, primer annealing at 55 °C and primer extension at 72 °C for 1 min each.

If it was necessary, e.g., in the case of organ homogenates, we increased the sensitivity of the test by an additional amplification. Two microliters of the PCR-product from the above described reaction were added to the 50 µl of the standard PCR-mix, as recommended by Perkin-Elmer Cetus. The primers used in this reaction were OBVD-7 and OBVD-8,

Table 1. Sequences of the synthetic oligonucleotides and their location along the BVDV genome

Primers[a]	Sequence	Location and direction
OBVD-5	GGTATGATGG ATGCAAGTGA G	1362–1382 >
OBVD-6	AAGCAGCGTA TGCTCCAAAC C	< 1860–1880
OBVD-7	ACTTCAACGC CATGAGTGGA ACAAGCATGG	1405–1434 >
OBVD-8	CTTTTTTCCT AGTATCCCGA GCTGCTTGCC	< 1808–1837

Probes	Sequence	Location and direction
OBVD-4	ACCCTTAACA GCTTGCAAGA A	1594–1614 >
OBVD-4a	GTTCTTTCCT TTCTTGCAAC C	< 1604–1624

[a] Functions of primers:
  OBVD-5 and OBVD-6: external primers
  OBVD-7 and OBVD-8: internal primers
  OBVD-6: also used as reverse transcriptase primer

30 base-long each. Each of the 32 cycles included two segments: 94 °C for 45 seconds for denaturation and 74 °C for 1 min for primer annealing as well as primer extension in the same segment.

Identification of the PCR products

To visualize the yield, 10 μl amounts of the PCR products were run on 2.5% agarose gels at 100 V for 45 minutes. The gels were stained with ethidium bromide as described elsewhere [14].

To control the specificity of the amplification, the PCR products were simultaneously tested by DNA dot-blot hybridization and by Southern-blot hybridization. In order to avoid reading errors, the dot-blot hybridization were made with three-fold dilutions of the PCR products [2]. PCR products of equine herpesvirus type 1, pseudorabies virus and bovine leukemia virus were also hybridized in order to estimate the specificity of the BVDV probes. These PCR products were amplified with their respective primers as described elsewhere [2, 3, Ballagi-Pordány and Belák, manuscript in preparation].

Filters were hybridized with the biotinylated BVDV oligonucleotide probes in the same way as previously described [2, 3, 9].

Results

Electrophoresis of the single PCR products

By applying the single PCR, a product of 520 bp was detected on the gels (Fig. 1). Such a fragment was yielded by each specimen of the cell cultures which had been inoculated with various BVDV strains, but not by specimens of cell cultures infected with heterologous viruses (Fig. 1).

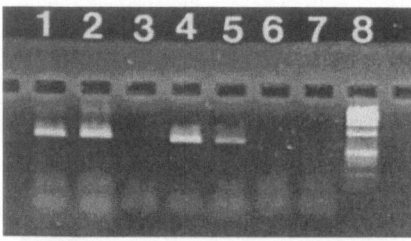

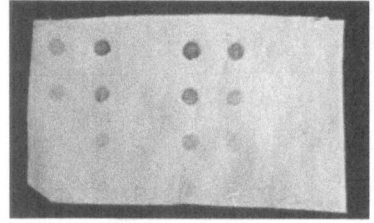

Fig. 2. PCR amplification of BVDV cDNA from specimens of acutely diseased and of persistently infected cattle, respectively. **A** Electrophoresis of the PCR products of serum specimens from persistently infected cattle positive at virus isolation (1 and 2) and negative at virus isolation (3). Spleen homogenates of acutely infected cattle positive at virus isolation (4 and 5) and negative (6 and 7) at virus isolation. The sera (positions 1 to 3) were examined in single PCR, whereas, the organ suspensions (positions 4 to 7) were tested by double PCR. DNA-size marker in position 8 is the same as in Fig. 1. **B** Dot-blot hybridization of three-fold dilutions of the PCR products with the OBVD-4a probe

When applying the single PCR, we were unable to detect the viral RNA in each of the virus-infected organ homogenates tested (not shown). However, in the serum samples of the persistently infected animals the virus was detected even by the single method of amplification (Fig. 2).

Electrophoresis of the double PCR products

By applying the double PCR, a band of 432 bp was seen on the gels. By the double amplification the virus signal was detected even in the organ homogenates (Fig. 2).

Nucleic acid hybridization of the PCR products

The Southern blot hybridization (not shown) as well as the dot-blot hybridization revealed that the bands had nucleic acid sequences specific for BVDV (Figs 1 to 3). The hybridization studies showed a considerable variability of the various BVDV strains in the regions from where the OBVD-4 probe was synthesized. At 44 °C hybridization temperature, calculated according to the theoretical formula of Davis et al. [9], the OBVD-4 probe recognized two local isolates and the NADL strain only (Table 2). By decreasing the temperature to 21°C, three local isolates, as well as the strains NADL, Singer and Ug59 were recognized by the probe (not shown). This probe did not react with the rest of the strains and with three non-cp local isolates. The OBVD-4a probe detected each BVDV strain and isolate at 44 °C, but did not hybridize to the PCR products of heterologous viruses (Figs 1, 3 and Table 2).

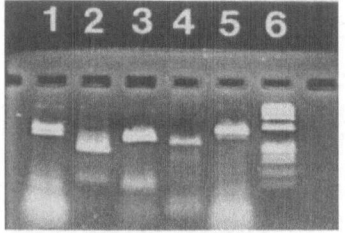
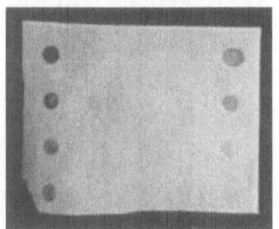

Fig. 3. Specificity test of the OBVD-4a probe on PCR products of heterologous viruses. The nucleic acid sequences of equine herpesvirus 1, pseudorabies virus and bovine leukemia virus were amplified with their respective primers. BVDV strains NADL and 0715, amplified with BVDV primers OBVD-5 and OBVD-6, were used as positive controls. **A** Electrophoresis of the PCR products of BVDV strain NADL (1), equine herpesvirus type 1 (2), pseudorabies virus (3), bovine leukemia virus (4) and BVDV strain 0715 (5). DNA-size marker at position 6 is the same as above. **B** Dot-blot hybridization of three-fold dilutions of the PCR products with the OBVD-4a probe

Table 2. Electrophoresis and hybridization of the PCR products of cell cultures infected with various BVDV strains. Single PCR was performed. The PCR products were identified by electrophoresis as well as hybridization at 44 °C

Virus strains and isolates	Cytopathic effect	Electrophoresis of PCR product	Hybridization of PCR product with probe...	
			A	B
A-1138/69	+	+	NT	+
NADL	+	+	+	+
New York-1	−	+	−	+
Oregon C24V	+	+	NT	+
Osloss 2482	+	+	NT	+
Singer	+	+	−	+
Ug59	+	+	−	+
0715/80	−	+	NT	+
S1*	−	+	+	+
S2	−	+	−	+
S3	−	+	−	+
S4	−	+	+	+
S5	−	+	−	+
S6	−	+	−	+

A, probe OBVD-4; B, probe OBVD-4a
S*, Swedish isolate; NT, not tested

Discussion

In this paper we present a specific and sensitive PCR assay for detection of BVDV nucleic acid sequences. By this approach BVDV infection can specifically and rapidly be diagnosed in cell cultures, in specimens of calf serum and in organ preparates of acutely diseased calves. The viral RNA is first converted into single stranded cDNA, which is then used for the enzymatic amplification. The diagnostic PCR products are detected by electrophoresis. The final identification is made by nucleic acid hybridization.

The aim of this project was to apply a "PCR plus hybridization system" which detects all the variants of BVDV. Such a "all-round" BVDV detection system would have a reliable diagnostic applicability. The composition of such a PCR panel, i.e., the selection of primers and probes from highly conserved regions of the viral genome was hampered by the poor information available concerning the nucleotide sequences of BVDV [8]. The sequence of the complete genome is not yet at hand and incomplete nucleotide sequences are available only for several BVDV strains [7, 18]. Two of these strains, NADL and Osloss, showed an approximate aligned sequence homology of 74% along the genome [8].

By using the sequencing data of Collett et al. [7] first we selected primers and probes from the highly conserved p80 region of the genome of the NADL strain. Unexpectedly, these primers and probes detected only the NADL strain (not shown). Subsequently, we selected primers and probes from the gp48 region of the same strain. By using these primers, a specific PCR product was amplified from all the BVDV-infected cell cultures as well as from the clinical specimens (Figs 1 and 2). This indicates an adequate selection of primer sequences from the gp48 region of the BVDV genome.

In contrast to the primers, the OBVD-4 probe, selected also from the gp48 region, proved to be too specific. This probe recognized the NADL strain and two local isolates only. This indicates that in the map regions from 1594 to 1616, the local isolates S1 and S3 are more closely related to the NADL strain than are the other local isolates. The high specificity of OBVD-4 was decreased by applying the lower stringency of 21°C, which allowed five mismatches out of the 21 nucleotides. However, the probe still did not possess the spectrum required for diagnostic use. Thus, the OBVD-4 probe can not be used as a "all-round" probe to detect BVDV infection.

The hybridization results of OBVD-4a show that the nucleic acid sequences of this probe, i.e., map regions from 1604 to 1624, are common (or contain only several mismatches) in the various BVDV strains tested in these experiments. We predict that the genomic regions of the primers and of OBVD-4a are conservative in the BVDV genome and will thus be useful components of a "all-round" diagnostic system to detect BVDV nucleic acid sequences.

Utilizing this "PCR plus hybridization system", we successfully detected the BVDV genome in the serum samples as well as in the organ specimens. In the serum the virus had been detected both by single and double PCR. However, in the organ suspensions only the double PCR gave reproducibly positive results. According to the results in Table 3, the double PCR had higher sensitivity of detection than the single amplification. It is likely, that the amount of the viral RNA was higher in the serum samples of persistently infected animals than in the organs of acutely diseased calves.

Table 3. Detection of BVDV genomic sequences in clinical specimens. Number of positive cases versus total number of specimens examined by virus isolation and by the PCR. The PCR products were identified by electrophoresis as well as hybridization with the OBVD-4a probe

Specimens	Virus isolation	PCR single	PCR double
From acute cases,			
Lungs	1/4	0/4	2/4
Spleen	2/5	0/5	2/5
Placenta	0/2	0/2	0/2
From chronic cases,			
Serum	5/8	4/8	5/8
Total	8/19	4/19	9/19

These results suggest that the double PCR assay should be chosen as a reliable diagnostic procedure. Recent data show that the use of double-nested primers enhances not only the sensitivity, but also the specificity of the PCR [19].

The PCR, together with the simple non-radioactive hybridization assay, provides a novel direct method for the detection of the BVDV infection. By this method a confirmed direct diagnosis of BVDV infection is made within two to three days. Correlation of the results with those obtained by the conventional diagnostic procedures (Table 2) showed that the detection assay of the BVDV genome by PCR amplification is both specific and sensitive. The "PCR plus hybridization system" will have a broad application for the rapid detection of BVDV in clinical specimens and in batches of fetal calf serum, as well as for the studies of BVDV pathogenesis.

Acknowledgements

The authors are grateful to Prof. Z. Dinter for the helpful discussions and constructive criticism of the manuscript. We thank Professors B. Liess and V. Moennig for providing the BVDV strains. The excellent technical assistance of Miss Annie Persson is appreciated. This work was supported by grant Nr 892002 of Foundation for Agricultural Research, Sweden and by grant 60102-1 of the National Veterinary Institute.

References

1. Baker JC (1987) Bovine viral diarrhea virus: a review. J Am Vet Med Assoc 190: 1449–1458
2. Ballagi-Pordány A, Klingeborn B, Flensburg J, Belák S (1990) Equine herpesvirus type 1: detection of viral DNA sequences in aborted fetuses with the polymerase chain reaction. Vet Microbiol (in press)

3. Belák S, Ballagi-Pordány A, Flensburg J, Virtanen A (1989) Detection of pseudorabies virus DNA sequences by the polymerase chain reaction. Arch Virol 108: 279–286

4. Bolin SR, McClurkin AW, Cutlip RC, Coria MF (1985) Response of cattle persistently infected with noncytopathic bovine viral diarrhea virus to vaccination for bovine viral diarrhea and to subsequent challenge exposure with cytopathic bovine viral diarrhea virus. Am J Vet Res 46: 2467–2470

5. Brock KV, Brian DA, Rouse BT, Potgieter LND (1988) Molecular cloning of complementary DNA from a pneumopathic strain of bovine viral diarrhea virus and its diagnostic application. Can J Vet Res 52: 451–457

6. Brownlie J, Clarke MC, Howard CJ (1984) Experimental production of fetal mucosal disease in cattle. Vet Rec 114: 535–536

7. Collett MS, Larson R, Gold C, Strick D, Anderson DK, Purchio AF (1988) Molecular cloning and nucleotide sequence of the pestivirus bovine viral diarrhea virus. Virology 165: 191–199

8. Collett MS, Moennig V, Horzinek MC (1989) Recent advances in pestivirus research. Review article. J Gen Virol 70: 253–266

9. Davis LG, Dibner MD, Battey, JF (1986) Basic methods in molecular biology. Elsevier, New York, p 75

10. Diderholm H, Dinter Z (1966) Infectious RNA derived from bovine viral diarrhea virus. Zentralbl Bakteriol I. Abt Orig 201: 270–272

11. Guitteny AF, Fouque B, Mougin C, Teoule R, Bloch B, (1988) Histological detection of messenger RNAs with biotinylated synthetic oligonucleotide probes. J Histochem Cytochem 36: 563–571

12. Juntti N, Larsson B, Fossum C (1987) The use of monoclonal antibodies in enzyme linked immunoadsorbent assay for detection of antibodies to bovine viral diarrhoea virus. J Vet Med 34: 356–363

13. Liess B, Frey HR, Orban S, Hafez SM (1983) Bovine Virusdiarrhoe (BVD) – "Mucosal Disease": Persistente BVD-Feldvirusinfektionen bei serologisch selektierten Rindern. DTW 90: 261–266

14. Maniatis T, Fritsch, EF, Sambrook J (1982) Molecular cloning. A laboratory manual. Cold Spring Harbor Laboratory, Cold Spring Harbor, New York, p 150

15. Potgieter LND, Brock KV (1989) Detection of bovine viral diarrhea virus by spot hybridization with probes prepared from cloned cDNA sequences. J Vet Diagn Invest 1: 29–33

16. Purchio AF, Larson R, Collett MS (1983) Characterization of virus-specific RNA synthesized in bovine cells infected with bovine viral diarrhea virus. J. Virol 48: 320–324

17. Renard A, Guiot C, Schmetz D, Dagenais L, Pastoret PP, Dina D, Martial JA (1985) Molecular cloning of bovine viral diarrhea viral sequences. DNA 4: 429–438

18. Renard A, Dino D, Martial J (1987) Vaccines and diagnostics derived from bovine diarrhea virus. European Patent Application number 86870095.6. Publication number 0208672, 14 January 1987

19. Rimstad E, Hyllseth B (1990) Double PCR and magnetic separation of DNA used in identification of infectious pancreatic necrosis virus, a dsRNA virus. In: Olsvik O, Bukholm G (eds) Application of molecular biology in diagnosis of infectious diseases. Oslo, March 16th and 17th, 1990, pp 94–98

20. Westaway EG, Brinton MA, Gaidamovich SY, Horzinek MC, Igarashi A, Kääriäinen L, Lvov DK, Porterfield JS, Russell PK, Trent DW (1985) Togaviridae. Intervirology 24: 125–139

Author's address: Dr. S. Belák, Department of Virology, The National Veterinary Institute, Biomedical Center, Box 585. S-751 23 Uppsala, Sweden.

Arch Virol (1991) [Suppl 3]: 191–197

cDNA probes for the detection of pestiviruses

Catherine Cruciere, L. Bakkali, Monique Gonzague and **E. Plateau**[1]

[1] Centre National d'Etudes Vétérinaires et Alimentaires,
Laboratoire Central de Recherches Vétérinaires, (C.N.E.V.A./L.C.R.V.),
France

Accepted March 20, 1991

Summary. Probes were prepared from genomic RNA of Hog Cholera Virus (HCV) after synthesis of cDNA and cloning. Six probes were selected according to their place on the viral genome determined by sequencing and comparison with BVDV sequence. These probes were hybridized with two strains of HCV (Alfort and Nord), two strains of Bovine Viral Diarrhea (BVDV) (NADL, New York) and four strains of Border Disease (BD) (Lyon 1, Lyon 2, Aveyron, IEMVT). This panel of six probes seem to be able to differentiate pestiviruses but some differences rely only on slight intensity of the hybridization.

Key words: cDNA probes, pestiviruses, diagnosis, distinction.

Introduction

Bovine viral diarrhea virus (BVDV) and border disease virus (BDV) are the cause of important diseases of large and small ruminants. These two viruses belong to the *Pestivirus* genus [8] and are genetically and antigenically related to hog cholera virus (HCV) of pigs.

The laboratory diagnosis of these diseases is traditionally based on virus isolation and/or serology [5, 6, 7]. Identification and differenciation between these three viruses is an important problem in pigs which can be infected by BVDV and BDV as well as by HCV.

The development of molecular biology gave a new approach to the comprehension of the biology of viruses, and new tools for the diagnosis such as cDNA probes. In this report we describe the preparation of hog cholera virus probes and their application for the detection and differentiation of pestiviruses.

Materials and methods

Cells and viruses

The HCV strain Alfort [1] twice cloned in 96 wells plates by end point dilution and HCV strain Nord (graciously given by Dr Leforban, SPP, Ploufragen) were amplified on porcine kidney (PK) cell line 15 cultured in modified Eagle's medium (MEM) supplemented with 8% foetal calf serum (FCS). The New York BVDV strain and BDV strains Lyon 1 and Lyon 2 were graciously given by Dr Chappuis (Rhone Merieux Lyon). The BDV strain IEMVT was graciously given by Dr Lefevre (IEMVT France). These strains were grown on sheep kidney (SK) cell line cultured in OPTIMEM supplemented with 10% FCS. The NADL BVDV (ATCC) strain was amplified on bovine nasal turbinate cell line (BT) cultured in MEM supplemented with 10% of basal medium supplement (BMS) (SEROMED, France). All stock cells and FCS were regularly tested for the absence of contamination by pestiviruses.

Production and purification of the virus

PK 15 cells were infected with a m.o.i. of 5 (as determined by immunofluorescence assay) of HCV (Alfort). After one hour of incubation, MEM supplemented with 5% horse serum was added. Twenty nine hours later the infected cells were freeze-thawed three times. The viral suspension was added with 0.5 mg/ml of trypsin and incubated 10 min at 37°C before centrifugation at 3,000 g during 30 min and at 10,000 g during 1 h. Then, the suspension was ultracentrifuged in a 45 Ti rotor at 44,000 RPM during 3 h. The pellet was purified with sucrose gradient 22%–44% ultracentrifuged at 39,000 RPM during 36 h in SW40.

Preparation of genomic RNA

The HCV virus pellet resuspended in TNE (200 mM Tris, pH 8, 200 mM NaCl, 2 mM EDTA) with 400 µg/ml of proteinase K was incubated 30 min at 37°C. After addition of a same volume of TNE 2 X with 2% SDS it was again incubated 5 min at 50°C and 30 min at 25°C, then genomic RNA was precipitated with 2.5 volumes of ethanol after phenol-chloroform extraction.

Synthesis of cDNA and cloning

Approximately 10 µg of RNA from purified virions and 1 µg of random hexanucleotide primer in 10 µl of H_2O were heated to 65°C for 5 min, chilled on ice and mixed with dATP, dCTP, dGTP, dTTP 1 mM each, 15 units of RNasin or RNAase inhibitor and 100 units of AMV reverse transcriptase in buffer (50 mM Tris-HCl pH 8.3 at 42°C, 8 mM $MgCl_2$, 50 mM KCl, 1 mM DTT) in a final volume of 100 µl. The mixture was incubated 30 min at 25°C, 30 min at 37°C and 2 h at 42°C.

After phenol-chloroform extraction, the heteroduplex was treated with 17 units of RNAase T_2 in 250 mM NaCl, 10 mM $NaCOOCH_3$ pH 4.5 during 15 min at 37°C, re-extracted with phenol-chloroform and filtered on sephadex G50 medium. Then the hetero-duplex and the vector PBR 322 cleaved with Pst I were added with complementary homopolymer tracts. Tails of about 20 deoxycitidines were added to 2 picomoles of 3′ termini in a final volume of 20 µl of buffer (25 mM Tris HCl pH 7, 100 mM potassium cacodylate, 0.2 mM DTT, 1 mM $CoCl_2$, 50 µg bovine serum albumim, 0.2 mM dCTP or dGTP) and 14 units of terminal transferase. The mixture was incubated 3 min at 37°C, extracted with phenol-chloroform and purified on a column of sephadex G 50 medium.

Tailed heteroduplex and vector were ligated by contact of 1 mole of heteroduplex with 2 moles of vector at a final concentration of vector of 5 ng/μl in buffer (100 mM Tris HCl pH 7.5, 100 mM NaCl, 1 mM EDTA). The mixture was heated at 65°C 10 min and the annealing was performed overnight by incubation in the water bath.

The DNA recombinant was introduced in bacterial cells strain RR1 after their treatment by calcium chloride [9].

Analysis of the inserts

After selection of transformed bacteria by tetracycline resistance and ampicillin susceptibility the plasmid DNA was obtained from minipreparations realized by alkaline lysis method [9].

After PstI digestion of the plasmid DNA, the inserts were analysed by 1% agarose electrophoresis. The inserts of a length above 500 pb were selected.

The inserts in PBR 322 were labelled with 32P dCTP by "multiprime DNA labelling systems" (Amersham, France). Their viral origin was confirmed by hybridization with dot blot of RNA extracted from infected and non infected cells according to the method described by Maniatis (1982) for the isolation of m RNA from mammalian cells.

The respective position of these inserts was determined by reciprocal hybridization and restriction enzyme cleavage.

Nucleotide sequencing

DNA sequencing was carried out by random subcloning of sonicated DNA [4] in DNA replicative form of M 13 phage.

The M 13 dideoxy sequencing was carried out according to Sanger technique and previously used for the sequence analysis of bovine enteritic coronavirus F 15 genome [3].

Preparation of RNA from infected and non infected cells, dot-blot and hybridization

After three freeze-thawings the cell suspension was clarified and ultracentrifugated as previously described. The pellet resuspended in 1:500 of the initial volume was treated with 2% of SDS 10 min at 20°C then RNA was extracted with phenol-chloroform and precipitated with sodium acetate 0.25 M and 2.5 volumes of ethanol.

The RNA from infected or non infected cells was denatured by mixing with the same volume of formaldehyde and heating at 50°C during 15 min. After ice chilling, a same volume of 10 X SSC (20 X SSC is 3M NaCl, 0.3 M Na citrate) was added and the RNA deposited on nitrocellulose filters (Schleicher and Schüell, Ceralabo, France) wetted by 5 X SSPE (20 X SSPE is 3M NaCl, 177 mM NaH_2PO_4, 20 mM EDTA pH 7.4).

Pre-hybridization was performed overnight at 42°C in 200 μl of 5 X SSPE by cm^2 of filter containing 50% formamide, 0.1% polyvinyl pyrrolidone (PVP), 0.1% ficoll, 0.1% bovine serum albumine (BSA), 0.1% SDS and 100 μg/ml of calf thymus DNA.

Hybridization was performed overnight at 55°C with 50 μl of buffer by cm^2 of filter with the same buffer 5 X SSPE containing 10 ng/ml of DNA probe labelled with 32P, dCTP (110 TBq/m mol, Amersham) by random priming (multiprime DNA labelling system kit, Amersham) and denatured 15 min at 100°C. Washing was initially performed 3 times at room temperature during 15 min in 2 X SSC, 0.1% SDS and then 2 times during 15 min at 52°C with 0.1% X SSC and 0.1% SDS.

For the discrimination of the pestiviruses those probes were prepared either from insert in PBR 322, either from insert in M13 DNA single strand or from insert purified from LMP agarose after electrophoresis.

Results

Among twelve different inserts six were selected according to their location on the HCV genome determined by total or partial sequencing and comparison with the genome of the NADL strain of BVDV and Alfort strain of HCV as published [2, 10].

The place of these inserts is presented in Fig. 1.

Comparative hybridization assays were performed with the different types of probes obtained from inserts in PBR 322, M13 DNA single strand or from purified inserts. The best results (data not shown) were obtained with the purified inserts which were used in the presented results.

After dot blot hybridization with the RNA extracted from non infected cells and cells infected with various strains of pestiviruses, different intensities in the hybridizations were observed (Fig. 2) and in comparison with the controls it is possible to classify the reactions in positive, weakly positive and negative (Table 1).

The results are summarized in Table 1. One of the inserts (784) recognized all the strains, while the five other inserts reacted only with some of the eight strains of pestiviruses tested.

All the inserts used as probes recognized the two strains of HCV (Alfort and Nord).

The sequences of the HCV genome corresponding to the inserts were compared with the homologous regions of the BVDV genome. The highest degree of homology (73%) was observed with the 784 insert which reacted with all the strains. A slightest homology (66.5%) was observed with the 779 fragment which recognized strongly the two strains of HCV, the NADL strain of BVDV and more weakly two BDV strains (Aveyron and Lyon 2). The lowest homology (about 63%) was obtained with two of the four other inserts (1610 and 1745) which did not recognize any of the two BVDV strain but only HCV and 3 strains of BDV (Aveyron, Lyon 1, Lyon 2). The degree of homology of the two last inserts is not yet determined.

Discussion

The purpose of this work was to detect and differentiate pestiviruses of different animal origin by cDNA probes. The six probes were tested against RNA extracts from undiluted and 1:10 cell extracts. When the probes reacted with undiluted dilution, a decrease of intensity of the reaction or a total extinction at the 10^{-1} dilution was observed.

This decrease is clear with HCV strain Nord (Fig. 2 lane H) and when the BVD strain Aveyron and Lyon 2 are tested against probe 784 (Lanes D and F). The total extinction at 10^{-1} dilution is explained for most of the other cases when the reaction is not very important with the undiluted extracts (lanes BCEG) but not for HCV Alfort (Lane A). In this case,

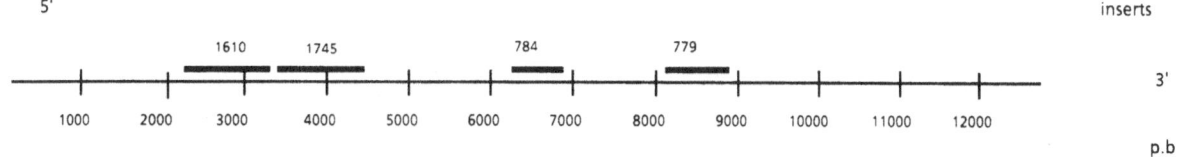

Fig. 1. Sites of the inserts 1610, 1745, 784, 779 on the genome of HCV. The sites of the inserts 2038 and 1331 have not been determined with precision, they are between 1745 and 784

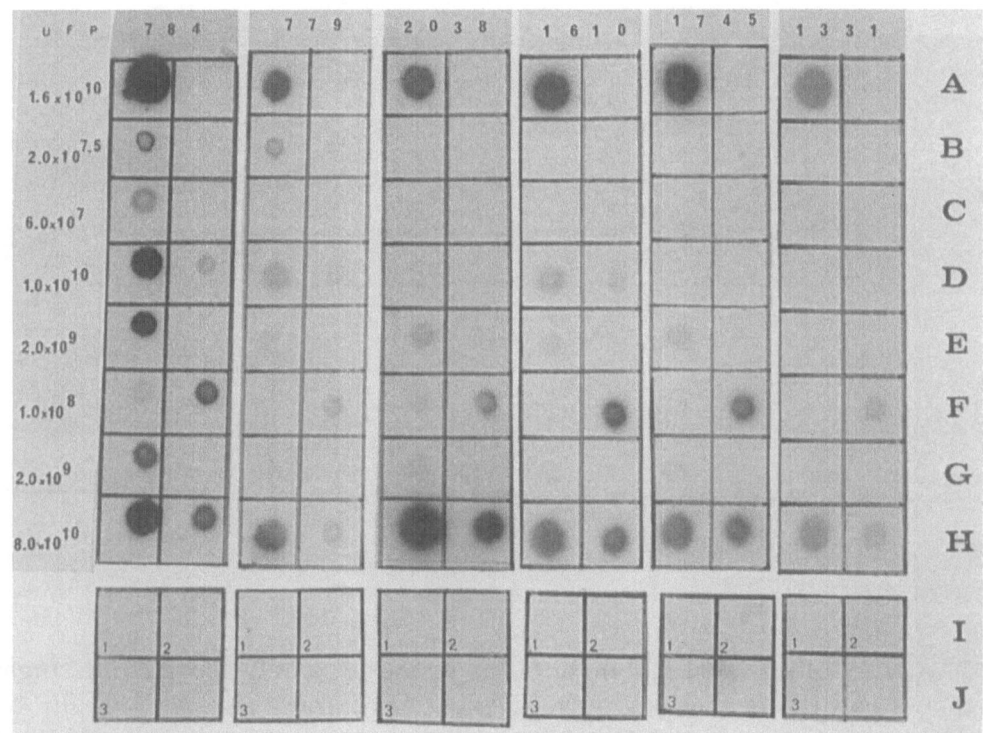

Fig. 2. Dot blot hybridization of 8 strains of pestiviruses with 6 different inserts. A: HCV strain Alfort; B: BVDV strain NADL; C: BVDV strain New York; D: BDV strain Aveyron; E: BDV strain Lyon 1; F: BDV strain Lyon 2; G: BDV strain IEMVT; H: HCV strain Nord. PFU column indicates the amount of virus deposited. Each insert was hybridized with 0.004 ml of initial dilution of each strain (Left column) and 0.004 ml of a 10^{-1} dilution of each strain (Right column). For strain F, the blots corresponding to the initial and 10^{-1} dilutions were unvoluntarily inverted. Lanes I and J: control hybridization with uninfected cells (1: PK; 2: BT; 3: SK)

despite the initial strong reaction no trace is observed at the 10^{-1} dilution. No explanation was given except an error in dilution manipulation.

By this panel of 6 cDNA probes it was possible to recognize and differentiate the various strains of pestiviruses. All the probes recognized the two HCV strains, but only one probe (784) recognized the whole range of

Table 1. Results of the dot blot hybridization expressed in positive weakly positive or negative between the HCV cDNA probes and various strains of pestiviruses

Virus Strains	Inserts					
	784	779	2038	1610	1745	1331
HCV (Alfort)	positive	positive	positive	positive	positive	positive
BVDV (NADL)	positive	positive	negative	negative	negative	negative
BVDV (New York)	positive	negative	negative	negative	negative	negative
BDV (Aveyron)	positive	weakly positive	negative	weakly positive	negative	negative
BDV (Lyon 1)	positive	weakly positive	positive	positive	weakly positive	negative
BDV (Lyon 2)	positive	weakly positive	positive	positive	positive	weakly positive
BDV (IEMVT)	positive	negative	weakly positive	negative	weakly positive	positive
HCV (Nord)	positive	negative	positive	positive	positive	positive
Homology (HCV/BVDV)	73.3%	66.5%	ND	63%	63.7%	ND

positive   weakly positive   negative

Lower line: degree of homology between the part of the HCV genome (Alfort) corresponding to the inserts and the corresponding part of the BVDV genome (NADL).
ND = Not determined

pestivirus strains tested and none of the probes was only specific of a single type of pestivirus. Besides, one of the BDV strain (Lyon 2) could be differentiated from the two HCV strains only by probe 779 which hybridized less with Lyon 2. In the same way the BVDV (New York) strain of pestivirus of ruminants could be differentiated from BDV (IEMVT) and BDV (Aveyron) only by a slight hybridization of probes 2038 and 1745, and probes 779 and 1610 with these two strains of BDV respectively. More probes should be necessary to avoid the difficulties due to weak hybridization and more strains of pestiviruses of different origin should also be tested to confirm these possibilities of differentiation. The highest specificity of the probes with HCV is not surprising as the probes were prepared from the Alfort strain of HCV. It can be presumed that more specific BDV and BVDV probes could be prepared from homologous pestiviruses.

The sensitivity of the probes was determined only by the titer of the supernatants and the determination of the number of viral particles corresponding to the volume deposited on the filters for hybridization. The quantity of viral RNA detected was not determined as the extracts were not purified and contained viral and cellular RNA. In spite of this, the high titers of virus in the undiluted samples and the rapid decrease of the reaction indicates that the sensitivity is not excellent and should be improved.

The yield of such high titers is possible after adaptation of the virus and the value of the probes is still interesting for the confirmation and distinction of the pestiviruses in reference laboratories.

An important step will be the adaptation of such techniques to material directly extracted from organs or samples of infected animals, combined with the preparation of non radioactive labelled probes for a wider use in conventional laboratories.

References

1. Aynaud JM, Galicher C, Lombard J, Bibard C, Mierzejewska M (1972) Peste porcine classique: les facteurs d'identification in vitro du virus en relation avec le pouvoir pathogène. Ann Rech Vét 3: 209–235
2. Collett MS, Larson R, Gold C, Strick D, Anderson DK, Purchio AF (1988) Molecular cloning and nucleotide sequence of the pestivirus bovine viral diarrhea virus. Virology 165: 191–199
3. Crucière C, Laporte J (1988) Sequence and analysis of bovine enteritic coronavirus (F15) genome. I. Sequence of the gene coding for the nucleocapsid protein; analysis of the predicted protein. Ann Inst Pasteur/Virol 139: 123–138
4. Deininger PL (1983) Random subcloning of sonicated DNA: application to shotgun DNA sequence analysis. Anal Biochem 129: 216–223
5. Fernelius A (1964) Non cytopathogenic bovine viral diarrhea viruses detected and titrated by immunofluorescence. Can J Comp Med Vet Sci 28: 121–126
6. Fernelius AL, Antower WC, Lambert C, Mc Clurikin AW, Matthews PJ (1973a) Bovine viral diarrhea virus in swine: characteristics of virus recovered from naturally and experimentally infected swine. Can J Comp Med 37: 13–20
7. Fernelius AL, Antower WC, Malmquist WA, Lambert G, Matthews PJ (1973b) Bovine viral diarrhea in Jioine: neutralizing antibody in naturally and experimentally infected swine. Can J Comp Med 37: 96–102
8. Horzinek MC (1981) Non arthropod borne Togaviruses. Academic Press Inc, New York
9. Maniatis T, Fritsch EF, Sambrook J (1982) Molecular cloning: a laboratory manual. Cold Spring Harbour Laboratory, Cold Spring Harbour, New York
10. Meyers G, Rumenapf T, Thiel HJ (1989) Molecular cloning and nucleotide sequence of Hog cholera virus. Virology 171: 555–567

Author's address: Dr. E. Plateau, Centre National d'Etudes Vétérinaires et Alimentaires, Laboratoire Central de Recherches Vétérinaires, (C.N.E.V.A./L.C.R.V.) 22, rue Pierre Curie, 94703 Maisons-Alfort Cedex. France.

Arch Virol (1991) [Suppl 3]: 199–208

Detection of persistent bovine viral diarrhea virus infections by DNA hybridization and polymerase chain reaction assay

K. V. Brock

Food Animal Health Research Program, Ohio Agricultural Research and Development Center, Wooster, OH, USA

Accepted March 20, 1991

Summary. The detection of persistent bovine viral diarrhea virus (BVDV) infections in cattle by DNA dot blot hybridization was done with a cloned cDNA probe prepared from noncytopathic BVDV (strain NY-1). Due to the variability of specific hybridization results, detection of BVDV by primer-directed polymerase chain amplification was done. Primers were chosen within a reported area of sequence conservation and amino acid homology of Pestiviruses. The amplification region extended from nucleotide 6322 to 7475 and was based on published BVDV sequence data (NADL strain). BVDV RNA was extracted by two methods (proteinase K and guanidinium isothiocyanate) from serum and white blood cell preparations collected from 3 persistently-infected heifers. cDNA was synthesized from extracted BVDV-genomic RNA using reverse transcriptase. Reaction conditions were optimized to amplify the 1153 base pair fragment from the cDNA preparation. Detection of BVDV in the samples by DNA dot blot hybridization using a nucleic acid probe corresponding to the amplified region (6322 to 7475) was compared with polymerase chain reaction assay. The increased sensitivity of the polymerase chain reaction assay provided clearer identification of persistently-infected animals than DNA hybridization under similar conditions.

Key words: Bovine viral diarrhea virus, hybridization, polymerase chain reaction, PCR, diagnostic assay, togaviridae.

Introduction

The prevention and control of bovine viral diarrhea virus (BVDV) infections presently centers around the detection and removal of animals persistently-infected with BVDV and the judicious use of vaccination programs [1, 4, 11]. The effectiveness of such control measures is dependent on maintaining closed herds and rigorous screening of animals for BVDV infection prior to entering the herds [1, 11]. At present, virus isolation, which is widely used, is the preferred method for detection of BVDV infections [10, 12]. Virus isolation is possible due to the persistent viremia present in persistently-infected animals. However, several disadvantages of BVDV isolation exist; difficulty encountered in maintaining cell cultures free of BVDV, time, and expense [10, 12]. Therefore; rapid, sensitive, and specific methods of virus detection are needed to improve herd screening efforts to control BVDV infections.

Due to the molecular cloning of several BVDV strains [2, 6, 15], the ability to detect BVDV by DNA hybridization and primer-directed amplification using the polymerase chain reaction (PCR) is available. Previous work has demonstrated that nucleic acid hybridization probes can be used to detect BVDV infections in cattle and may be applicable for herd screening [2, 3]. The recent development of PCR amplification provides another rapid and extremely sensitive method of detecting specific nucleic acid sequences with diagnostic as well as research applications. Due to sequence specific primer-directed amplification, PCR assays can detect as little as 1 to 2 copies of a specific nucleotide sequence in a sample. Therefore, the application of PCR reaction assay to provide amplification of the target BVDV sequences was investigated and compared with nucleic acid hybridization.

Materials and methods

Animals and sampling

Whole blood samples were collected from 3 animals persistently-infected with BVDV. Two (#1 and #2) of the persistently-infected animals were obtained from private herds infected with BVDV. The third (#3) animal was produced by intravenous inoculation of a seronegative dam at 90 days of gestation [13] with 20 ml of whole blood collected from persistently-infected animal #1.

Cell cultures and quantitation of virus

The titers of BVDV were determined in serum obtained from the 3 persistently-infected animals by virus isolation. A 1.0 ml volume of each serial ten-fold dilution of serum was inoculated in replicates of 3 onto BVDV negative secondary bovine turbinate cells passaged 5 to 10 times in Dulbecco's minimum essential medium containing 10% horse serum. The inoculum was removed after 1 hour and replaced with fresh medium. Cell cultures were incubated for 5 days at 37 °C in 5% CO_2. Noncytopathic BVDV was detected by indirect fluorescent antibody assay (IFA) using an anti-BVDV (NADL and New York-1) polyclonal serum prepared in a gnotobiotic calf [3].

Extraction of BVDV RNA from serum samples

Method 1: RNA extraction was done by guanidinium isothiocyanate and phenol chloroform extraction [5]. Serum (1 ml) was diluted with 9 ml of guanidinium solution (4 M guanidinium isothiocyanate, 25 mM sodium citrate, pH 7.0, 0.5% sarcosyl, and 0.1 M 2-mercaptoethanol) and the following were sequentially added: 0.1 ml 2 M sodium acetate, pH 4.0; 10 ml of Tris-buffered phenol; and 2 ml of chloroform/isoamyl alcohol (24:1). The mixture was agitated, placed on ice for 10 minutes, and centrifuged at 10,000 g for 20 minutes at 4 °C. The aqueous phase was removed and precipitated with an equal volume of isopropanol and placed at − 20 °C overnight. The extracted RNA was pelleted by centrifugation at 10,000 g for 30 minutes at 22 °C. The RNA pellet was dried and then resuspended in diethylpyrocarbonate (DEPC) treated H_2O. In addition to serum, RNA was also extracted from white blood cells (WBC). The WBC from 5.0 ml of whole blood were pelleted and RNA was extracted using 1/10 the volumes described above for serum.

Method 2: A 1.0 ml volume of serum or a 1.0 ml suspension of WBC from 5.0 ml of whole blood were adjusted to a concentration of 1.0% SDS and 10 mg of proteinase K was added [2]. The mixture was incubated at 37 °C for 3 hours and then phenol/chloroform/isoamyl alcohol (25:24:1) extracted followed by a chloroform/isoamyl alcohol (24:1) extraction. The aqueous phase was removed and precipitated with an equal volume of isopropanol and placed at − 20 °C overnight. The extracted RNA was pelleted by centrifugation at 10,000 g for 30 minutes at 22 °C. The RNA pellet was dried and then resuspended in DEPC treated H_2O.

Primer-directed amplification

The NADL BVDV sequence data [6] and HCV sequence data[14] were compared. Oligonucleotide primers (24-mers) were designed within an identified area of high conservation of amino acid sequence within the two Pestiviruses which extends approximately from nucleotide 5324 to 9716, Fig. 1. Primer selection was made using a computer program (Oligo, National Biosciences, Hamel, MN) designed to construct optimal oligonucleotide primers for use in PCR assays [16]. Two primers, one upstream and one downstream primer were selected: primer 6322, 5′-CTGCCAAATGCCTCAACCAAAGCT-3′; primer 7451, 5′-GGACAACCCGGTCACTTGCTTCAG-3′, Fig. 1. The theoretical melting temperature of both primers was calculated to be 60.6 °C. Optimized PCR reaction conditions were previously determined using purified genomic BVDV RNA (strain NY-1) (unpublished results).

The PCR assay was done using reagents supplied in a kit, (GeneAmp; Perkin-Elmer Cetus Corp., Norwalk, Conn) and a DNA thermal cycler (Coy Laboratory Products Inc., Ann Arbor, MI). First strand cDNA synthesis was done using MLV-reverse transcriptase (Bethesda Research Laboratories, Bethesda, MD) and random primers. The first strand reaction contained 1X PCR buffer (10X PCR buffer contains 500 mM KCl, 100 mM Tris-HCl [pH 8.3], 15 mM $MgCl_2$, and 0.1% [wt/vol] gelatin), 1 mM of each deoxynucleotide (dATP, dGTP, dTTP, dCTP), 12.5 units of RNasin (Promega, Madison, WI), 300 ng random hexanucleotides, 200 units M-MLV reverse transcriptase, and extracted BVDV RNA in a reaction volume of 20 μl. The reaction mixture was incubated at 42 °C for 45 minutes, heat inactivated at 95 °C for 5 minutes, and 80 μl of the following reaction mixture was added: 1X PCR buffer, 1.25 μM of primer 6322, 1.25 μM of primer 7452, and 2.5 units of Amplitaq polymerase (Perkin-Elmer Cetus Corp., Norwalk, Conn). The reaction mixture was initially denatured at 94 °C for 2 minutes followed by 30 cycles of the following reaction parameters: template denaturation at 94 °C for 1 minute, primer annealing at 55 °C for 1.5 minutes, and extension at 72 °C for 3 minutes. An additional incubation was done at 72 °C for 7 minutes to complete the extension of open (5′ overhangs) templates.

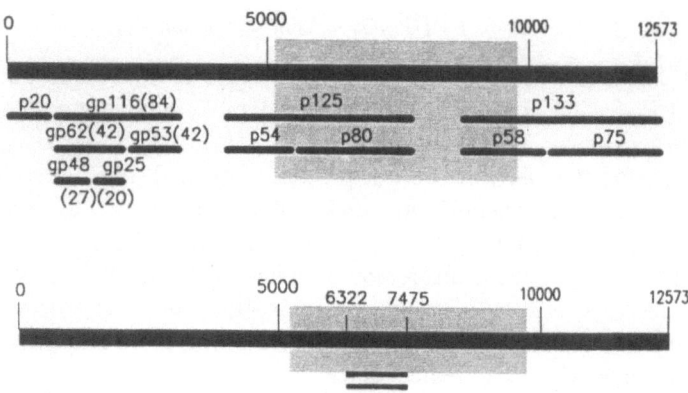

Fig. 1. Specific region of primer-directed amplification (below) within the genome of BVDV in relation to the genomic organization of BVDV (above). Grey area represents the area of amino acid conservation of BVDV and HCV [Collett et al., 1988] [Meyers et al., 1989]. Genomic organization based on figure published by Collect et al. [Collett et al., 1989]. Note amplification region (below) extends from nucleotides 6322 to 7475

Detection of amplified product

Following amplification, 4 µl of the 100 µl PCR reaction mixture was examined by 1% agarose gel electrophoresis using a 123 bp ladder as a molecular weight size standard. Southern blots of the agarose gels were done according to the methods of Southern [19]. Hybridization of Southern blots was done essentially as described for dot blot hybridization. Additionally, Southern blots were hybridized with a probe prepared from a cDNA clone (pBV4-p80) from the corresponding p80 region of NADL BVDV (kindly provided by Dr. Marc Collett, Bethesda MD).

Dot blot hybridization assay

Hybridization blots were prepared, hybridized, and washed following hybridization as described previously with the following modifications [2, 3]. BVDV cDNA (strain NY-1), corresponding to the PCR amplification region, was prepared from purified genomic RNA amplified by PCR using primer 6322 and primer 7475 as described above, C-tailed using terminal deoxytransferase, and cloned into G-tailed, Pst-I cut pUC9 (unpublished results). Purified insert DNA was nick translated to incorporate $dCT^{32}P$ to a specific activity of approximately 10^7 cpm/µg. Dot blots were prepared with total RNA extracted from 1.0 ml, 0.5 ml, and 0.1 ml of serum and WBC from 5.0 ml of whole blood using methods 1 and 2 as previously described [2].

Results

The 50% endpoints of cell culture infective doses/ml ($CCID_{50}$/ml) were calculated from the results of IFA from ten-fold serum dilutions. The $CCID_{50}$/ml of serum from the 3 persistently-infected animals ranged from 10^4 to $10^{6.5}$. Table 1.

Table 1. Titers of BVDV ($CCID_{50}$/ml of serum) detected in cell culture by an indirect immuno-fluorescence assay

Animals	$CCID_{50}$/ml[a]
#1	10^6
#2	10^4
#3	$10^{6.5}$

[a] Values represent the mean titers of 3 replicates

Results of the dot blot hybridization assay were positive for all serum and WBC samples collected from the 3 persistently-infected animals with the probe prepared with previously amplified and cloned cDNA from NY-1 BVDV RNA. A hybridization signal was clearly detected when dot blots prepared from 1.0 ml and 0.5 ml of serum and the WBC preparation from animals #1 and #2 were hybridized with the probe prepared with previously amplified and cloned cDNA from NY-1 BVDV RNA, Fig. 2. The minimum volume of serum required to obtain a positive hybridization signal from the persistently-infected animals in this study ranged from greater than 0.1 ml to 0.5 ml. Specific hybridization was not detected in the 0.1 ml serum extracts. The quality of hybridization signals were reduced below 0.5 ml of serum and were difficult to interpret when dot blots were prepared from RNA extracted from less than 0.5 ml of serum.

The product length of primer-directed amplification using primers 6322 and 7451 based on published sequence data from NADL BVDV was calculated to be 1153 bp, Fig. 1. The linear electrophoretic migration of the amplified PCR products in 1% agarose was similar between all samples and comigrated between molecular weight standard segments 9 and 10, 1107 and 1230 bp (1160 bp) respectively, Fig. 3. Amplified product was obtained from all 3 animals using extraction methods 1 and 2. Following PCR amplification, the total amount of DNA in the reaction mixture was approximately 15 to 25 µg, quantitated by spectrophotometric absorbance. The quantity of amplified cDNA from RNA extracted using method 1 was greater than from RNA extracted using method 2. Southern blot analysis of the agarose gel (Fig. 3) demonstrated that the amplified product was BVDV specific by hybridization with probe prepared from amplified cDNA from NY-1 BVDV, Fig. 4. The specificity of the amplified products were also confirmed by a similar pattern of hybridization (Fig. 4) with a cDNA clone (pBV4-p80) from the corresponding region of the p80 of NADL BVDV.

The detection limit of BVDV in serum samples using the PCR assay was determined. Serum obtained from one persistently infected animal (#2) was diluted ten-fold in TE buffer (10 mM Tris-HCl and 1 mM EDTA, pH 7.4)

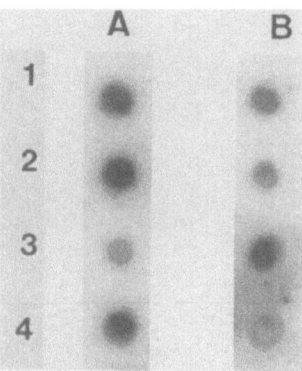

Fig. 2. Dot blot hybridization of serum and white blood cell preparations (WBC) hybridized with cloned cDNA amplified from purified BVDV genomic RNA (strain New York 1). Column A: row 1, BVDV NY-1 infected cell culture supernatant; row 2 and 3, border disease virus isolates; row 4, BVDV NADL infected cell culture supernatant; Column B: row 1, GITC extracted serum from animal #1; row 2, GITC extracted serum from animal #2; GITC extracted WBC from animal #1; row 4, GITC extracted WBC from animal #2

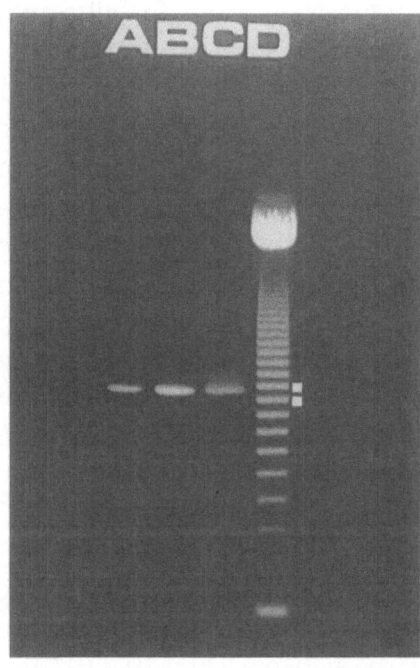

Fig. 3. Agarose gel electrophoresis of amplified cDNA of genomic RNA extracted from serum of 3 persistently-infected animals stained with ethidium bromide. Column A, PI animal #1; column B, PI animal #2; column C, PI animal 3; column D, 123 bp ladder. Bands in columns A, B and C comigrate between notated bands 9 and 10 of the molecular weight standard (1107 to 1230 bp)

from 10^{-1} to 10^{-6}. The PCR assay was done with RNA extracted (method 1) from a 100 µl fraction of the ten-fold serum dilutions as previously described. Following amplification the products were examined by 1% agarose gel electrophoresis. A detection limit of 10 µl of serum (10^4 $CCID_{50}$/ml) from animal #2 was determined to be sensitivity level for the PCR assay.

Discussion

The results of primer-directed amplification indicate that the PCR assay can be used to amplify a conserved region of the BVDV genome from RNA extracted directly from samples collected from persistently-infected animals.

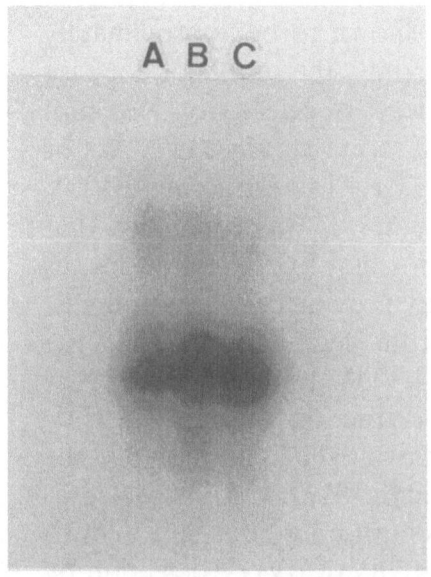

Fig. 4. Southern blot prepared from the agarose electrophoresis gel in Fig. 2 and hybridized with cloned cDNA amplified from purified BVDV genomic RNA (strain New York 1). Hybridization conditions were as previously described. Note specific hybridization with comigrating fragments in columns A, B, and C

Similar amplification products were consistently obtained from all 3 animals sampled in this study. Likewise, specific amplification of cDNA from various samples collected from acutely (serum, WBC, and milk) and persistently-infected (serum, WBC, and embryo transfer flush fluids) animals has also been done (unpublished results). The difference in hybridization signal intensity of samples from animals #1 and #2 appears to reflect the differences in the levels of viremia, Table 1 and Fig. 2.

Detection of BVDV by dot blot hybridization has been done in a variety of samples collected from acutely (serum, WBC, nasal swabs, vaginal swabs, and milk) and persistently-infected animals (serum, WBC, nasal swabs, vaginal mucus, and embryo transfer flush fluids) (unpublished results) [2]. The limits of BVDV RNA detection by the DNA dot blot hybridization assay using infected cell culture fluids are similar to virus isolation [2]. The detection of BVDV by dot blot hybridization in animal samples is more difficult than detection in infected cell culture fluids. Based on the results of this study and previous work, the detection of BVDV by dot blot hybridization assay in serum and white blood cell preparations has practical application [3]. However, difficulty may be encountered in evaluating the intensity of the hybridization signals and thus true positive and negative reactions in samples with low levels of BVDV RNA may be difficult to distinguish. The successful application of a diagnostic assay requires that the margin of identity between positive and negative samples be distinguishable. Therefore, due to amplification of specific cDNA sequences the use of PCR may increase the margin of identity between negative and positive samples not available by standard hybridization assays.

The results of dot blot hybridization indicate that the sensitivity of detection was less than expected by standard methods. The threshold level of sensitivity in cell culture has previously been shown to be approximately 10–20 pg of purified BVDV RNA [2]. In this study, the results of dot blot hybridization with serum did not reach this level of sensitivity. Dot blot hybridization signals with RNA extracted from serum volumes of 0.1 ml or less were weak or undetectable. Considering the virus titers in serum, BVDV RNA should have been detected in lower volumes. It was concluded that sample preparation was the limiting step for dot blot hybridization in this study. Dilution of serum prior to extraction may provide better de-proteinization and help protect the genomic RNA from degradation during the extraction procedure. Although high levels of BVDV are present in persistently-infected animals, the processing and extraction of intact BVDV genomic RNA remains difficult and currently may be the limiting step in the hybridization and PCR assays. The susceptibility of BVDV RNA to degradation may effectively lower the sensitivity of detection by dot blot hybridization and PCR assay. The comparative sensitivity of BVDV detection by dot blot hybridization (0.1 ml to 0.5 ml of serum) and PCR amplification (0.01 ml of serum) in serum containing approximately 10^4 $CCID_{50}$/ml of BVDV was less than expected, based on other published reports [17, 18], and may have been considered to be due to incomplete recovery of BVDV RNA in extracted samples. Therefore, one important consideration during sample preparation and RNA extraction must be to prevent contaminating RNAses which may result in false negative reactions.

As expected, PCR amplification was more sensitive than detection by dot blot hybridization. Amplified cDNA product could easily be detected from RNA extracted from 10 µl of serum (10^4 $CCID_{50}$/ml) which corresponds to approximately 10^2 $CCID_{50}$/ml. Theoretically, PCR requires only a few genomic copies to allow amplification. Due to the difficulties of extracting RNA from serum and the possibility of decreased first strand synthesis due to secondary RNA structure, this level of sensitivity may not be practically achieved.

The results of PCR primer-directed amplification support the assumption that a high degree of nucleic acid conservation does exist between strains of BVDV corresponding to the selected amplification region, Fig. 1 [7, 8, 9]. Meyers et al. noted a region of amino acid similarity between BVDV and hog cholera virus extending between positions 1550 and 2300 (nucleotides 5024 to 7274) which overlaps with the region of PCR amplification used in the present study, Fig. 1 [14]. The conserved nature of this region is supported by the hybridization and PCR results of the amplified region (6322 to 7451 bp) between several cytopathic and noncytopathic BVDV strains (NADL, NY-1, and 2 uncharacterized strains from persistently-infected animals). Using the nucleic acid hybridization probe prepared from cloned cDNA amplified from NY-1 BVDV, specific hybridization has been obtained with BVDV strains; Singer, NADL, NY-1, BVDV-72; and 2 field isolates of border disease virus (unpublished results). It will be interesting to determine if this region of sequence conservation exists between most if not all BVDV strains.

There are several advantages of using PCR amplification to detect BVDV. The most important is the speed and sensitivity of the assay. Due to the sensitivity available using PCR, amplification can be done directly from samples obtained from persistently-infected animals. This technique can be a valuable tool for diagnostic detection of BVDV in samples collected from animals as well as providing research applications. The time required to obtain specific results from the time of sample collection could be reduced to 24 hours. Due to the automation of the DNA thermal cycler, the machine can be programmed to run the incubation profiles overnight and hold at 4 °C allowing results to be obtained the next day. Amplification of BVDV cDNA directly from samples collected from infected animals provides the ability to characterize virus without the requirement of virus isolation in cell culture. Therefore, the possibility that in vitro biotypic differences may or may not be present in vivo may be investigated further using PCR methodology. If a genetic component to biotypic variation is demonstrated in vitro it will be important to identify the same mechanism existing in vivo. PCR amplification followed by sequencing of the amplified product may provide the methodology to identify genomic sequences which influence biotypic differences. Although the detection of BVDV using molecular techniques provides specific identification, the ability to distinguish noncytopathic and cytopathic strains will depend on the identification of BVDV genomic differences that determine biotypic strain differences. Additionally, the use of PCR amplification coupled with detection by in situ hybridization may provide the sensitivity required to investigate the possibility of latent BVDV infections [18].

In conclusion, the detection of BVDV can be done in serum samples and white blood cell preparations collected from persistently-infected animals by dot blot hybridization assay and PCR amplification. The application of PCR BVDV cDNA amplification provides a more sensitive method of virus detection than the hybridization assay. PCR amplification may serve as a primary method of BVDV detection or to complement dot blot hybridization or virus isolation.

Acknowledgements

Salaries and research support were provided by state and federal funds appropriated to the Ohio Agricultural Research and Development Center, The Ohio State University. This work was supported in part by Cooperative Research Agreement 88-34116-3653 from the U.S. Department of Agricultural Science and Education Administration. The author would like to thank Dr. M. S. Collett for kindly providing the pBV4-p80 BVDV clone and Jerry Meitzler and Sylva Riblet for their technical assistance.

References

1. Bolin SR (1990) Control of bovine virus diarrhoea virus. Rev Sci Tech Off Int Epiz 9: 163–171.
2. Brock KV, Brian DA, Rouse BT, Potgieter LND (1988) Molecular cloning of complementary DNA from a pneumopathic strain of bovine viral diarrhea virus and its diagnostic application. Can J Vet Res 52: 451–457
3. Brock KV, Potgieter LND (1990) Detection of bovine viral diarrhea virus in serum from cattle by dot blot hybridization assay. J Vet Microbiol (in press)
4. Brownlie J, Clarke MC (1990) Bovine virus diarrhoea virus: speculation and observations on current concepts. Rev Sci Tech Off Int Epiz 9: 223–230
5. Chomczynski P, Sacchi N (1987) Single-step method of RNA isolation by acid guanidinium thiocyanate-phenol-chloroform extraction. Anal Biochem 162: 156–159
6. Collett MS, Larson R, Gold C, Strick D, Anderson DK, Purchio AF (1988) Molecular cloning and nucleotide sequence of the pestivirus bovine viral diarrhea virus. Virology 165: 191–199
7. Collett MS, Larson R, Belzer SK, Retzel E (1988) Proteins encoded by bovine viral diarrhea virus: the genomic organization of a pestivirus. Virology 165: 200–208
8. Collett MS, Anderson DK, Ritzel E (1988) Comparisons of the pestivirus bovine viral diarrhoea virus with members of the flaviviridae. J Gen Virol 69: 2637–2643
9. Deacon NJ, Lah M (1989) The potential of the polymerase chain reaction in veterinary research and diagnosis. Aust Vet J 66: 442–444
10. Edwards S (1990) The diagnosis of bovine virus diarrhoea-mucosal disease in cattle. Rev Sci Tech Off Int Epiz 9: 115–130
11. Harkness JW (1987) The control of bovine viral diarrhoea virus infection. Ann Rech Vet 18: 167–174
12. Heuschele WP (1975) BVD virus strain variation and laboratory diagnostic problems. Proc Am Assoc Vet Lab Diagn 18: 91–111
13. McClurkin AW, Littledyke ET, Cutlip RC, Frank GH, Coria MF, Bolin SR (1984) Production of cattle immunotolerant to bovine viral diarrhea virus. Can J Comp Med 48: 156–161
14. Meyers G, Rumenapf T, Thiel HJ (1989) Molecular cloning and nucleotide sequence of the genome of hog cholera virus. Virology 171: 555–567
15. Renard A, Guiot C, Schmetz D, Dagenais L, Pastoret P-P, Dina D, Martial JA (1985) Molecular cloning of bovine viral diarrhea viral sequences. DNA 4: 429–438
16. Rychlik W, Rhoads RE (1989) A computer program for choosing optimal oligonucleotides for filter hybridization, sequencing and in vitro amplification of DNA. Nucleic Acids Res 17: 8543–8551
17. Saiki RK, Gelfand DH, Stoffel S, Scharf SJ, Higuchi R, Horn GT, Mullis KB, Erlich HA (1988) Primer-directed enzymatic amplification of DNA with a thermostable DNA polymerase. Science 239: 487–491
18. Salimans MMM, Vanderijke FM, Raap AK, Vanelsacker-niele AMW (1989) Detection of parvovirus B19 DNA in fetal tissues by in situ hybridisation and polymerase chain reaction. J Clin Pathol 42: 525–530
19. Southern EJ (1975) Detection of specific sequences among DNA fragments separated by gel electrophoresis. J Mol Biol 98: 503–517

Author's address; Dr. Kenny V. Brock, Food Animal Health Research Program, 1680 Madison Ave., Wooster OH 44691, U.S.A.

Arch Virol (1991) [Suppl 3]: 209–215

Differentiation of pestiviruses by a hog cholera virus-specific genetic probe

Brief Report

Ch. Schelp, J. Dahle, Th. Krietsch, O.-R. Kaaden[1], and **B. Liess**

Institute for Virology, Hannover Veterinary School, Hannover, Federal Republic of Germany

Accepted March 20, 1991

Summary. A hog cholera virus (HCV)-specific genetic probe has been generated after cloning of the genomic viral RNA. This probe distinguished between HCV and the closely related bovine viral diarrhoea virus (BVDV). Furthermore, it detected a broad spectrum of HCV strains and isolates which differ in their phenotype such as virulence.

Key words: Pestiviruses, BVDV, HCV, genetic probe.

*

Classical swine fever (CSF) or hog cholera (HC) is a contagious and economically important disease of pigs and is caused by hog cholera virus (HCV). HCV belongs to the genus pestivirus of the family Togaviridae [1]. This genus is composed of three antigenically closely related viruses: hog cholera virus of pigs, bovine viral diarrhoea virus (BVDV) of cattle and border disease virus (BDV) of sheep. These viruses show strong structural and serological similarities. BVDV and BDV are indistinguishable by serological techniques, whereas HCV is considered to be a more separate antigenic entity [2, 3].

Beside the classical form of HC atypical symptoms can be observed, which render the diagnosis more difficult. A number of HCV variants have been isolated, which are genetically related but apparently different in their

_____

[1] Corresponding author

virulence. Moreover, pestiviruses are able to cross species barriers. BVDV naturally infects pigs, sheep, goats and many wild ruminants [3, 4]. Both virus infections induce homologous antibodies. To differentiate among pestiviruses investigations about the different degrees of neutralization of infectivity, virulence assays or monoclonal antibodies are necessary [2]. These techniques are based on differences of the viral gene products. More recently, genetic probes were considered to distinguish those closely related viruses at the genetic level by hybridization with a specific probe.

For that purpose the HCV strain Alfort 187 [5] was cloned into the lambda gt 11 vector (Promega) according to standard procedures [6]. Briefly, HCV virions were harvested from the tissue culture supernatant and the genomic RNA was extracted. cDNA synthesis was primed with calf thymus DNA oligonucleotides, and after methylation and linker addition the double stranded cDNA was cloned into the Eco RI-site of lambda gt 11. HCV-specific recombinants hybridized in Northern blots specifically with the viral genomic RNA of about 12 kb [7, 8]. In Western blots fusion proteins could be detected with polyclonal serum from infected pigs (data not shown).

The viral cDNA was further subcloned into the plasmid vector pTZ 18 (Pharmacia) and sequenced [9]. The sequences were aligned to recently published data for BVDV [10] and HCV [11], respectively. For further hybridization experiments an Eco RI-fragment of the clone lambda HC 6 was chosen. Figure 1 shows the complete nucleotide sequence and the deduced amino acid sequence of the Eco RI-fragment in comparison to the corresponding sequences of BVDV strains NADL [10] and Osloss [12]. The comparison revealed a 79.4% homology at the nucleotide level and a 92.6% homology with regard to the amino acid sequence for BVDV-NADL. The values for BVDV-Osloss were 79.8% and 95.1%, respectively. There was one amino acid exchange and 8.6% mismatches at the nucleotide level in comparison to the published HCV-Alfort 187 sequence. These differences may have resulted from the high mutation rate of RNA viruses and the passage history in different cell lines [13]. In analogy to the BVDV genome organization these sequences probably code for a part of the nonstructural protein p125 [11]. p125 is known to be a highly conserved protein within the genus pestivirus [4, 10, 11, 14]. Therefore, it should be possible to detect many HCV variants by hybridization experiments. Variation of the hybridization conditions may give informations about the relatedness of the tested strains at the genetic level.

The sensitivity of the genetic probe was first tested in the homologous system. Cytoplasmic RNA from PK-15 cells infected with the HCV strain Alfort 187 was isolated [15], spotted on nitrocellulose membranes and hybridized. Virus-specific RNA could be detected in extracts from 2×10^4 cells as starting material (Fig. 2). Then, the RNA from nine different HCV strains and two BVDV strains was investigated. PK-15 cells were infected with the HCV strains Alfort 187 [5], CAP [16], ALD [17], Eystrup

```
HC6        GAATTCACCTGTGTGACAGCATCAGGAACTCCGGCCTTCTTTGATCTCAAGAACCTCAAAGGCTGGTCAGGGCTACCGATATTT
HCV  5551                        C                C              T G
NADL 5843      G   C C        C A     T     C  C A A  T G     A          CT G  T
Osl. 5800      A   T T   G      A        C  G A TT G  A          T       C
HC6        E   F   T   C   V   T   A   S   G   T   P   A   F   F   D   L   K   N   L   K   G   W   S   G   L   P   I   F
HCV  1730
NADL 1820
Osl. 1807
```

```
HC6        GAGGCATCAAGTGGAAGGGTAGTCGGCAGGGTCAAGGTCGGGAAGAATGAGGACTCTAAACCAACCAAGCTTATGAGTGGAATA
HCV             A                    A                C   T C           C         G
NADL            A C C C G       G T       A     A A        A G         T    A AA A          C
Osl.              T T    C    G         A T A A A         A C G C   A AT A          T C
HC6        E   A   S   S   G   R   V   V   G   R   V   K   V   G   K   N   E   D   S   K   P   T   K   L   M   S   G   I
HCV
NADL                                                           E                   I
Osl.                                                           E
```

```
HC6        CAAACAGTCTCCAAAAGTACCACAGCATTGACAGAAATGGTAAAGAAAATAACGACCATGAACAGGGGAGAATTC
HCV            G T T     CG          G    G   G      G                     G        5793
NADL           G C     A    AC AG     C    C G   C      G     C G              C     6085
Osl.             C     A     C AG C T     A   G   C      G     C G              C T   6042
HC6        Q   T   V   S   K   S   T   T   D   L   T   E   M   V   K   K   I   T   T   M   N   R   G   E   F
HCV                            A                                                              1810
NADL                        N   A                                  S                   D      1900
Osl.                            A                                  S                   D      1887
```

Fig. 1. Homology between HCV and BVDV. The cDNA sequences of HC6, HCV, BVDV-NADL and BVDV-Osloss (first to fourth line) and the deduced amino acid sequences (fifth to eighth line, respectively) are shown. For HCV and BVDV strains only differences from HC6 are specified. Numbers refer to the published BVDV and HCV sequences

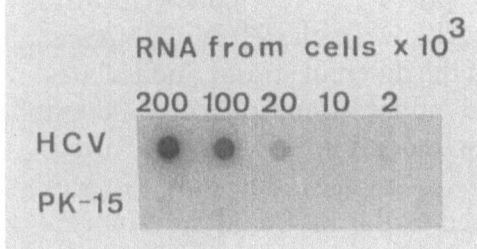

Fig. 2. Detection of HCV-RNA. RNA from different numbers of cells (as indicated $\times 10^3$) was spotted onto the nitrocellulose membrane, fixed and hybridized to 2×10^6 cpm/ml ^{32}P-labeled cDNA at 42°C in 50% formamide. The membrane was washed three times at room temperature in $2 \times$ SSC/0.1% SDS and three times at 50°C in $0.1 \times$ SSC/0.1% SDS

[18], Glentorf [19], Bergen [20], Thiverval [21], 331 [22] and Osterode (Veterinäramt Osterode, 1983). The cytopathogenic BVDV strain NADL and the noncytopathogenic BVDV strain 7443 were propagated in MDBK cells. The cytoplasmic RNA from 2×10^5 and 1×10^5 cells was hybridized to ^{32}P-labeled HCV-cDNA. Duplicates of these dot blots were hybridized with ^{32}P-labeled BVDV-NADL-cDNA. The genomic localization of the BVDV-cDNA provided by Dr. M. S. Collett (plasmid pBV-KPB) is unknown. It just served as a positive control for BVDV-RNA. Under standard

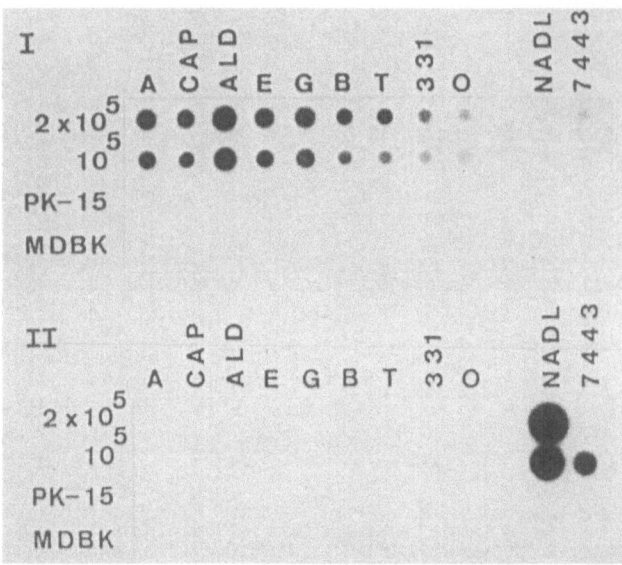

Fig. 3. Dot blot hybridization with RNA from different HCV and BVDV strains. RNA from 2×10^5 or 10^5 cells infected with different HCV and BVDV strains was spotted onto nitrocellulose membranes and hybridized to 2×10^6 cpm/ml ^{32}P-labeled HCV-cDNA (I) or 4×10^6 cpm/ml ^{32}P-labeled BVDV-cDNA (II). RNA from mock-infected PK-15 and MDBK cells served as negative controls. Hybridization and washing conditions were as mentioned in Fig. 2. The dot with RNA from 2×10^5 cells infected with strain 7443 was not hybridizing due to a technical problem. (Abbreviations of HCV strains: A = Alfort 187, E = Eystrup, G = Glentorf, B = Bergen, T = Thiverval and O = Osterode)

reaction conditions seven out of nine HCV strains could be easily distinguished from the tested BVDV strains (Fig. 3). Only HCV strains Osterode and 331 could not be detected. The BVDV probe also reacted specifically.

To investigate the degree of homology between the HCV-cDNA sequence and the unknown RNA sequences of the different strains and isolates hybridization experiments were performed under conditions of different stringencies. RNA from 1×10^5 infected or mock-infected cells was used. Cell-bound virus was titrated prior to RNA extraction, and only samples with comparable titers were further subjected to dot blot analysis. Hybridization was carried out at 37 °C in 20, 30 and 50 per cent formamide. The membranes were washed for several hours in $5 \times$ SSC/0.1% SDS at 49 °C, 56 °C and 71 °C, respectively (the equivalent effective temperatures calculated from the salt and formamide concentrations, 23). After drying, the dots were punched out and counted in a liquid scintillation counter. These experiments revealed a different degree of homology among the tested strains to HCV-Alfort 187 (Table 1). The virulent strains ALD and Eystrup, as well as the field strain Glentorf hybridized to the Alfort 187 probe to a greater extent than the avirulent strains CAP and Bergen. Even the vaccine strain Thiverval showed a reduced degree of homology, although Thiverval originated from the virulent strain Alfort [21]. The virulent field strain

Table 1. Ratio of cpm bound to pestivirus-RNA dot/cpm bound to Alfort 187-RNA dot

RNA	Formamide concentration				
	20%	30%	50%	Average ($\pm$S.D.)	
Alfort 187	1	1	1	1	(virulent)
ALD	1.023	0.945	0.925	0.964 ($\pm$0.052)	(virulent)
Glentorf	0.898	0.869	0.808	0.858 ($\pm$0.045)	(virulent)
Eystrup	0.931	0.756	0.699	0.795 ($\pm$0.121)	(virulent)
CAP	0.974	0.717	0.636	0.776 ($\pm$0.176)	(avirulent)
Thiverval	0.959	0.704	0.630	0.764 ($\pm$0.172)	(vaccine)
Bergen	0.837	0.671	0.568	0.692 ($\pm$0.136)	(avirulent)
Osterode	0.860	0.653	0.507	0.673 ($\pm$0.177)	(virulent)
331	0.798	0.658	0.520	0.659 ($\pm$0.139)	(virulent)
7443 (BVDV)	0.746	0.718	0.525	0.663 ($\pm$0.120)	(noncytopathogenic)
NADL (BVDV)	0.701	0.618	0.523	0.614 ($\pm$0.089)	(cytopathogenic)
PK-15	0.225	0.083	0.034	0.114 ($\pm$0.099)	
MDBK	0.279	0.086	0.052	0.139 ($\pm$0.122)	

Osterode and 331 resembled the BVDV strains concerning the degree of homology within this particular part of the genome. In case of the HCV strain 331, this is in good agreement with findings based on serological techniques, which showed a closer relatedness of this strain to BVDV strains than other HCV strains [24, 25]. The degree of homology to Alfort 187 among the tested HCV strains seemed to reflect their different phenotypes. Virulent strains hybridized to a greater extent than avirulent strains.

This HCV-Alfort 187 fragment detected all pestiviruses tested so far. Standard hybridization conditions were stringent enough to exclude the examined BVDV strains. Lowering stringency the reactivity was expanded to all tested pestiviruses. These properties make this cDNA fragment suitable for diagnostic hybridization tests. Until now HCV diagnosis suffers from time consuming virus cultivation in cell culture. Beside the monoclonal antibodies [26, 27] diagnostic hybridization may be helpful to circumvent this problem.

Acknowledgements

The authors wish to thank Dr. M. S. Collett for providing the BVDV-cDNA and Ch. Gentzsch for excellent technical assistance. The work was supported by Intervet International BV and the Commission of the European Communities within the activities of the EEC/FAO Liaison Laboratory for Hog Cholera/Classical Swine Fever.

References

1. Westaway EG, Brinton MA, Gaidamovich SY, Horzinek MC, Igarashi, Kääriäinen L, Lvov DK, Porterfield JS, Russel PK, Trent DW (1985) Togaviridae. Intervirology 24: 125–139

2. Carbrey EA (1989) Diagnostic procedures. In: B Liess (ed) Classical swine fever and related viral infections. Martinus Nijhoff Publishing, pp 99–114

3. Wensvoort G, Terpstra C, De Kluyver EP (1989) Characterization of porcine and some ruminant pestiviruses by cross-neutralisation. Vet Microbiol 20: 291–306

4. Collett MS, Moennig V, Horzinek MC (1989) Recent advances in pestivirus research. J Gen Virol 70: 253–266

5. Dahle J, Liess B, Frey HR (1987) Neutralizing antibody development following sequential inoculation of pigs with strains of bovine viral diarrhea virus and hog cholera virus. J Vet Med B 34: 729–739

6. Maniatis T, Fritsch EF, Sambrooks S (1982) Molecular cloning: a laboratory manual. Cold Spring Harbour Laboratory, Cold Spring Harbour, New York

7. Rümenapf T, Meyers G, Stark R, Thiel HJ (1989) Hog Cholera virus-characterization of specific antiserum and identification of cDNA clones. Virology 171: 18–27

8. Moormann RJM, Hulst MM (1988) Hog Cholera virus: identification and characterization of the viral RNA and the virus-specific RNA synthesized in infected swine kidney cells. Virus Res 11: 281–291

9. Sanger F, Nicklen S, Coulson AR (1977) DNA sequencing with chain-terminating inhibitors. Proc Natl Acad Sci USA 74: 5463–5467

10. Collett MS, Larson R, Gold C, Strich D, Anderson DK, Purchio AF (1988) Molecular cloning and nucleotide sequence of the pestivirus bovine viral diarrhea virus. Virology 165: 191–199

11. Meyers G, Rümenapf T, Thiel HJ (1989) Molecular cloning and nucleotide sequence of the genome of hog cholera virus. Virology 171: 555–567

12. Renard A, Dino D, Martial J (1987) Vaccines and diagnostics derived from Bovine Diarrhea virus. European Patent Application number 86870095.6. Publication number 0208672, 14 January 1987

13. Smith DB, Inglis SC (1987) The mutation rate and variability of eukaryotic viruses: an analytical review. J Gen Virol 68: 2729–2740

14. Collett MS, Larson R, Belzer SK, Retzel E (1988) Proteins encoded by bovine viral diarrhea virus: the genomic organization of a pestivirus. Virology 165: 200–208

15. Gough NM (1988) Rapid and quantitative preparation of cytoplasmic RNA from small numbers of cells. Anal Biochem 173: 93–95

16. Laude H (1978) Virus de la peste porcine classique: isolement d'une souche cytolytique a partir de cellules IB-RS2. Ann Microbiol 129A: 553–561

17. Sato U, Nishimura Y, Hanaki T, Nobuto K (1964) Attenuation of hog cholera virus by means of continuous cell-virus propagation (ccvp) method. Arch Ges Virusforsch 14: 394–403

18. Dräger K, Kamphans S, Maass W (1964) Use of suiferin a live attenuated swine fever vaccine for the control of the disease in an infected herd in north Germany. Tierärztl Umschau 19: 405–414

19. Pittler H, Brack M, Schulz LCL, Rohde G, Witte K, Liess B (1968) Untersuchungen über die Europäische Schweinepest. I. Ermittlungen zur gegenwärtigen Seuchensituation in Norddeutschland. DTW 75: 537–542

20. Van Oirschot JT (1980) Ph.D. Thesis, State University of Utrecht, pp 120–125

21. Launais M, Aynaud JM, Corthier G (1972) Swine fever virus: properties of a clone (Thiverval strain) isolated in cell culture at low temperature. Use in vaccination. Rev Med Vet 123: 1537–1554

22. Mengeling WL, Packer RA (1969) Pathogenesis of chronic hog cholera: host response. Am J Vet Res 30: 409–417

23. Howley PM, Israel MA, Law MF, Martin MA (1979) A rapid method for detecting and mapping homology between heterologous DNAs. J Biol Chem 254: 4876–4883

24. Corthier G, Aynaud JM, Galicher C, Gelfi J (1974) Activite antigenique comparee de deux togavirus: le virus de la peste porcine et la virus de la maladie des muqueuses. Ann Rech Vet 5: 373–393
25. Neukirch M, Liess B, Frey HR, Prager D (1980) Serologische Beziehungen zwischen Stämmen des Europäischen Schweinepest-Virus und dem Virus der Bovinen Virus-diarrhoe. Fortschritte der Veterinärmedizin 30: 148–153
26. Wensvoort G, Terpstra C, Boonstra J, Bloemraad M, Van Zaane D (1986) Production of monoclonal antibodies against swine fever and their use in laboratory diagnosis. Vet Microbiol 12: 101–108
27. Hess RG, Coulibaly COZ, Greiser-Wilke I, Moennig V, Liess B (1988) Identification of hog cholera viral isolates by use of monoclonal antibodies to pestiviruses. Vet Microbiol 16: 315–321

Author's address: Dr. O.-R. Kaaden, Institute for Virology, Hannover Veterinary School, Bünteweg 17, D-W-3000 Hannover 71, Federal Republic of Germany.

Arch Virol (1991) [Suppl 3]: 217–224

Lesions in aborted bovine fetuses and placenta associated with bovine viral diarrhoea virus infection

R. D. Murray

Department of Veterinary Clinical Science, University of Liverpool Veterinary Field
Station, Leahurst, NESTON, South Wirral, U.K.

Accepted March 20, 1991

Summary. Abortions in dairy cattle were investigated on 55 dairy farms sited in North West England, using a multi-level diagnostic technique. After pathological examination of fetal and placental tissues collected at the time of abortion, possible causes for these abortions could be identified, supported by bacteriological and serological laboratory findings.

Of 150 abortions investigated, Bovine Viral Diarrhoea (BVD) virus infection was related to 40 episodes (27% of the total), often accompanied by evidence of concurrent infections.

Lesions associated with BVD abortions were found in fetal eyelid, lung, and occasionally myocardium. Lesions in the lung were most consistent, characterized by mononuclear inflammatory cell infiltration of peribronchiolar and inter-alveolar tissues. Placental lesions were non-specific.

It is concluded that the lesions observed are insufficient to be the primary cause of abortion. However, the pathological changes associated with BVD infection in the placenta may allow secondary opportunist pathogens to cross the feto-maternal barrier, thereby threatening the health of the fetus and the physiological and endocrinological functions of the placenta which maintain pregnancy.

Key words: Bovine viral diarrhoea virus, lesions, fetus, placenta.

Introduction

Bovine Virus Diarrhoea (BVD) infections of susceptible pregnant cattle are usually subclinical, but there is a high probability of transplacental spread of virus to the fetus. The fetus responds to such infection in a variety of ways including abortion, or the birth of weak and undersized calves with or without congenital malformations. Clinically normal calves are also born. The principal determinant of the outcome of BVD infection in pregnant cattle is the age of the fetus when viral challenge occurs; the immune status of the dam may also be important [7].

Infection in susceptible cattle during the first 90 days of pregnancy often results in persistently infected fetuses, or abortion which frequently goes unnoticed and is perceived as an infertility problem [16]. Abortion has been reported after pregnant heifers were exposed to experimental BVD infection at 29–41 gestation; fetal death followed infection by 12–27 days, and abortion occurred 30–50 days later [4]. Abortions in late pregnancy have been reported in field outbreaks of BVD, but often accompanied by evidence of recent challenge with other infectious agents such as *Leptospira hardjo*, *Coxiella burnetti* [15], and *Corynebacterium pyogenes*, heamolytic Staphylococci, and heamolytic Streptococci [13].

Fetal lesions associated with BVD infection have been widely described. Following experimental injection of virus into pregnant heifers, lesions of cerebellar hypoplasia alone or with hydrancephaly, cataracts, retinal degeneration, optic neuritis, mummification [2, 17, 19], and thymic hypoplasia [14] have all been recorded. Those associated with natural infections have included mummification and the presence of growth arrest lines in tibias [15]. This paper describes histopathological findings in fetuses and placenta from abortions which were associated with natural BVD infection on dairy farms, spanning a two year period.

Materials and methods

Abortion cases on dairy farms located in North West England were investigated according to the statutory requirements of the Brucellosis (England and Wales) Order 1978. The cause of these episodes was investigated by obtaining additional samples at the time of the initial farm visit, namely the fetus, placenta, and placentome; maternal blood samples were also taken but a further sample was obtained not less than four weeks later. Diagnosis was based upon a systematic, multi-level laboratory investigation as described by Kradel [11]. Post mortem examinations were performed on all fetuses, and routine sampling provided abomasal contents, thoracic fluid, liver and kidney for laboratory analysis.

Bacteriology was carried out on abomasal contents, lung, and placentomes; material was plated out on blood agar and McConkeys's media, incubated under aerobic and partial anaerobic conditions, and colonies identified after 24 hrs and 48 hrs. Pure cultures were defined as >30 similar colonies growing on one plate.

Serology was carried out on thoracic fluid samples. Infectious Bovine Rhinotracheitis (IBR) and BVD antibodies were identified using serum neutralization (SN) tests, and microscopic agglutination (MAT) was used to detect *L. hardjo* antibody. Titres of >1/10 were considered to be significant. Fluorescent antibody tests (FAT) were performed on liver and kidney tissue homogenates, using antibody against *L. hardjo* as described by Ellis and others [8].

Antibody titres to BVD and IBR viruses in paired maternal serum samples were determined using either SNT or Enzyme-linked Immuno-assay (EIA) methods; four-fold changes in titres or consistent titres of >1/128 were considered significant. Titres of antibody to *L. hardjo* were measured by MAT after the method of Wolff [20]; values >1/400 were considered as significant when assessed in relation to abortion episodes. Formalin-fixed specimens of fetal eyelid, lung, heart, liver, placentome, and chorio-allantoic membrane (CAM) were obtained; sections 6–8 μ thick were cut, mounted, and stained with heamatoxylin and eosin. They were all examined by the author. The histopathological findings for each case were recorded and related to the results of bacteriological and serological examinations. A diagnosis was made only where pathological changes could be explained from laboratory methods used and their results.

Results

From an estimated breeding population of 3800 cattle on 55 dairy farms, 150 abortions were investigated. Serological evidence of previous BVD infection was found in 40 cases; of these, histological lesions were found in 27 cases, and only in these cases was BVD considered to be associated with the abortion. The crown/rump lengths of 26 fetuses were 46–92 cm (170–270 days gestation), there being one of 36 cm (155 days gestation). Fetal immunoglobulins IgG and IgM and SN antibodies to BVD virus were found in 13 out of 27 fetuses examined; significant titres of SN antibody were found in paired maternal sera in only 8 cases. The distribution of lesions found in these cases is shown in Fig. 1.

Lesions in eyelid

Mononuclear inflammatory cell infiltration in the sub-conjunctival connective tissue was present in twelve cases. In one further case, severe congenital bilateral conjunctivitis was characterized by epithelial necrosis, congestion and oedema in sub-epithelial tissue, together with a severe mixed inflammatory cell infiltration; SN titres to BVD were found at 1/80 in thoracic fluid and >1/2000 in maternal serum samples, but no virus was detected either by electron microscopy or immunoperoxidase (IPx).

Lesions in lung

The most common lesion was that of mononuclear inflammatory cell infiltration in peribronchiolar and inter-alveolar tissue, found in 14 out of 24 cases examined. Viewed at low power, the lung parenchyma appeared to hypercellular; under higher magnification, this increased cellularity was seen to significantly increase the thickness of inter-alveolar tissue by surrounding capillaries within it. Numerous plasma cells were present. Congestion was also present. Where concurrent bacterial infection was present, polymorpho-nuclear inflammatory cells were also found at this site as well as within airways along with cellular debris. Inter-lobular congestion and oedema was also a feature in 11 cases.

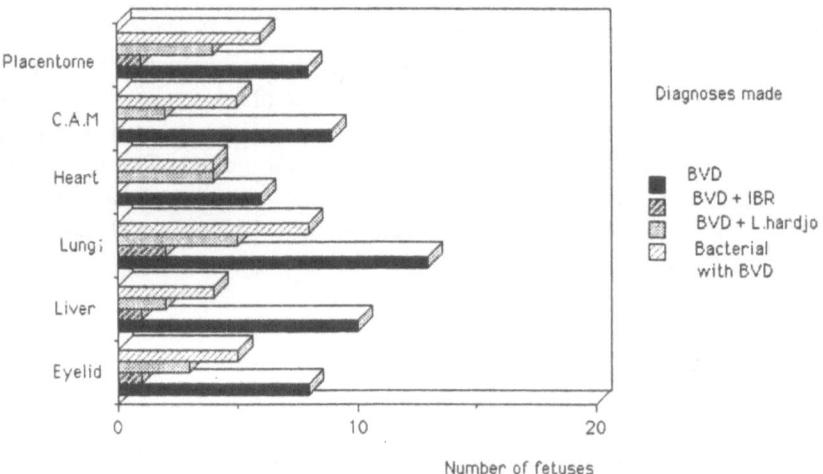

Fig. 1. Distribution of histological lesions in 27 abortions associated with BVD infection

Lesions in heart

These were found in six cases and presented as a non-specific myocarditis; focal areas of degenerative or necrotic change, which included swelling and loss of striation in myocardial fibres, were surrounded by mononuclear inflammatory cells.

Lesions in liver

Of 12 cases where lesions were present, a periacinar accumulation of immature leucocytes was found in 11 cases. Occasional areas of coagulative necrosis were recorded in two cases, but there was no indication from other results as to the pathogenesis of the lesion.

Lesions in placentome

These were found in 14 cases. As would be expected, degenerative or necrotic lesions were always found, and in themselves were not considered significant. Extensive, discrete areas of mineralization were found in six cases (out of a total of 16 cases exhibiting this lesion). Vasculitis accompanied by oedema, congestion, or heamorrhage, was a feature in three cases where IBR virus and two cases where *L. hardjo* were concurrent infections; significant mixed inflammatory cell infiltrations were found in septal villous connective tissue adjacent to blood vessels. Of the remainder, five cases were associated with concurrent bacterial infection and a diffuse placentitis was present with accompanying moderate to severe, inflammatory cell infiltration. Where no other pathogens were found, inflammatory changes were minimal.

Lesions in chorio-allantois

These were recorded in eight cases where it was not possible to obtain a placentome. Necrosis was always present. Fibrosis and oedema were found in two cases, and mineralization in four. Surprisingly, inflammatory cell infiltration within the chorionic villi—though present—was often minimal, except when concurrent bacterial infections occurred.

Discussion

Fetuses that are younger than 150 days gestation when challenged by BVD virus infection may produce specific lesions in various organs [5]; sequential development of cerebellar lesions in this age of calf have been well documented [2]. By comparison older fetuses present only mild, generalized lesions, although specific lesions described include vasculitis [18], necrotizing bronchiolitis, nonsuppurative meningitis and dermatitis [5].

The present study offers no additional observations to those already quoted, with one exception. Changes which have been described simply reflect an active immune response to viral challenge by bovine fetuses, older than 170 days gestation, an opinion supported by increased concentrations of immunoglobulins IgG and IgM or positive titres to BVD found in nearly half the fetuses examined. Since bovine fetuses are capable of producing serum-neutralising antibody to BVD by 200 days gestation [1], not all fetuses examined in this study would be capable of mounting a SN antibody response. One calf, of eight months gestation, presented specific lesions of bilateral conjunctivitis and cataracts. The conjunctival lesion was characterised by epithelial necrosis, suppuration, with a heavy mixed inflammatory cell infiltration. Since there are no reports in the literature of congenital conjunctivitis associated with BVD infection, this finding ought to be examined from two stand-points;

    (a) is the lesion primarily associated with BVD infection, or
    (b) has the lesion developed as a result of concurrant fetal infection?

It is generally recognised that virus can be isolated from cattle where active lesions associated with clinical cases of BVD-mucosal disease are found [3]. Detectable SN antibodies are found only in animals that survive infection, and are virus free. Clinical conjunctivitis is often found in cattle suffering from mucosal disease, and it was therefore surprising that virus was not isolated from this calf. Instead, a moderate fetal SN titre of 1/80 and high maternal titre of 1/2000 indicated recent BVD infection, followed by a good immunological response. The presence of cataract in the fetus suggests that it was infected around 150 days gestation. Scott and others [17] have shown that high serum neutralisation titres in both dam and fetus are present following maternal infection with BVD at 150 days gestation and the birth of full-term calves; in these same calves, histological evidence of previous degenerative and inflammatory changes in the eyes were still evident. Clinical experience has shown us that some clinical mucosal disease cases can eventually produce antibody and survive, the lesions found in the acute stages of disease slowly disappearing. This may well be the explanation of conjunctival lesions found in this calf.

Necrotic conjunctivitis is not a lesion associated with other common fetal infections; no similar lesions have been found with bacterial or fungal infections of the fetus [13]. The only conjunctival inflammatory lesions found with abortions are those of sub-conjunctival inflammatory cell infiltrations. Epithelial hyperplasia and goblet cell formation has been previously reported, but no associated aetiology has been described [13]; however, similar lesions have been described after experimental infection with *Ureaplasma diversum* into the amniotic cavity of pregnant cattle by Miller [12].

Casaro and others [5], Kendrick [9], and Done and others [6] have all stated that no significant lesions are present in placental material obtained from cattle which have been experimentally infected with BVD virus during pregnancy. Only non-specific lesions of plasma cell infiltration and mild, non-suppurative arteritis at the base of caruncles have been described. It is recognized that specific placental lesions do not occur after BVD infection, and that viral transfer across the feto-maternal boundary does not require lesions to be present. This study is further evidence of that opinion. The pathology described is essentially that of non-specific degenerative and necrotic changes in maternal epithelium and adjacent fetal trophoblast cells, accompanied variously by oedema, congestion, and focal inflammatory lesions in the caruncle. Extensive mineralisation was seen in epithelial cells of 25% of sections examined, a change usually associated with tissue hypoxia; this would not be unusual in placental tissue recovered from cattle which had aborted and consequently retained fetal membranes. However, if this finding is part of the normal pattern of degenerative changes found in placenta post-partum, why was the lesion not more frequently observed? Of the 150 cases examined in this study, widespread mineralisation was found in less than 20% of placental tissues [13].

Extensive and significant lesions were found, associated with evidence of concurrent infections with opportunist pathogens as well as BVD virus; they were probably not directly caused by BVD virus infection. In particular, concurrent *L. hardjo* infection produced severe lesions of congestion, oedema, and haemorrhage which was evidence of a major vascular crisis in the placentome; this would have been sufficient to totally disrupt the physiological function of the placenta, and particularly that of hormonal support to maintain the corpus luteum of pregnancy.

It is concluded that BVD virus infection during pregnancy may cause degenerative lesions in chorionic and endometrial epithelium, either directly through intra-cellular infection or indirectly as a result of caruncular congestion or oedema. These lesions do not necessarily threaten the well-being of the fetus nor the continuing maintenance of pregnancy, and are seen as extensive, focal areas of mineralisation. However, these breaches of the feto-maternal barrier allow opportunist pathogens access to the fetus via the transplacental or haematogenous routes, thereby infecting the fetus which may cause its death and subsequent abortion.

References

1. Braun RK, Osburn BI, Kendrick JW (1973) Immunologic response of bovine fetus to bovine viral diarrhoea virus. Am J Vet Res 34: 1127–1132
2. Brown TT, Bistner SI, De Lahunter A, Scott FW, McEntee K (1975) Pathogenetic studies of infection of the bovine fetus with bovine viral diarrhoea virus. Vet Pathol 12: 394–404
3. Brownlie J (1985) Clinical aspects of the bovine virus diarrhoea/mucosal disease complex in cattle. In Practice 19: 195–201
4. Carlsson U, Fredriksson G, Alenius S, Kindahl H (1989) Bovine virus diarrhoea virus, a cause of early pregnancy failure in the cow. J Vet Med 36: 15–23
5. Casaro APE, Kendrick JW, Kennedy PC (1971) Response of the bovine fetus to bovine diarrhoea-mucosal disease virus. Am J Vet Res 32: 1543–1562
6. Done JT, Terlecki S, Richardson C, Harkness JW, Sands JJ, Patterson DSP, Sweasey D, Shaw IG, Winkler CE, Duffell SJ (1980) Bovine virus diarrhoea-mucosal disease virus: pathogenicity for the fetal calf following maternal infection. Vet Rec 106: 473–479
7. Duffell SJ, Harkness JW (1985) Bovine virus diarrhoea-mucosal disease infection in cattle. Vet Rec 117: 240–245
8. Ellis WA, O'Brien JJ, Neill SD, Ferguson HW, Hanna J (1982) Bovine leptospirosis; microbiological and serological findings in aborted fetuses. Vet Rec 110: 147–150
9. Kendrick JW (1971) Bovine viral diarrhoea-mucosal diseases virus infection in pregnant cows. Am J Vet Res 32: 533–544
10. Kent Lloyd KC (1985) Bovine viral diarrhoea in a newborn calf. J Am Vet Med Assoc 186: 600–601
11. Kradel DC (1978) Abortion storms—projection for the future. Cornell Vet 68 [Suppl 7]: 195–199
12. Miller RC (1984) Personal communication
13. Murray RD (1989) A field study of placental and fetal lesions associated with bovine abortion. DVM&S thesis, University of Edinburgh
14. Ohmann HB (1982) Experimental fetal infection with bovine viral diarrhoea virus II morphological reactions and distribution of viral antigen. Can J Comp Med 46: 363–369
15. Pritchard GC, Borland ED, Wood L, Pritchard DG (1989) Severe disease in a dairy herd associated with acute infection with bovine virus diarrhoea virus, *Leptospira hardjo* and *Coxiella burnetti*. Vet Rec 125: 625–629
16. Roeder PL (1985) Viral abortion. British Cattle Veterinary Association Proceedings 1984/85, 275–282
17. Scott FW, Kahrs RF, De Lahunta A, Brown TT, McEntee K, Gillespie JH (1973) Virus induced congenital anomalies of the bovine fetus following experimental infection with bovine viral diarrhoea-mucosal disease virus. Cornell Vet 63: 536–560
18. Shoey A (1979) Symposium: clinical aspects of bovine viral diseases. Bovine Practitioner 14: 128–129

19. Trautwein G, Hewicker M, Liess B, Orban S, Peters W (1987) Cerebellar hypoplasia and hydrancephaly in cattle associated with transplacental bovine virus diarrhoea infection. DTW 94: 588–590
20. Wolff JW (1954) Laboratory diagnosis of Leptospirosis. Thomas Springfield, Illinois.

Author's address: Dr. R. D. Murray, BVM&S. DBR. DVM&S. MRCVS., Department of Veterinary Clinical Science, University of Liverpool Veterinary Field Station, Leahurst, NESTON, South Wirral, L64 7TE, U.K.

Arch Virol (1991) [Suppl 3]: 225–230

Immunological reactivity of bovine viral diarrhea virus proteins after proteolytic treatment

Brief Report

S. Kamstrup, L. Roensholt and **K. Dalsgaard**

Animal Biotechnology Research Centre, State Veterinary Institute for Virus Research,
Lindholm, Kalvehave, Denmark

Accepted March 20, 1991

Summary. The immunological reactivity of bovine viral diarrhea virus proteins after proteolytic treatment is described. The results indicate that the epitopes detected are very dependent on conformation of the protein. A partially protease resistant 22 kD fragment of the biotype-specific p80 is identified.

Key words: Bovine viral diarrhea virus, mucosal disease, proteolytic treatment.

Bovine viral diarrhea virus (BVDV) is currently classified as a member of the family togaviridae, genus pestivirus [10]. The virus is the causative agent of the bovine viral diarrhea/mucosal disease (BVD/MD) syndrome in cattle and is of great economic importance [1]. The pathogenesis of mucosal disease (MD) is generally complex and involves persistence of a non-cytopathogenic biotype virus in animals specifically immunotolerant to the persisting biotype. Contemporary presence of an immunologically matching cytopathogenic biotype virus arisen either by superinfection or mutation usually causes the MD syndrome including haemorrhagic diarrhea and death [2]. In order to prevent the disease it is important to avoid intra-uterine infection of the calf fetus and the subsequent birth of BVDV persistently infected calves. One way of obtaining this could be prophylactic immunization of heifers before insemination.

The recent molecular cloning and sequencing of the genome of BVDV [3, 15] and hog cholera virus [11] has directed much of the research on BVDV towards molecular biological methods, i.e. expression of defined fragments of the viral genome and screening of resulting proteins, or peptide synthesis of predicted amino acid sequences. However, we have chosen a different approach. To evaluate the potential usefulness of fragments of BVDV proteins for immunization purposes, we have examined the reactivity of an immune serum against BVDV proteins derived from virus infected cell cultures, fragmented by amino acid specific proteinases.

The present investigation was done with the Danish cytopathogenic BVDV strain Ug59 and a corresponding antiserum.

The antiserum was raised in a pig by initial inoculation with BVDV (Ug59 strain) followed by inoculation with hog cholera virus (Alfort strain) 3 weeks later.

BVDV strain Ug59 was propagated in primary calf kidney monolayer cell cultures (in 1 liter roller bottles, area approximately 850 cm^2) grown in Eagles MEM (Glasgow modification) supplemented with streptomycin (0.1 mg/ml), neomycin (100 U/ml) and 10% bovine serum (free from BVDV and BVDV antibodies). At confluency of cells, the medium was replaced with 50 ml serum-free medium and each bottle was inoculated with 1 ml medium containing $10^{6.3}$ TCID$_{50}$ of BVDV (Ug59 strain). After adsorption for 2 hours at 37 °C an equal volume of medium containing 10% serum was added.

A cytopathic effect was observed after 48 hours, with approximately 20% of the cells released from the glass surface. Remaining cells were scraped off into the medium and the cells pelleted by low speed centrifugation.

The antigen content of the cellular fraction was estimated by ELISA (L. Roensholt, manuscript in preparation) to be approx. 90% of the total antigenic mass. Therefore, the supernatant was discarded. Briefly, the ELISA was based on a swine anti BVDV antiserum as catching antibody and a rabbit anti BVDV antiserum as detecting antibody. The rabbit antibodies bound were detected by peroxidase conjugated anti rabbit antibodies and OPD as chromogen. The staining intensity of different dilutions of antigen was taken as a measure of antigen content.

The cell fraction was lysed in 1 ml per roller bottle of a 5% solution of the non-ionic detergent Mega-10 (Iscotec AB, Sweden) in isotonic NaCl, and subsequently ultrasonicated (3 × 10 seconds, on ice). The lysate was clarified by high speed centrifugation (30.000 g, 5 °C, 1 h) and used for protein characterization.

SDS polyacrylamide gel electrophoresis was performed according to standard protocols [8]. Whole proteins and fragments were separated in gels of 10 and 15% acrylamide, respectively. The reducing agent dithiothreitol was included in the sample buffer for separation of cell lysates. Semi-dry electrotransfer to nitrocellulose (NC), (0.2 μm, Schleicher and

Schuell) was performed using 40 mM glycine, 50 mM Tris, 1.3 mM SDS, 20% v/v methanol as buffer. Transfer was done with 200 mA for 2 h at room temperature. Immunoblotting was performed with the following steps: Blocking for 30 min in 10% bovine serum (free from BVDV and BVDV antibodies) in dilution buffer (phosphate buffered saline containing 0.5 M NaCl and 1% Triton X-100 (Serva, FRG)). After blocking 1/200 vol of pig BVDV antiserum was added and incubation continued for 2 h. Washing in dilution buffer 3×5 min. Incubation for 2 h with peroxidase-conjugated rabbit anti-swine antibodies (Dakopatts, Denmark) diluted 1:500 in dilution buffer containing 10% serum. Washing as above. Staining by hydrogen peroxide and o-dianisidin as chromogen. All incubations were carried out at room temperature on a rotary shaker.

Western blots from extracts of infected cells (Fig. 1A) showed at least 4 distinct bands with apparent molecular masses of 115 kD, 80 kD, 56 kD and 45–50 kD. The most prominent bands are the 80 and 115 kD bands, while the 56 kD band is more diffuse. The appearance and apparent molecular mass of all bands correspond to those published by others [4, 6, 7, 9, 12, 13, 14]. In order to make the nomenclature of this report consistent with other published data, we will use the designations proposed by Collett et al. [5], i.e. p125 for the 115 kD band, p80 for the 80 kD band, gp53 for the 56 kD band and gp48 for the 45–50 kD band.

By calculating the mobility of each protein relative to a stain (pyronine Y), it was possible to cut out unstained gel cubes containing the different BVDV proteins. In order to investigate the proteolytic breakdown pattern of individual proteins, the gel cubes were transferred to the wells of a 15% gel, cast with a stacking gel of 50 mm's length, in contrast to the normal length of 20 mm. The cubes were overlaid with 20% glycerol sample buffer in order to cover the gel piece. Then the desired amount of enzyme, dissolved in 10% glycerol sample buffer, was added. The difference in glycerol concentration of the overlay and enzyme buffer ensured that no mixing of the protease and protein occurred during sample application. When applying voltage, the mixing took place simultaneously and uniformly for all samples. Pyronine Y was used as a marker of the electrophoresis. When the migration front had passed 2/3 through the stacking gel, the proteins were concentrated into a narrow zone ("stack"), and the enzymatic reaction took place for 30 min. without applying voltage to the gel. After this treatment the proteins were electrophoresed into the separation gel, transferred to NC and probed with the antiserum.

We have applied this procedure to the four BVDV proteins described using gel fragments from similarly electrophoresed proteins of uninfected cells as a control. The enzymes used were trypsin, Staph. aureus V8 protease, papain (Sigma, USA), thermolysin and alpha-chymotrypsin (Serva, FRG).

The general finding is a quick loss of detectable immunological reactivity of all detected BVDV proteins, pointing to conformation-dependent epi-

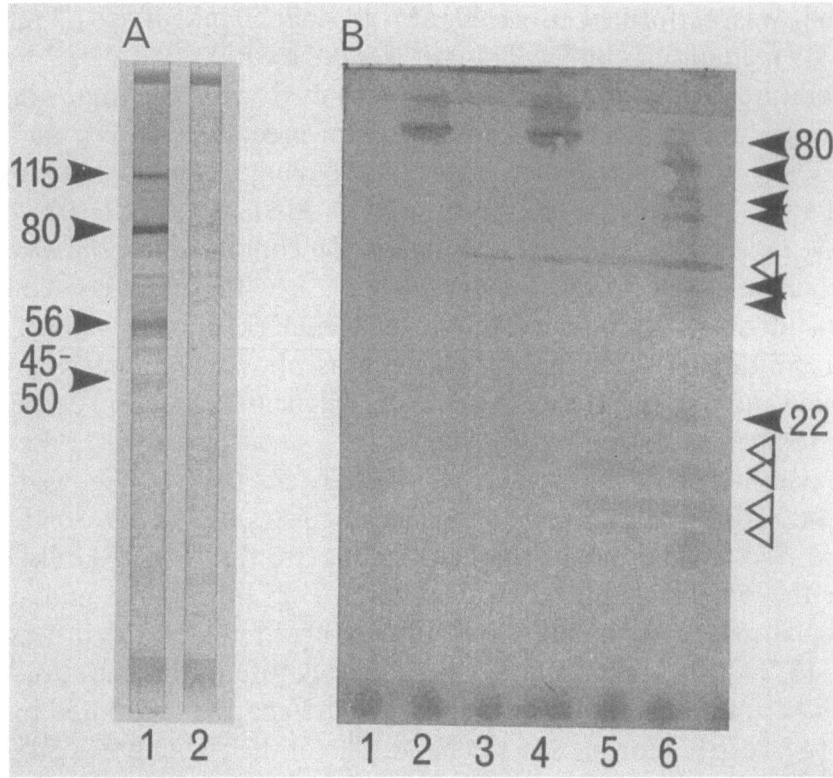

Fig. 1. A Western blots of cell lysates of (1) BVDV-infected and (2) uninfected cell cultures. The four bands of molecular weight 115 kD, 80 kD, 56 kD and 45–50 kD are marked by arrows. These bands correspond to p125, p80, gp53 and gp48, respectively. **B** Western blot of p80, digested with thermolysin; lanes 1, 3 and 5: uninfected cell extract, lanes 2, 4, 6: BVDV-infected cell extract. Lanes 1 to 6 contain 0, 0, 0.01, 0.01, 1 and 1 unit of thermolysin, respectively. Numerous specific bands of different sizes are seen for BVDV infected cells (arrows), treated with 1 U of protease. Unspecific bands are seen as well (unfilled triangles)

topes. However, we did find a general pattern of a 22 kD stable fragment cleaved from p80, obtained with thermolysin, papain and Staph. aureus V8 protease. The exact size varied slightly depending on the protease used. By further increasing the protease concentration this fragment could be cleaved to non-reacting peptides. This may simply reflect the presence of cleavage sites in the reactive amino acid sequence. The inability to demonstrate the same size fragment in p125, the precursor of p80 [4], is surprising but may be a matter of sensitivity or conformation. In the initial electrophoresis p80 had a stronger reaction than p125 and may therefore better maintain detectable reactivity after proteolytic digestion.

By the present method we have not been able to show any minor fragments of gp53 which can be considered as obvious candidates for vaccination purposes. However, this method has certain limitations: 1. Immunoreactivity in western blots does not necessarily detect the possible neutralizing epitopes in the viral proteins. Therefore, there might be fragments with neutralizing epitopes, that do not show up in our assay. 2. The background staining hampers simultaneous detection of many differently sized fragments – the signal disappears by dilution.

These investigations confirm and further support our general finding (data not shown) that immunological reactivity of BVDV proteins is highly dependent on the conformation of the proteins. The presence of linear epitopes should in our assay have shown up as low molecular weight bands, migrating close to the front of the electrophoresis.

To obtain increased sensitivity, we have tried to immunoprecipitate radiolabelled BVDV proteins prior to proteolytic treatment. However, we have not been able to precipitate any BVDV-specific proteins following labelling with ^{3}H-leucine or ^{3}H-mannose. Attempts to inhibit cell protein synthesis by using high concentrations of sodium chloride in the labelling medium as described by Donis and Dubovi [6, 7] apparently also blocked viral protein synthesis, as no specific bands were observed for any of the tested concentrations of sodium chloride (0, 50, 150, 200 and 250 mM above normal). This may be due to the application of a different strain of BVD virus or differences in propagation conditions.

Acknowledgements

The excellent technical assistance of Ms. Inge Nielsen is greatly appreciated. This work was supported in part by the Danish Biotechnology Research Programme/Animal Biotechnology Research Centre.

References

1. Baker JC (1987) Bovine viral diarrhea virus: a review. JAVMA 190: 1449–1458
2. Brownlie J, Clarke, MC, Howard CJ (1984) Experimental production of fatal mucosal disease in cattle. Vet Rec 114: 535–536
3. Collett MS, Larson R, Gold C, Strick D, Anderson DK, Purchio AF (1988) Molecular cloning and nucleotide sequencing of the pestivirus bovine viral diarrhea virus. Virology 165: 191–199
4. Collett MS, Larson R, Belzer SK, Retzel E (1988) Proteins encoded by bovine viral diarrhea virus: the genomic organization of a pestivirus. Virology 165: 200–208
5. Collett MS, Moennig, V, Horzinek MC (1989) Recent advances in pestivirus research. J Gen Virol 70: 253–266
6. Donis RO, Dubovi EJ (1987) Characterization of bovine viral diarrhea-mucosal disease virus-specific proteins in bovine cells. J Gen Virol 68: 1597–1605
7. Donis, RO, Dubovi EJ (1987) Glycoproteins of bovine viral diarrhea-mucosal disease virus in infected bovine cells. J Gen Virol 68: 1607–1616
8. Laemmli UK (1970) Cleavage of structural proteins during the assembly of the head of bacteriophage T4. Nature 227: 680–685
9. Magar R, Minocha HC, Lecomte J (1988) Bovine viral diarrhea virus proteins: heterogeneity of cytopathogenic and non-cytopathogenic strains and evidence of a 53K glycoprotein neutralization epitope. Vet Microbiol 16: 303–314

10. Matthews REF (1982) Classification and nomenclature of viruses: fourth report of the international committee on taxonomy of viruses. Intervirology 17: 97–101
11. Meyers G, Rümenapf T, Thiel HJ (1989) Molecular cloning and nucleotide sequence of the genome of hog cholera virus. Virology 171: 555–567
12. Ohmann HB (1988) BVD virus antigens in tissues of persistently viremic, clinically normal cattle: implications for the pathogenesis of clinically fatal disease. Acta Vet Scand 29: 77–84
13. Pocock DH, Howard CJ, Clarke MC, Brownlie J (1987) Variation in the intracellular polypeptide profiles from different isolates of bovine virus diarrhea virus. Arch Virol 94: 43–53
14. Purchio AF, Larson R, Collett MS (1984) Characterization of bovine viral diarrhea virus proteins. J Virol 50: 666–669
15. Renard A, Guiot C, Schmetz D, Dagenais L, Pastoret PP, Dina D, Martial JA (1985) Molecular cloning of bovine viral diarrhea viral sequences. DNA 4: 429–438

Author's address: S. Kamstrup, Animal Biotechnology Research Centre, State Veterinary Institute for Virus Research, Lindholm, DK-4771 Kalvehave, Denmark.

Arch Virol (1991) [Suppl 3]: 231–238
© Springer-Verlag 1991

Polymerase chain reaction amplification of segments of pestivirus genomes

P. M. Roehe and **M. J. Woodward**

Central Veterinary Laboratory, New Haw, Weybridge, Surrey, U.K.

Accepted March 20, 1991

Summary. Reverse transcription followed by polymerase chain reaction (PCR) amplification of a region of the viral genome at the 3′ end of the glycoprotein(s) gene was employed with the aim of determining its applicability as a diagnostic tool for pestiviruses. Candidate primers were designed from homologous segments detected by comparison between the sequences of strains NADL, Osloss and Alfort. A segment of 634 base pairs on the glycoprotein gene was targeted for amplification. Segments of five pestivirus strains of bovine viral diarrhoea virus, two of border disease virus and the Alfort 187 strain of hog cholera virus were amplified successfully.

Key words: Pestivirus, glycoprotein gene, polymerase chain reaction.

Introduction

The polymerase chain reaction (PCR) [16] has considerable potential as a diagnostic technique for the detection of pestiviruses [6], providing high sensitivity without the need for virus replication. The objective of this study was to determine suitable oligonucleotide primer sequences with potential for PCR diagnostic use. We report here preliminary results on the amplification of segments within the glycoprotein region of pestivirus genomes, using strains multiplied in vitro.

Materials and methods

Viruses and cells: Pestivirus strains and isolates used in the present experiments were: bovine viral diarrhoea virus (BVDV) laboratory strains NADL, C24V, Osloss, and field isolates C455, and 7206; border disease virus (BDV) field isolate 137/4 and pig isolate 87/6; hog cholera virus (HCV) laboratory strains Baker "A" and Alfort 187. The field isolates were mostly obtained from routine diagnostic submissions. Ruminant pestiviruses were grown in primary calf testis (CTs) cells. HCV strains were grown in porcine kidney (PK15) cells. For RNA extractions, cell monolayers were infected by adsorption for 1 hour at 37 °C with 0.5 ml of a virus suspension containing 10^3 to 10^6 $TCID_{50}$/50 μl. The monolayers were then washed, covered with fresh MEM with 10% fetal calf serum and incubated for four days at 37 °C. After incubation, the medium was removed and cells processed for RNA extraction as described below.

RNA extraction

A number of RNA extraction procedures were tested [1, 7, 8, 18]. All methods yielded viral RNA from which the targeted sequence were amplified successfully, provided that the RNA was extracted from intact cells. An adaptation of the method of Stallcup and Washington [18] was selected because of its ease and rapidity. With the exception of cell culture media, all solutions, plastics and glassware that came in contact with RNA-containing suspensions were treated with 0.1% diethyl-pyrocarbonate (DEPC, Sigma) and subsequently auto-claved. Gloves were used throughout and changed frequently to minimize RNAase contamination. Cells were scraped off the flasks with a rubber policeman, washed in MEM without serum and pelleted by centrifugation at 1000 g for 10 minutes. The cells were washed in ice-cold saline solution, pelleted for 5 seconds at 12000 g and the supernatant removed. Packed cell volume was approximately 100 to 150 μl, of which 50 μl were used for each extraction. To that cell volume, 500 μl of 10 mM EDTA (pH 8.0), 0.5% SDS were added, followed by an equal volume of 0.1 M sodium acetate (pH 5.2). The mixture was vortexed briefly, passed three times through an 18-gauge syringe needle, and the nucleic acid extracted with an equal volume of water-equilibrated phenol. The aqueous phase containing the RNA was transferred to fresh tubes, adjusted to 0.1 M Tris/HCl (pH 8.0), 0.2 M NaCl, followed by ethanol precipitation for at least 30 minutes in an ice bath. The RNA was collected by centrifugation in a refrigerated microcentrifuge at 12000 g for 10 minutes. The ethanol was removed and the RNA was resuspended in 200 μl of TE buffer (10 mM Tris/HCl pH 7.6, 1 mM EDTA). The solution was again made 0.2 M NaCl and precipitated with ethanol. The RNA was pelleted by centrifugation, the supernatant removed and the pellet resuspended in 100 μl of PCR buffer (10 mM Tris/HCl, 50 mM KCl, 1.5 mM KCl, 0.001% gelatin, pH 8.3).

Primers

The primers used for cDNA synthesis and amplification of segments of viral genomes were synthesized based on the BVDV NADL nucleotide sequence (2). Primer "G" (5′-TCAGCGAAGTAATCCCGGTG -3′, 58% G-C content) was complementary to nucleotides (5′ to 3′) 3485-3466; primer "F" (5′-CATATGGTCTGCAAGGCATAGG -3′, 50% G-C) was complementary to nucleotides (5′ to 3′) 3304-3283; primer "E" (5′-TCTTGCAAGTGGTGTGGT -3′, 53% G-C) was homologous with nucleotides (5′-3′) 3070-3086 and primer "D" (5′-ATAGGATGGACAGGAACTGT -3′, 45% G-C) was homologous with nucleotides (5′-3′) 2855-2874. For the amplification reactions the primer concentration was adjusted to 20 pmol/μl.

Reverse transcription

Reverse transcription was performed essentially as described [17] in a final volume of 20 μl. Viral RNA was boiled for 1 min, and 5 μl of the RNA suspension were added to a reaction mixture composed of 2 μl of 10 × PCR buffer, 1 μl of the appropriate primer ("G" or "F") dilution, 2.5 mM dNTPs, 35 units of RNA-Gard (Pharmacia), 2.5 mM MgCl$_2$, and 200 units of murine reverse transcriptase (USB). The reaction was incubated for 30 minutes at 37 °C and stopped by heating at 95 °C for 5 min.

Amplification of cDNA

Amplification of the cDNA was performed on the total reverse transcription reaction. In some experiments, the RNA preparations were tested neat and diluted 1/10 in PCR buffer. PCR was performed by the addition of 1 µl of the second primer ("D" or "E"), 79 µl of PCR buffer and 2.5 units of Taq polymerase (Cetus). The samples were overlaid with 100 µl of mineral oil (Sigma) to prevent evaporation. Temperatures for the reaction were 95 °C for 5 min, followed by 30 cycles of 95 °C for 1 min, 40 °C for 2 min and 70 °C for 4 min, and by a final period at 70 °C for 4 mins. After amplification, products were cooled to 4 °C and stored at this temperature until required.

Agarose gel electrophoresis

Gel electrophoresis was performed in 1.4% agarose (Sigma) or in 1% low melting point agarose (Nusieve), as described [11]. Five microlitre samples of each reaction were run in each slot.

Hybridization procedures

The 634 bp segment amplified from the NADL genome was used as probe. The fragment was excised from 1% low melting point agarose gel, boiled for 5 min and radiolabelled with [^{32}P]CTP (Amersham) as precursor using a commercially available labelling kit (Multiprime DNA labelling System, Amersham), following the instructions recommended by the manufacturers. Gels were transferred onto "Hybond-N" nylon membranes (Amersham) as described (17). Pre-hybridization was carried out for 1 hour in 6 × SSC, 5 × Denhardt's solution, 1% SDS and 10 µg/ml of denatured salmon sperm DNA (Sigma) at 42 °C. The labelled probe was added and after overnight incubation at 42 °C the membranes were washed twice with 2 × SSC at 42 °C, twice with 2 × SSC, 0.1% SDS at 65 °C and twice with 0.2 × SSC, 0.1% SDS at 65C (15). Membranes were then autoradiographed (Fuji X-ray film).

Results

The analyses by gel electrophoresis of the amplification products of the C24V, NADL and the BDV field isolate 137/4 are shown in Fig. 1a. (lanes 1, 2 and 3, respectively). Ethidium bromide staining revealed discrete bands in the region corresponding to the expected 634 bp fragment size, in all three viruses. However, the BDV isolate 137/4 (Fig. 1, lane 3) gave rise to a faint band in the expected 634 bp region, together with some other bands of different sizes. No amplification product was obtained from uninfected cells (Fig. 1a, lane 4). The results of the hybridization with the NADL probe of the same gel are shown in Fig. 1b. A strong signal was obtained with the NADL strain, mostly at the expected region (Fig. 1b, lane 2). The autoradiograph was deliberately overexposed to highlight the faint signal detected at

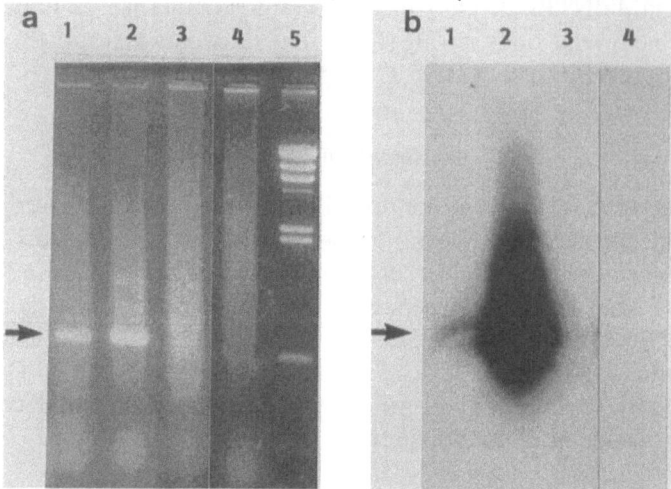

Fig. 1. Agarose gel (1.4%) electrophoresis of PCR amplification products derived from pestivirus genomes. **a** Ethidium bromide stained gel. BVDV strains C24V (1), NADL (2), BDV 137/4 (3), uninfected CT cells (4), molecular weight marker (5). The arrow represents the region of migration of a DNA fragment of approximately 634 bp. Amplification was performed with primers G–D (see text). **b** Hybridization of the gel shown above. Probe was a 634 bp fragment amplified from the strain NADL. Lanes labelled as above. Refer to text for methods

the equivalent region of C24V (Fig. 1b, lane 1). No hybridization was observed with the BDV isolate 137/4, although after longer exposure a very faint smear could be detected in the region where the 634 bp was expected (data not shown).

Figure 2 shows the results of the amplification carried out with other ruminant pestivirus field isolates grown in cell culture. The three viruses tested previously were included as standards. The tests were performed under the same conditions as above except that primer combination "G–E" was employed for the amplification of three field isolates (Fig. 2a, lanes 4, 5 and 6), two of which (7206 and 87/6) were refractory to amplification with the primer combination "G–D" (Fig. 2a, lanes 8 and 9). Hybridization of the

Fig. 2. Agarose gel (1.4%) electrophoresis oʃ PCR amplification products derived from pestivirus genomes of field isolates of ruminant pestiviruses. **a** Ethidium bromide stained gel (see text). BVDV strain NADL (1, 10), BDV 137/4 (2, 4), BVDV C24V (3), BVDV isolate 7206 (5, 8), BDV isolate 87/6 (6, 9), BVDV field isolate C455 (7), control uninfected cells (11), molecular weight marker (12). Lanes (5, 6, 7) were amplified with primers E–G. All others were amplified with primers G–D (see text). The arrow represents the region of migration of a DNA fragment of approximately 634 bp. **b** Hybridization of the gel shown above. Probe was a 634 fragment amplified from the strain NADL (see text). Lanes labelled as above. **c** Amplification of HCV strains Baker A (1), Alfort diluted 1/10 (2), Alfort neat (3). BVDV Osloss strain (4) was included as positive control. Lanes (5) control uninfected cells and (6) molecular weight markers. Refer to text for methods

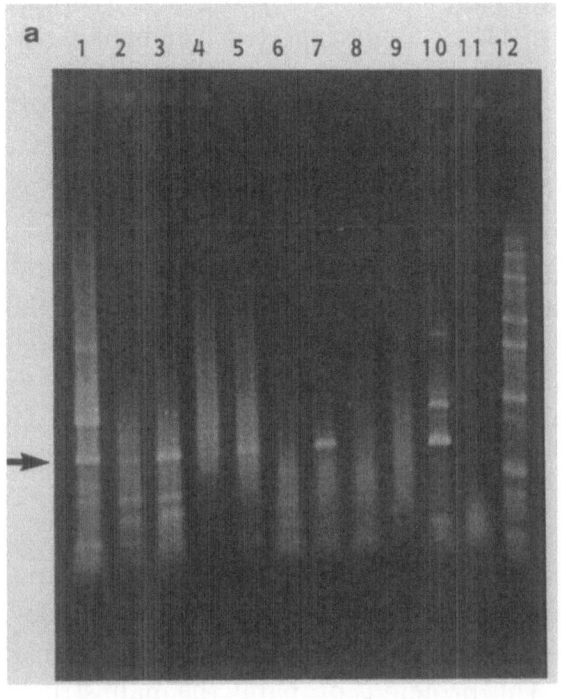

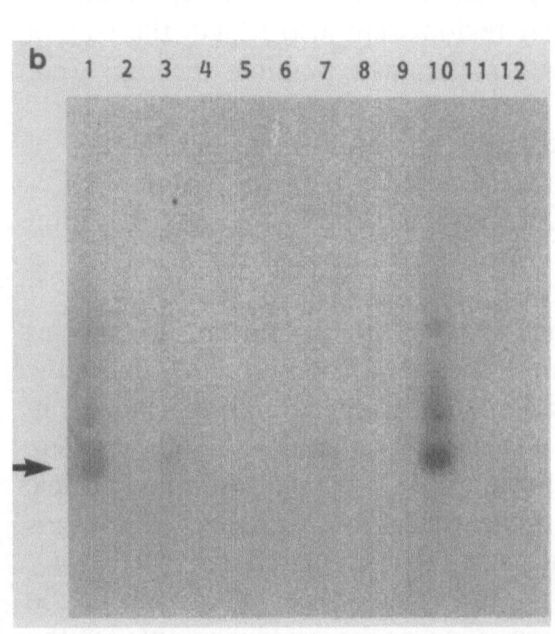

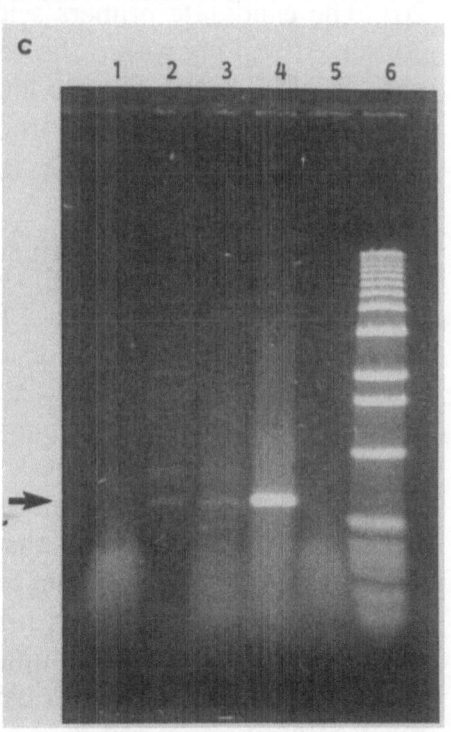

same gel (Fig. 2b) showed that the NADL probe only hybridized to strains NADL (lanes 1 and 10), C24V (lane 3) and the BVDV isolate C455 (lane 8) at the region of the expected 634 bp. The probe also hybridized to the NADL strain at regions of larger than expected fragment sizes (Fig. 2b, lanes 1 and 10).

The results of the agarose gel electrophoresis of the PCR products with cDNA derived from standard laboratory strains of HCV tested under the same conditions are shown in Fig. 2c. The BVDV strain Osloss was included for comparison (lane 4). Amplified products represented by bands at the expected size of approximately 634 bp were obtained from the strain Alfort, from both neat (lane 3) and diluted RNA preparations (lane 2). The Baker A strain was refractory to amplification under the conditions employed. Control uninfected PK15 cells did not give rise to any amplification products (data not shown).

Discussion

The segment which was targeted for amplification corresponded to a 634 bp segment of the pestivirus genome which probably encodes a portion of the glycosylated protein of molecular mass 53 kDa (gp53) in BVDV [3, 5] and HCV strains [13]. This protein is probably the main structural glycoprotein of the envelope of pestiviruses. It has been shown to be the major protein involved in neutralization of both BVDV and HCV strains [3, 4, 5, 13, 14, 20]. The candidate primers selected appear to be representative of short conserved regions within the highly variable region putatively coding for gp53 [3, 14]. Although two other primers (E and F) corresponding to oligonucleotide sequences within that region were also tested, the pair of primers expected to amplify the largest segment possible (634 bp) were preferentially used.

PCR amplification was achieved with all five strains and isolates of BVDV tested, with the two BDV isolates and with the Alfort strain of HCV (Figs 1 and 2). With isolates BVDV 7206 and BDV 87/6 amplification required a different set of primers ("G–E", Table 1) to give rise to an amplified product of the approximate expected size (Fig. 2a, lanes 6 and 7). It appears that some undetermined factor led to the generation of a shorter amplifiable fragment. The HCV strain Baker A was not amplified under the same conditions that led to amplification of the fragment of predicted size in strain Alfort. The reason for that is unclear and remains to be investigated. The HCV strain Alfort was successfully amplified, despite the limited nucleic acid homology with BVDV [3, 14]. Annealing at low stringency conditions (45 °C) probably favoured amplification in those circumstances [10].

The generation of cDNA using the same "downstream" primer for the reverse transcription and PCR amplification was the strategy adopted in the present study, since it bypasses the addition of random oligonucleotides in the cDNA generation. However, cDNA generation with random hexamers has been found more consistent and capable of generating higher yields of the targeted product [14]. This approach will be investigated further in pestivirus diagnosis.

We have successfully employed the PCR amplification of the segments of the genome of some pestivirus strains grown in vitro. The lack of hybridization of the NADL probe with the majority of the strains tested (Figs 1b, 2b) could be interpreted as a sign of sequence variability within the targeted region. This variability might provide the basis for further characterization of strains which could be done by restriction enzyme analysis of the particular segment or by nucleotide sequencing [7, 8, 19]. Sequencing of nucleotides of the amplified fragments shall provide definitive proof of their specificity, and it is being presently pursued in our laboratory. Although the system here described is useful for the production of large amounts of the targeted sequence, the application of the method in clinical specimens should bypass the need for virus growth in vitro in order to become a rapid diagnostic tool. The results of such application will be presented in a future publication.

Acknowledgements

Thanks are due to Dr. M. Collett for providing a copy of the NADL sequence. Equally the authors thank Dr. M. Moormann for the HCV Brescia sequence, and for the access to information before publication. Also thanks to Sharon Rankin and Gareth Sullivan for their excellent technical assistance. This paper was carried out while PMR was in receipt of grants from The British Council and the Brazilian National Research Board (CNPq).

References

1. Chirgwin JM, Przybyla AE, MacDonald RJ, Rutter WJ (1979) Isolation of biologically active ribonucleic acid from sources enriched in ribonuclease. Biochemistry 18: 5294–5299
2. Collett MS, Larson R, Gold C, Strick D, Anderson DK, Purchio AF (1988) Molecular cloning and nucleotide sequence of the pestivirus bovine viral diarrhea virus. Virology 165: 191–199
3. Collett MS, Larson R, Belzer SK, Retzel E (1988) Proteins encoded by bovine viral diarrhea virus: the genomic organization of a pestivirus. Virology 165: 200–208
4. Collett MS, Moennig V, Horzinek MC (1989) Recent advances in pestivirus research. J Gen Virol 70: 253–266
5. Donis RO, Corapi W, Dubovi EJ (1988) Neutralizing monoclonal antibodies to bovine viral diarrhoea virus bind to the 56K to 58K glycoprotein. J Gen Virol 69: 77–86
6. Edwards S (1990) The diagnosis of bovine virus diarrhoea-mucosal disease in cattle. Sci Tech Rev Off Int Epiz 9: 115–130
7. Gama RE, Horsnell PR, Hughes PJ, North C, Bruce CB, Al-Nakib W, Stanway G (1989) Amplification of rhinovirus specific nucleic acids from clinical samples using the polymerase chain reaction. J Med Virol 28: 73–77
8. Hyypia T, Auvinen P, Maronen M (1989) Polymerase chain reaction for human picornaviruses. J Gen Virol 70: 3261–3268

9. Innis MA, Gelfand DH (1990) Optimization of PCRs. In: Innis MA, Gelfand DH, Sninsky JJ, White TJ (eds) PCR protocols: a guide to methods and applications. Academic Press, London pp 3–12

10. Kawasaki ES (1990) Amplification of RNA. In: Innis MA, Gelfand DH, Sninsky JJ, White TJ (eds) PCR protocols: a guide to methods and applications. Academic Press, London pp 21–27

11. Maniatis T, Fritsch EF, Sambrook J (1982) Molecular cloning: a laboratory manual. Cold Spring Harbour Laboratory, Cold Spring Harbour, New York

12. Meyers G, Rumenapf T, Thiel H-J (1989) Molecular cloning and nucleotide sequence of the genome of hog cholera virus. Virology 171: 555–567

13. Moormann RJM, Hulst MM (1988) Hog cholera virus: identification and characterization of the viral RNA and virus-specific RNA synthesized in infected swine kidney cells. Virus Res 11: 281–291

14. Moormann RJM, Warmerdam PAM, Van der Meer B, Schaaper WMM, Wensvoort G, Hulst MM (1990) Molecular cloning and nucleotide sequencing of hog cholera virus strain Brescia and mapping of the genomic region encoding envelope protein E1. Virology 176: (accepted for publication)

15. Renard A, Dina D, Martial JA (1987) Complete nucleotide sequence of bovine viral diarrhoea virus genome and its fragment, useful for making antigenic proteins useful in therapy and diagnosis. European Patent Application No. 0208672

16. Saiki RK, Scharf S, Faloona F, Mullis KB, Horn GT, Erlich HA, Arnheim N (1985) Enzymatic amplification of β-globin genomic sequences and restriction site analysis for diagnosis of sickle cell anemia. Science 230: 1350–1354

17. Sambrook J, Fritsch EF, Maniatis T (1989) Molecular cloning: a laboratory manual 2nd ed. Cold Spring Harbour Laboratory, Cold Spring Harbour, New York

18. Stallcup MR, Washington LD (1983) Region-specific initiation of mouse mammary tumour virus RNA synthesis by endogenous RNA polymerase II in preparation of cell nuclei. J Biol Chem 258: 2802–2807

19. Torgersen H, Skern T, Blaas D (1989) Typing of human rhinovirus based on sequence variations in the 5′ non-coding region. J Gen Virol 70: 3111–3116

20. Wensvoort G (1989) Topographical and functional mapping of epitopes on hog cholera virus with monoclonal antibodies. J Gen Virol 70: 2865–2876

Author's address: P. M. Roehe, Central Veterinary Laboratory, New Haw, Weybridge, Surrey, England KT15 3NB, U.K.

Arch Virol (1991) [Suppl 3]: 239–244

Production of monoclonal antibodies to study the molecular biology of bovine viral diarrhea virus

Brief Report

F. Drèze, A. Collard and **Ph. Coppe**

Division immunologie animale, Centre d'Economie Rurale, Marloie, Belgium

Accepted March 20, 1991

Summary. Five monoclonal antibodies produced against bovine viral diarrhea virus were characterized for some of their biological activities. All of them bound to varying degrees to pestivirus strains but failed to neutralize the virus. One of the antibodies immunoprecipitated four polypeptides presumably involved in viral envelope organization.

Key words: BVD, pestivirus, monoclonal antibodies.

*

Bovine Viral Diarrhea Virus (BVDV) is the causative agent of fatal mucosal disease in cattle. Like hog cholera virus (HCV) and border disease virus (BDV), BVDV is a member of the genus Pestivirus. All three agents are serologically related [5]. Although they are currently classified as members of the family Togaviridae, their morphological and genomic organization resembles that of the Flaviviridae [2]. Our understanding of the molecular biology of pestiviruses still has several gaps. In this report, we describe the preparation of a panel of monoclonal antibodies (MoAbs) suitable for the analysis of viral structural organization. The MoAbs were characterized by an indirect immunoperoxidase assay using a number of pestivirus strains, virus neutralization and immunoprecipitation.

With one exception the viral strain used for the immunization of mice for the production of MoAbs was NADL propagated in bovine testicle cells. MoAb AM2G5 was raised against a non-cytopathic strain originating from a permanently contaminated bovine turbinate cell line. The antigens used for immunization were either infected cell extracts, pelleted virus [9] or virus

Table 1. Immunization procedures used to prime Balb/C mice

MoAbs	Inoculum	Days	Adjuvant	Route
X1A9	NADL strain	0	C	Intraperitoneal
	Virus pelleted	30	I	Intraperitoneal
		51	—	Intraperitoneal
		54[a]		
AE3E2	NADL strain	0	—	Intravenous
	Infected cell extracts	1	—	Intraperitoneal
		2	—	Intraperitoneal
		3	—	Intraperitoneal
		4[a]		
AM2G5	Non cytopathic BVDV strain	0	—	Intravenous
	Infected cell extracts	1	—	Intraperitoneal
		2	—	Intraperitoneal
		3	—	Intraperitoneal
		4[a]		
BT4B2	NADL strain	0	C	Intraperitoneal
BT6G9	Virus purified on	33	C	Intraperitoneal
	sucrose gradient	263	C	Intraperitoneal
		277	I	Intraperitoneal
		290	—	Intraperitoneal
		293[a]		

[a] shows the day of fusion
C (Complete Freund's Adjuvant) and
I (Incomplete Freund's Adjuvant)

purified on 25–60% sucrose gradients. Immunization procedures and time periods varied and are summarized in Table 1. Spleen cells of immunized mice were fused with SP2/0 Ag14 myeloma cells following a standard procedure. Hybridoma clones producing BVDV specific MoAbs were selected by an indirect immunoperoxidase assay performed on infected and non-infected cells fixed with isopropanol. The neutralizing effect of the MoAbs was assessed applying a standard test with 100 $TCID_{50}$ of NADL propagated in bovine testis cells. For the radioimmunoprecipitation (RIP), bovine testis cells were infected with NADL at a multiplicity of infection of 1. After 24 hours, infected cells were pulsed for four hours with L-(^{35}S) methionine (Amersham, 800 Ci/mM). Thereafter cells were lysed using the following buffer: 0.15 M NaCl, 1% sodium deoxycholate, 1% Triton X-100, 0.1% SDS, 10 mM Tris/HCl pH 7.2, PMSF (1 mM). Lysates were ultracentrifuged at 90000 × g for one hour. They were then mixed with MoAb X1A9 bound to sheep anti-mouse immunoglobulins coated magnetic beads (Dynal). Electrophoretic analysis was performed on 10–15% polyacryl-

Table 2. Reactivity of monoclonal antibodies against pestiviral isolates

Strains	X1A9	AE3E2	AM2G5	BT4B2	BT6G9
NADL	+ + +	+	+ +	+ + +	+ + +
Oregon C24V	+ +	−	+	+ +	+
Osloss nc	−	−	−	ND	ND
Singer nc	−	−	−	ND	ND
8875	−	−	+	ND	ND
R1350/90	−	ND	ND	−	−
Osloss c	−	−	+	ND	ND
Singer	+ + +	ND	ND	+ +	+
Lamspringe	−	−	+	ND	ND
0321	−	−	−	ND	ND
New-York	−	−	−	ND	ND
7443	−	−	−	ND	ND
Nebraska nc	+ +	−	−	ND	ND
Strain nc 1	+ +	ND	ND	ND	ND
Strain nc 2	−	ND	ND	ND	ND
Strain nc 3	+ +	ND	ND	ND	ND
Strain nc 4	+ +	ND	ND	−	−
Strain nc 5	−	ND	ND	ND	ND
Aveyron	+ + +	ND	+ +	ND	ND
BD nc	−	ND	−	ND	ND
Alfort	+ +	ND	+	ND	ND
331	−	ND	−	ND	ND

+ Reaction in indirect immunoperoxidase assay
− Failure to react in immunoperoxidase assay
ND not done

amide gradient gels in the presence of SDS (Pharmacia) and was followed by fluorography and autoradiography.

Five hybridoma lines secreting MoAbs against BVDV antigens were obtained from four fusions. Two of them (X1A9, BT4B2) secreted immunoglobulins of the IgG2a isotype, whereas the other three MoAbs (AM2G5, AE3E2, BT6G9) were IgM.

The binding pattern of the five MoAbs was partially determined using an immunoperoxidase assay with different pestiviruses. As shown in Table 2, all antibodies reacted differently. None of them reacted with all viruses tested. So far, within the limits of our tests, AE3E2 seems to react with its homologous strain NADL only. Although antibodies X1A9 and AM2G5 failed to react with all BVDV strains tested, they recognized the HCV strain Alfort and the BDV strain Aveyron. None of the MoAbs displayed neutralizing activity against NADL.

RIP analysis of the MoAb X1A9 using (^{35}S)methionine-labelled extracts of infected cells led to the identification of four polypeptides (Fig. 1). The

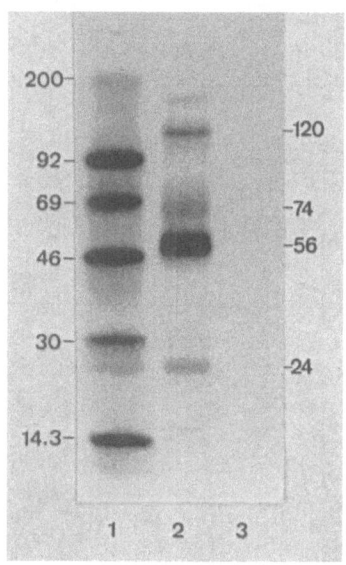

Fig. 1. Autoradiogram of the polypeptides immunoprecipitated with X1A9. Lane 1 shows molecular weight
marker proteins ($\times 10^{-3}$). X1A9 reacted with infected
(Lane 2) and non infected (Lane 3) lysates. Numbers on
right show the molecular weight values of polypeptides
precipitated with X1A9 ($\times 10^{-3}$)

bands were absent with lysates of uninfected cells. Their molecular weights
were 120kD (P120), 74kD (P74), 56kD (P56) and 24kD (P24). Corresponding results were obtained using NADL propagated in a different cell line and
two other viral strains (Oregon and a non-cytopathic strain, results not
shown). Under non-reducing conditions, the electrophoretic pattern of the
four polypeptides appeared completely modified: P24 was no longer visible,
the quantity of P56 decreased, whereas intensity of P74 and P120 increased
(Fig. 2).

As unanimously described elsewhere [7, 8, 12], we observed a very low
yield in the production of BVDV specific hybridomas when using antigen
from mice immunized with crude antigen preparations or semi-purified
antigen. BVDV antigen appears to be poorly immunogenic for Balb/C mice.
In addition, virions purified by ultracentrifugation on sucrose gradients
remained somewhat associated with cellular antigens.

The reactivity of the five MoAbs with different pestiviruses provided
additional proof of the antigenic diversity existing among members of the
group. However, the results obtained with X1A9 and AM2G5 also empha-
sized that conserved epitopes exist. Antigenic analysis of the 53kD viral
envelope glycoprotein of BVDV with MoAbs revealed variations between
strains [1, 4]. It is feasible that some of the epitopes, subject to the observed
antigenic changes, are involved in the virus attachment to cells and/or in its
penetration [1, 4, 6]. Our antibodies seemed to react with epitopes not
relevant for the above functions.

X1A9 identified four polypeptides using the RIP technique. One of them
(P56) appeared to be the viral envelope glycoprotein GP53 [3, 4, 6]. The
presence of the other three polypeptide bands in the electrophoretic pattern
could be explained by either antigenic relationships or coprecipitation. The

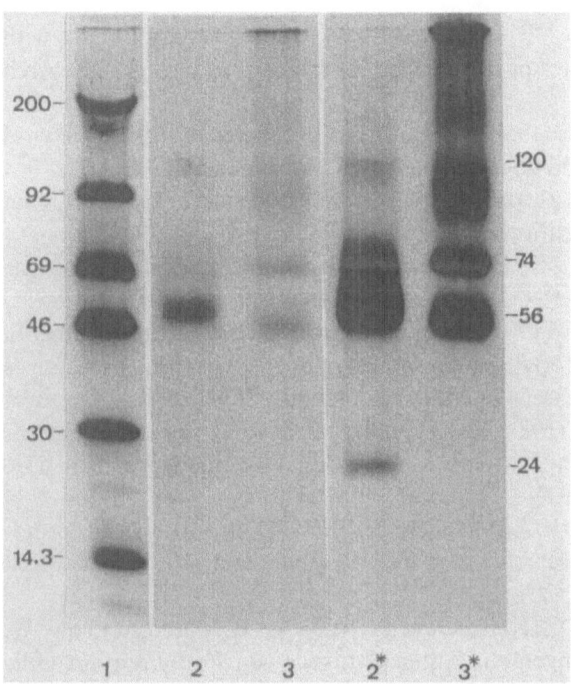

Fig. 2. Comparison of electrophoretic pattern of X1A9 immunoprecipitated polypeptides under reducing (Lane 2, 2*) and non-reducing conditions (Lane 3, 3*). All lanes are from the same gel, but for lanes 1–3 the time of exposure was 16 hours and 5 days for lanes 2*–3*. Lane 1 shows molecular weight marker proteins ($\times 10^{-3}$)

latter cannot be ruled out since the molecular weights of the four polypeptides were in accordance with those of the viral glycoproteins (GP53 and GP25) described earlier by Collett and coworkers and those of the glycoprotein precursors (GP62 and GP118) [3]. All four molecules could share a common epitope. In this case, X1A9 would confirm published results [3]. Another possibility is that the coprecipitated polypeptides constituted complexes which remained intact under the conditions used for immunoprecipitation. Changes which appeared when analyzing the polypeptides under non-reducing conditions, support this possibility, since they indicated the involvement of disulfide bonding. The coprecipitation of two glycoproteins GP55 and GP33 had been described for HCV [11]. The envelope protein of Flaviviruses seems to be linked with disulfide bonding to the viral M protein (derived from prM) [13], which is suggested to be an anchor protein [10]. In a similar manner, P56 could be linked by disulfide bonding to P24, which also is a hydrophobic protein [3] and thus a good candidate for an anchor protein. The complex formed by P56 and P24 would amount to a molecular weight of 74kD. We actually observed an increase of the concentration of P74 when electrophoretic analysis was carried out without reduction. Homopolymeric associations of proteins as described for the E protein of Flaviviruses could also play a role [13]. The P120 could have been a dimer of P56. However, the above hypotheses still have to be confirmed by more experimental data in order to give a complete view of pestiviral envelope organization.

244 F. Drèze et al.: Molecular biology of BVDV

References

1. Bolin S, Moennig V, Kelso Gourley NE, Ridpath J (1988) Monoclonal antibodies with neutralizing activity segregate isolates of bovine viral diarrhea virus into groups. Arch Virol 99: 117–123
2. Collett MS, Anderson DK, Retzel E (1988) Comparisons of the pestivirus bovine viral diarrhoea virus with members of the flaviviridae. J Gen Virol 69: 2637–2643
3. Collett MS, Larson R, Belzer SK, Retzel E (1988) Proteins encoded by bovine viral diarrhea virus: the genomic organization of a pestivirus. Virology 165: 200–208
4. Donis RO, Corapi W, Dubovi EJ (1987) Neutralizing monoclonal antibodies to bovine viral diarrhoea virus bind to the 56K Glycoprotein. J Gen Virol 69: 77–86
5. Howard CJ, Brownlie J, Clarke MC (1987) Comparison by the neutralization assay of pairs of non-cytopathogenic and cytopathogenic strains of bovine virus diarrhoea virus isolated from cases of mucosal disease. Vet Microbiol 13: 361–369
6. Magar R, Minocha HC, Lecomte J (1988) Bovine viral diarrhea virus proteins: heterogeneity of cytopathogenic and non cytopathogenic strains and evidence of a 53K Glycoprotein neutralization epitope. Vet Microbiol 16: 303–314
7. McHugh PH, Mackie DP, McNulty MS, McFerran JB (1988) Production of monoclonal antibodies to bovine virus diarrhoea virus and their reactivity with selected pestivirus isolates. J Vet Med B 35: 207–213
8. Moennig V, Bolin SR, Coulibaly COZ, Kelso Gourley NE, Liess B, Mateo A, Peters W, Greiser-Wilke I (1987) Untersuchungen zur Antigenstruktur von Pestiviren mit Hilfe monoklonaler Antikörper. DTW 94: 572–576
9. Peters W, Greiser-Wilke I, Moennig V, Liess B (1986) Preliminary serological characterization of bovine viral diarrhea virus strains using monoclonal antibodies. Vet Microbiol 12: 195–200
10. Rice CM, Lenches EM, Eddy SR, Shin SJ, Sheets RL, Strauss JH (1985) Nucleotid sequence of yellow fever virus: implications for flavivirus gene expression and evolution. Science 229: 726–733
11. Stark R, Rümenapf T, Meyers G, Thiel H-J (1990) Genomic localization of hog cholera virus glycoproteins. Virology 174: 286–289
12. Teyssedou E, Magar R, Justewicz DM, Lecomte J (1987) Cell-protective monoclonal antibodies to bovine enterovirus-3 and partial or no activity against other serotypes. J Virol 61: 2050–2053, 16: 303–314
13. Wengler G, Wengler G (1989) Cell-associated west nile flavivirus Is covered with E + Pre-M protein heterodimers which are destroyed and reorganized by proteolytic cleavage during virus release. J Virol 63(6): 2521–2526

Author's address: F. Drèze, Division immunologie animale, Centre d'Economie Rurale, rue du Carmel, 1, B-5406 Marloie, Belgium.

Arch Virol (1991) [Suppl 3]: 245–251

Determination of level of antibodies to bovine virus diarrhoea virus (BVDV) in bulk tank milk as a tool in the diagnosis and prophylaxis of BVDV infections in dairy herds

R. Niskanen[1], S. Alenius[2], B. Larsson[1] and S-O. Jacobsson[1]

[1] Swedish University of Agricultural Sciences, Faculty of Veterinary Medicine, Department of Cattle and Sheep Diseases, Uppsala, Sweden
[2] National Veterinary Institute, Uppsala, Sweden

Accepted March 20, 1991

Summary. An indirect ELISA has been evaluated for determination of the level of antibodies to BVDV in individual milk samples and recently in bulk tank milk from dairy herds. As part of an epidemiological study, bulk milk and individual milk samples from all cows in 15 dairy herds were analysed for antibodies to BVDV two times one year apart. There was an excellent correlation between the level of antibodies in the bulk tank milk and the prevalence of BVDV antibody positive cows. The mean prevalence of BVDV antibody positive cows in the 15 dairy herds was 45.5% (188/413) at the first sampling and 46.2% (191/413) one year later. Seven of the herds had no, or only a low number of, antibody positive cows. In contrast, between 52 to 100% of the cows in seven other herds were antibody positive to BVDV. In the 15th herd all cows without antibodies at the first sampling were antibody positive to BVDV one year later, indicating a recently introduced BVDV infection in this herd.

Analysis of bulk milk samples for BVDV antibodies is now routinely used in Sweden as a tool in diagnosis and prophylaxis of BVDV infections in dairy herds. The importance and advantages of this diagnostic technique, that has made it possible to establish BVDV-free dairy herds, is discussed.

Key words: Bovine virus diarrhoea virus, antibodies, bulk tank milk, enzyme-linked immunosorbent assay, prophylaxis.

Introduction

Advances in the understanding of the epidemiology of bovine virus diarrhoea virus (BVDV) and improved diagnostic methods have led to a highly increased rate of publications from all over the world that describes disease outbreaks associated with the introduction of BVDV into dairy herds. Different clinical manifestations that are commonly described include abortions, malformed or stillborn calves, retarded growth rate in calfhood, a high incidence of diarrhoea and pneumonia in persistently infected (PI) calves, or deaths in the typical mucosal disease form of BVDV [1]. All these different clinical manifestations are primarily caused by a transplacental BVDV infection of pregnant susceptible heifers or cows [6]. However, there are only a few publications in the literature that discuss the possibility to establish BVDV free dairy herds and keep them free in the future. We have shown by using an enzyme-linked immunosorbent assay (ELISA) that there is a good correlation between the level of serum antibodies to BVDV and the level of antibodies in milk during the whole lactation [4]. There is also a good correlation between the level of antibodies in bulk tank milk and the prevalence of antibody positive cows [5].

The purpose of this paper is to show how bulk milk samples can be used to monitor the BVDV status in dairy herds and to discuss how this ELISA for detection of BVDV antibodies in milk, can be used as a tool in the diagnosis and prophylaxis of BVDV infections and make it possible to establish BVDV-free dairy herds.

Materials and methods

Dairy herds and milk and blood samples

Bulk tank milk samples, as well as individual milk samples from all lactating cows were obtained twice one year apart (November 1987 and November 1988) from 15 randomly selected dairy herds. No vaccines against BVDV were used in these herds. The dairy herds were located in the county of Kopparberg and the number of lactating cows per herd sampled varied between 16 and 55. During 1989 blood samples were obtained from all animals, with the exception of previously tested antibody positive cows, in four of the herds shown to be infected with BVDV. Seronegative viraemic animals that were shown to be persistently infected (PI) with BVDV were subsequently removed from these herds. In May 1990 bulk tank milk samples were again collected from all the 15 dairy herds and analysed for antibodies to BVDV. The blood samples were collected in vacutainer tubes (Becton-Dickinson). Milk from individual lactating cows and bulk tank milk was collected in 10 ml plastic tubes containing 1.5 mg of the preservative bronopol (2-brom-2 nitropropane-1,3-diol). The skim milk was collected from below the fat layer after centrifugation of whole milk for 10 min at $3000 \times g$. Blood samples were centrifugated in the same manner and the serum was removed. Skim milk and sera were stored at $-20°C$ until analysis.

Detection of antibodies to BVDV in skim milk and blood serum

An indirect ELISA[1] was used for detection of antibodies to BVDV in both sera [3] and skim milk [4]. Briefly, microtitre plates were coated with detergent – solubilized BVDV

---

[1] SVANOVA, Biotech AB, Uppsala, Sweden.

antigens and stored at $+4°C$ until analysis. Immediately before analysis the plates were washed twice with phosphate-buffered saline containing 0.05% Tween (PBS-T). Undiluted skim milk and blood serum, diluted 1:10 in PBS-T containing 5% horse serum, were tested in volumes of 100 µl per well in duplicate. Following an incubation period of one hour at 37°C, the plates were washed three times with PBS-T and a monoclonal antibody to bovine IgG, conjugated with horse-radish peroxidase, was added in an optimal dilution in PBS-T. The plates were incubated and then washed as described above. A substrate to the enzyme was added in a volume of 200 µl per well and the reaction was interrupted after 10 minutes by adding 50 µl H_2SO_4. The absorbance value was measured at 450 nm. Serum with a mean absorbance value above 0.20 was regarded as positive for antibodies to BVDV [3]. Skim milk was considered positive for antibodies to BVDV when the mean absorbance value exceeded 0.04 [4].

Positive and negative control samples were always run in parallel with the test samples. The positive and negative control samples of skim milk had a mean absorbance value of 0.93 (S.D. = 0.16) and 0.01 (S.D. = 0.01), respectively.

Interpretation of ELISA absorbance values in bulk tank milk

In dairy herds without detectable antibodies to BVDV in bulk tank milk (ELISA absorbance value ≤ 0.04) all individual lactating cows are in the majority of herds also negative for BVDV antibodies. Herds with a low level of antibodies to BVDV in bulk tank milk (ELISA absorbance value between 0.05 and 0.20) have a low prevalence of antibody positive lactating cows (10.3% ± 6.5, mean ± S.D.). In contrast, 90 to 100% of lactating cows are antibody positive to BVDV in herds with a high level of antibodies in their bulk tank milk (ELISA absorbance value > 0.70) [5].

Virus isolation

Virus isolation was performed by inoculation of 0.1 ml serum from animals without antibodies to BVDV (ELISA absorbance value ≤ 0.20) on coverslip cultures of embryonic bovine turbinate (BTB) cells. The presence of BVDV was determined by an indirect immunofluorescence technique, applying a monoclonal antibody to BVDV.

Information to the dairy herd owners

At the end of 1988 all 15 herd owners in this study received general information about BVDV and more specific about the BVDV status in their dairy herds. After the information four herd owners (dairy herd number 12–15 in Table 1) wanted to have their herd screened for PI animals in an effort to try to establish a BVDV-free herd in the future. All 15 herd owners were also informed at this time about the risk of buying animals without testing for both virus and antibodies to BVDV.

Results

Prevalence of BVDV antibody cows

The mean prevalence of BVDV antibody positive cows in the 15 dairy herds was 45.5% (188/413) at the first sampling and 46.2% (191/413) one year later (Table 1).

Table 1. Prevalence of BVDV antibody positive cows in 15 dairy herds and level of antibodies to BVDV in bulk tank milk

Herd number	Number of BVDV antibody positive cows/total number of cows		Level[a] of antibodies in bulk tank milk		
	1987	1988	1987	1988	1990
1	0/55	0/54	0.00	0.00	0.00
2	0/26	0/27	0.00	0.00	0.00
3	0/19	0/17	0.00	0.00	0.00
4	0/18	0/21	0.00	0.00	0.00
5	4/28	0/20	0.40	0.00	0.00
6	2/41	2/39	0.10	0.16	0.00
7	9/23	2/23	0.65	0.09	0.23
8	21/23	16/22	0.45	0.57	0.56
9	27/27	12/23	0.82	0.58	0.24
10	16/16	13/20	0.79	0.53	0.41
11	24/24	27/27	0.89	0.82	0.89
12	24/26	28/28	0.94	0.53	0.37
13[b]	18/18	16/16	1.08	0.65	0.18
14[b]	42/44	49/50	0.81	0.66	0.58
15[b]	1/25	26/26	0.12	1.20	0.44
Total	188/413	191/413			

[a]The antibody level to BVDV was measured by an ELISA and the results are expressed as absorbance values.

[b]Persistently BVDV infected animals were removed from the herd during 1989.

As seen in Table 1 all cows were antibody negative in herd number 1 to 4 in 1987 and 1988 and only a low percentage of the cows were antibody positive to BVDV in herd number 5 to 7. In contrast, between 52 to 100% of the cows in the herd number 8 to 14 had antibodies to BVDV. In herd number 15 all cows without antibodies at the first sampling, were antibody positive to BVDV one year later indicating a recently introduced BVDV infection in this herd (Table 1). The owner of this dairy herd had bought pregnant heifers, without testing for BVDV, and kept them in close contact with the dairy cows during 1987 and 1988.

Level of antibodies to BVDV in bulk tank milk

In 1987 all lactating cows were free from antibodies to BVDV in herd number 1 to 4 and in 1988 in herd number 1 to 5. As can be seen in Table 1 no detectable antibodies to BVDV (absorbance ≤ 0.04) were found in the bulk milk samples analysed from these herds during the same time period. Neither were antibodies to BVDV found in the bulk milk samples collected from the herd number 1 to 6 in May 1990 (Table 1). In contrast all the other dairy herds with antibody positive cows had a detectable level of antibodies to BVDV in the bulk tank milk. In the herds with a high level (ELISA absorbance value >0.70) of antibodies to BVDV in the bulk milk, the majority of the cows were also antibody positive to BVDV.

Identification and removal of persistently infected animals

Attempts to isolate BVDV were performed during 1989 from serum of all animals that were seronegative to BVDV in the four herds 12, 13, 14 and 15 in Table 1. Animals that were found to be viraemic were blood sampled again approximately 3 to 4 weeks after the first blood sampling. In herd number 12 none of the animals was viraemic or persistently infected (PI). However, in herd number 13 two pregnant heifers were found to be PI and in herd number 14 one cow and her calf were both found to be PI with BVDV. These animals, apparently healthy, were subsequently removed from the herds and slaughtered. In herd 15 two one-year old heifers were found to be PI. In this herd two heifers in the same age group had died in mucosal disease, one month before the first blood sampling was performed. The two PI heifers now identified had also to be emergency slaughtered, due to mucosal disease, one week after the second blood sample was collected. Based on the serology data and the age of the PI animals identified it was decided that no further blood testing were necessary in these herds in order to detect PI animals. The level of antibodies to BVDV in the bulk tank milk from these four herds analysed in May 1990 were substantially lower than those determined in 1987 or 1988 (Table 1). This indicates a decreased herd immunity to BVDV among the cows in these four herds and points towards a BVDV-free status in the future.

Discussion

In this study we have shown that determination of the level of antibodies to BVDV in bulk tank milk by using an ELISA, is an easy, cheap and sensitive method to monitor the BVDV status in dairy herds (Table 1). We have in Sweden, during the last five years diagnosed BVDV to be the cause of a high incidence of abortions and an increased calf mortality in more than a hundered dairy herds. These dairy herds have a high level of antibodies to BVDV in the bulk tank milk, at the time when PI calves are identified. In a recent epidemiological study of BVDV, we used this ELISA to monitor the BVDV status in the herds, and it was shown that BVDV-free dairy herds have a general better fertility and less treatments for other diseases compared to dairy herds with active BVDV infections (Alenius, manuscript, 1990).

The studies performed have shown that BVDV is introduced to susceptible dairy herds, almost invariably by the introduction of, or contact with, primary or persistently BVDV-infected animals from other herds, or even more often, after purchase of antibody positive pregnant cattle from herds with active BVDV-infections. These healthy pregnant seropositive cattle, have subsequently given birth to persistently BVDV-infected calves, and thereby at parturition secured an effective spread of BVDV among susceptible pregnant heifers or cows. This has often resulted in the birth of several new PI calves, usually born 6 to 10 months after the first PI calf was born. Such PI heifer-calves also can become pregnant and give birth to new PI infected calves in the herd during a time period of several years [2]. If these calves, born after PI infected animals survive, the infection can be maintained in such a way in a closed dairy herd for decades. According to our experience BVDV is however often a self limiting infection in cattle herds. Even large dairy herds with active BVDV infections may become free from BVDV in a natural way if not reinfected with purchased PI or pregnant cattle with high antibody titres to BVDV, which indicate that they can carry a PI fetus. A large dairy herd became BVDV-free in such a natural way. This occurred over a time period of six years after PI calves were born in the herd. This dairy herd has now been BVDV-free for four years as monitored by analysing both individual and bulk milk samples for BVDV antibodies. The reason why infected cattle herds often becomes free from BVDV, is that PI cattle are commonly culled early in life due to either disease or an inadequate production.

An effective prophylaxis against BVDV could in the future be based on the prevention of dairy herds from becoming infected or reinfected with BVDV. We therefore consider the prophylaxis against BVDV infections in dairy herds, more or less exclusively, as an information problem. The knowledge of the epidemiology and diagnostic and virological techniques for an effective control program against BVDV are available [2].

About six hundred dairy farms in Sweden have now been analysed twice, one year apart, for antibodies to BVDV in the bulk tank milk. The dairy herd owners have also received general information about BVDV and more specific information about the BVDV status in their dairy herds. Several herd owners have understood, after they received the information, that many health problems in their herds have been caused by BVDV and are now taken measures to get rid of PI animals. Many owners of BVDV-free dairy herds, now only purchase animals directly from other free herds, in order to minimize the risk of an introduction of BVDV.

We are confident that the use of this ELISA in such a way will be an effective prophylaxis of BVDV infections in dairy herds in the future, with or without the use of safe and effective vaccines. The ELISA used in this study for detection of antibodies to BVDV in both milk and serum, is to our experience an excellent method to use in the diagnosis and prophylaxis of

BVDV-infections. It also can be used to monitor and validate a BVDV-free status in a large number of dairy herds in an easy way, by regular analysis of bulk milk samples for antibodies to BVDV.

Acknowledgements

This work was supported by grants from the Farmers Research Council for Information and Development. The excellent technical help from Maj Hjort and Eva Blomqvist is gratefully acknowledged. We thank Elisabeth Sjöberg for preparation of the manuscript.

References

1. Barber DML, Nettleton PF, Herring JA (1985) Disease in a dairy herd associated with the introduction and spread of bovine virus diarrhoea virus. Vet Rec 117: 59–64
2. Duffel SJ, Harkness JW (1985) Bovine virus diarrhoea-mucosal disease infection in cattle. Vet Rec 117: 240–245
3. Juntti N, Larsson B, Fossum C (1987) The use of monoclonal antibodies in enzyme-linked immunosorbent assays for detection of antibodies to bovine viral diarrhoea virus. J Vet Med B 34: 356–363
4. Niskanen R, Alenius S, Larsson B, Juntti N (1989) Evaluation of an enzyme-linked immunosorbent assay for detection of antibodies to bovine virus diarrhoea virus in milk. J Vet Med B 36: 113–118
5. Niskanen R (1991) The level of antibodies to bovine virus diarrhoea virus in bulk tank milk from dairy herds reflects the herd immunity to this virus. (Submitted for publication)
6. Oirshot JT van (1983) Congenital infections with nonarbo togaviruses. Vet Microbiol 8: 321–361

Author's address: S. Alenius, National Veterinary Institute, Box 7073, S-750 07 Uppsala, Sweden.

Arch Virol (1991) [Suppl 3]: 253–256

BVD-virus infection in goats – experimental studies on transplacental transmissibility of the virus and its effect on reproduction

K. Depner[1], **O. J. B. Hübschle**[2] and **B. Liess**[1]

[1]Institute of Virology, Hannover Veterinary School, Hannover, Germany
[2]Central Veterinary Laboratory, Windhoek, Namibia

Accepted March 20, 1991

Summary. Eight groups of altogether 25 goats without neutralizing antibodies against BVD virus, were inoculated either intranasally or intranasally and subcutaneously with two different BVD virus isolates during different stages of gestation. In all 18 goats inoculated within the first 78 days of gestation an abortion and foetal death rate of approximately 100% occurred. Only one goat gave birth to a clinically healthy kid. The other seven goats which were inoculated after the 78th day of gestation showed also a high foetal death rate. Only two of them gave birth to clinically healthy kids. Neutralizing antibodies against BVD virus could be detected in blood samples drawn from 14 kids born at normal term including stillborn and non-viable offsprings. BVD virus was reisolated from different organs taken from seven foetuses. It was not possible to isolate BVD virus from any of the normal offsprings.

Key words: BVD virus, goats, transplacental transmission, reproduction.

Introduction

Serological evidence of natural bovine virus diarrhea virus (BVDV) infection in goats was obtained in many countries [2, 3, 5, 8]. The existence of persistently BVDV-infected goats similar to cattle and sheep has never been reported. However, only two cases of clinically manifested pestivirus infections (Border Disease) in goats have been reported so far [4, 11].

The purpose of the present investigation was to attempt induction of viral persistence by inoculation of pregnant goats at various gestational stages with BVDV and to study the effect of transplacental transmission of BVDV in relation to the age of the fetus at the time of infection.

Materials and methods

Eight groups of altogether 25 oestrus-synchronised and seronegative goats were inoculated with two different pestivirus isolates at different stages of gestation (between day 30 and day 110) after the first service. Eleven goats were challenged with the third cell culture passage of the BVDV isolate Paulinenhof/89/Win, originally derived from an acute case of mucosal disease in an ox. This virus produced cytopathic effects in tissue cultures of bovine kidney cells. The other 14 goats were inoculated with the second cell culture passage of the BD/2109/Han/81 noncytopathic BVDV, originally isolated from a sheep [9]. Each goat was inoculated either intranasally or intranasally and subcutaneously with 10^6 to 10^7 $TCID_{50}$. Blood samples were taken from the mother goats and their offsprings at regular intervals for serology and attempts to isolate BVDV from buffy coat leucocytes. Attempts were also made to reisolate BVDV from organs of aborted fetuses and stillborn or non-viable offsprings. A direct neutralizing peroxidase-linked antibody (NPLA) assay as described by Hyera et al. [7] was used and BVDV cultural isolation performed on fetal calf kidney cell cultures.

Results

All of the mother goats produced neutralizing antibodies against BVDV within three weeks after inoculation. Ten goats aborted, nine goats gave birth to stillborn or non-viable kids, three goats gave birth to clinically healthy kids and three goats had no offsprings (Fig. 1). Neutralizing antibodies against BVDV could also be detected in blood samples taken before the first uptake of colostrum from 14 kids born at normal term including stillborn and non-viable offsprings. The serum antibody titer profiles ranged from 1/7,5 to 1/2560. These kids were all born to mother goats inoculated after the 64th day of gestation. BVDV was isolated from different organs taken from seven fetuses whose mothers had been inoculated between the 50th and the 90th day of gestation. In contrast it was not possible to isolate BVDV from any of the healthy offsprings exhibiting neutralizing antibody titres of various levels.

Discussion

The BVDV transmission studies in goats showed that the infection resulted in a high rate of presumably embryonic death, abortion, stillbirth and birth of non-viable kids. In 18 goats inoculated before day 78 of gestation abortion and fetal death rate was approximately 100%. Only one goat which was inoculated at day 70 of gestation gave birth to a clinically healthy kid. Seven goats which were inoculated after day 78 of gestation showed a high fetal death rate, too, with only two of them (inoculation at day 90 and 110) which gave birth to viable offsprings. This high frequency of abortion and fetal death has also been seen in previous experiments [6, 12]. The main reason for this outcome seems to have been a severe placentitis which occurred after challenging goats with pestivirus [1].

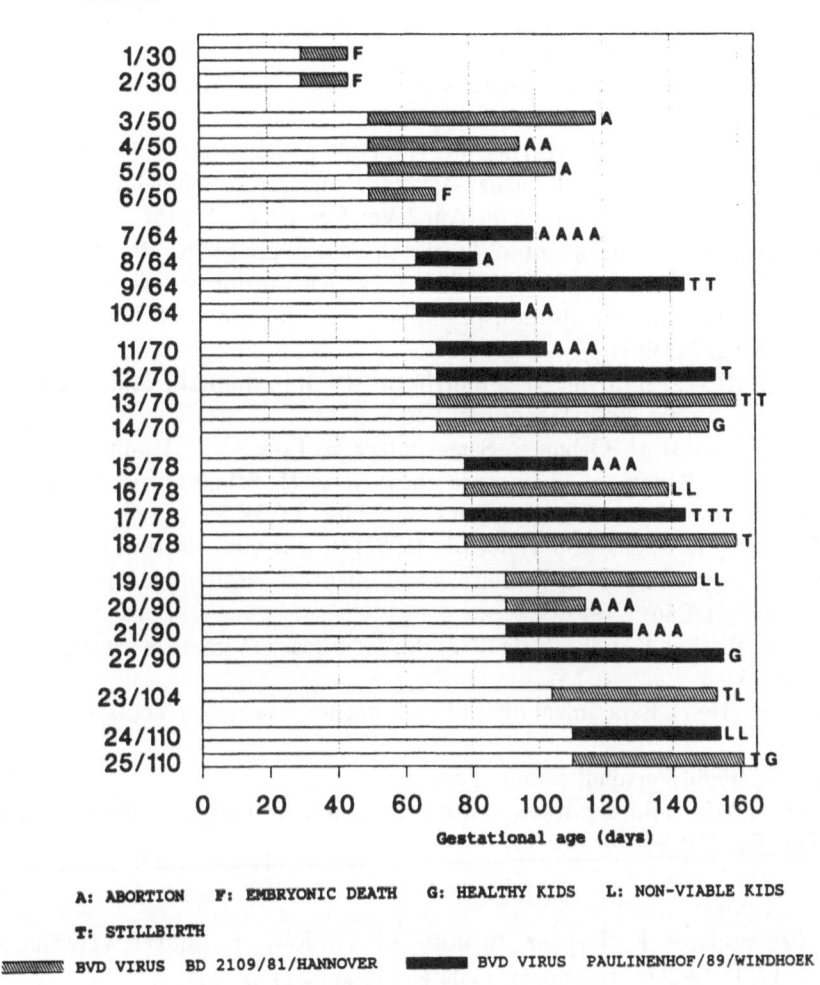

Fig. 1. Development of gestation after inoculation with two different pestivirus isolates at different gestational stages in goats

Persistent infections following intrauterine infection of immune incompetent fetuses as observed in cattle and sheep [10, 14], could not be verified for goats in the present experiment. It appears as if BVDV infections of seronegative pregnant goats result only rarely in persistently infected viable offspring [13]. Lack of spreading of BVDV by such animals through contact or vertical transmission in goat herds would therefore explain the low prevalence of antibody carriers against BVDV in goats.

References

1. Barlow RM, Rennie JC, Keir WA, Gardiner AC, Vantsis JT (1975) Experiments in border disease: VII. The disease in goats. J Comp Pathol 85: 291–297
2. Brown CC, Olander HJ, Castro AE, Behymer DE (1989) Prevalence of antibodies in goats in north-eastern Brazil to selected viral and bacterial agents. Trop Anim Health Prod 21: 167–169

3. Elazhary MASY, Silim A, Dea S (1984) Prevalence of antibodies of bovine respiratory syncytial virus, bovine diarrhoea virus, bovine herpesvirus-1 and bovine parainfluenza-3 virus in sheep and goats in Quebec. Am J Vet Res 45: 1660–1662

4. Fraser GC, Littlejohns JR, Moyle A (1981) The isolation of aprobable pestivirus of goat. Aust Vet J 57: 197–198

5. Fulton RW, Downing MM, Hagstad HV (1982) Prevalence of bovine herpesvirus-1, bovine viral diarrhea, parainfluenza-3, bovine adenovirus-3 and 7 and goat respiratory syncytial viral antibodies in goats. Am J Vet Res 43: 1454–1457

6. Huck RA (1973) Transmission of border disease in goats. Vet Rec 92: 151

7. Hyera JMK, Liess B, Frey H-R (1987) A direct neutralizing peroxidase-linked antibody assay for detection and titration of antibodies to bovine viral diarrhoea virus. J Vet Med B 34: 227–239

8. Hyera LMK (1989) Bovine viral diarrhoea (BVD) in domestic and wild ruminants in northern Tanzania. Diss. Hannover, Vet. School

9. Liess B, Blindow H, Orban S, Sasse-Patzer B, Frey H-R, Timm D (1982) Aborte, Totgeburten, Kümmern, Lämmersterben in zwei Schafherden Nordwestdeutschlands—"Border Disease" in der Bundesrepublik? DTW 89: 6–11

10. Liess B, Frey H-R, Orban S, Hafez SM (1983) Bovine virusdiarrhoe (BVD)—"Mucosal Disease": persistente BVD-Feldvirusinfektionen bei serologisch selektierten Rindern. DTW 90: 261–266

11. Loken R, Bjerkas I, Hyllseth B (1982) Border disease in goats in Norway. Res Vet Sci 33: 130–131

12. Loken R (1987) Experimentally-induced border disease in goats. Res Vet Sci 33: 130–131

13. Loken R (1990) Personal communication

14. Terpstra C (1981) Border disease: virus persistence, antibody response and transmission studies. Res Vet Sci 30: 185–191

Author's address: K. Depner, Institute of Virology, Hannover Veterinary School, Bünteweg 17, D-W-3000 Hannover, Federal Republic of Germany.

Arch Virol (1991) [Suppl 3]: 257–260

BVD virus isolation techniques for routine use in cattle herds with or without previous BVD history

H.-R. Frey, K. R. Depner, C. C. Gelfert and **B. Liess**

Institute of Virology, Hannover Veterinary School, Hannover,
Federal Republic of Germany

Accepted March 20, 1991

Summary. Buffy coats of 1074 cattle were tested for BVD virus using the usual longterm-cultivation (LTC) in bovine kidney monolayer cell cultures (7 days) whereby 268 BVD virus carriers could be detected. Serum samples collected simultaneously from the same animals were examined by means of a shortterm-cultivation (STC) procedure of only two days in stationary macroplate cell cultures. Using this method only 172 amongst the former 268 BVD virus carriers were found. Of the remaining 96 serum samples from animals positive in buffy coat leucocytes by LTC and negative in sera by STC, further 19 cattle were found to be viraemic when the sera were additionally tested by LTC. These results are discussed with regard to the antibody level and the age of the animals. The reduced sensitivity of STC of sera is considered in relation to the favourable time and cost factor. STC of serum samples in connection with the serological results on a herd basis proved to be valuable for the examination of cattle of more than 6 months of age but not for calves below 6 months. This was particularly true in cattle herds with no previous BVD history.

Key words: BVD virus, persistent infection, diagnosis.

Introduction

Cattle persistently infected with BVD virus are usually detected by time-consuming and costly virus isolation procedures. These include incubation periods of about one week in rolled fetal calf kidney cell (FCK) cultures (longterm-cultivation: LTC) and final detection by a direct immunofluorescent assay [4]. Buffy coat leucocytes separated from peripheral heparinized blood samples are preferred for virus isolation. Blood samples are usually derived from animals found seronegative by preceding serological tests in cattle herds with or without previous BVD anamnesis. This procedure proved to be safe but time-consuming and hardly applicable when large numbers of blood samples are to be tested. Therefore a virus isolation procedure was introduced to select viraemic animals without too much time loss and at reasonable costs.

Materials and methods

In order to select persistently infected animals within a cattle herd the standard virus isolation technique (LTC) was applied in order to examine buffy coat leucocytes from 1074 cattle of any age for BVD virus. In parallel a more simplified technique was used to test the sera of the same animals for the same purpose. The virus isolation procedure was changed to a one-passage modification in macroplates with only a two days incubation period of inoculated FCK cultures. For final detection of viral antigen a direct peroxidase-linked antibody assay [2] was used (shortterm-cultivation: STC). The same serum samples were also submitted to standard virus neutralization tests [1].

Results

By application of the standard virus isolation procedure (LTC) to buffy coat leucocytes blood samples from 268 cattle were found to contain BVD virus. When the serum samples of the same animals were tested by the modified procedure (STC) only 172 animals could be found positive (Fig. 1). Thus the latter procedure missed 96 blood samples (36%) in which BVD virus had been detected by the standard technique (LTC). Amongst these 96 samples

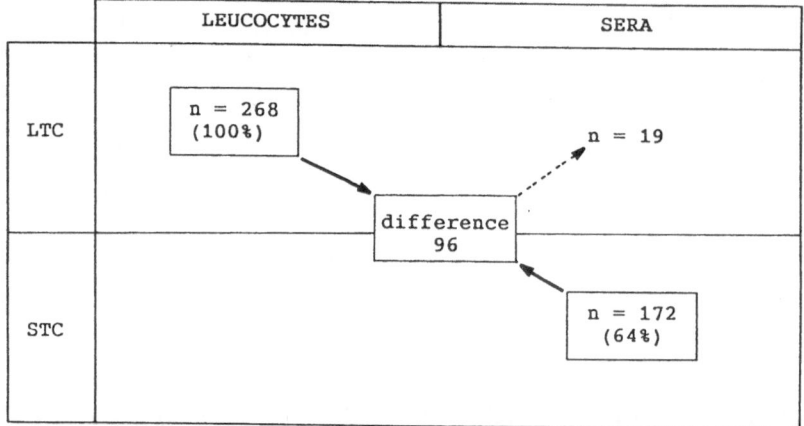

Fig. 1. Frequency of BVD virus isolation from leucocytes and sera using longterm- and shortterm-cultivation, respectively, in BVD viraemic cattle
Abbreviations:
LTC: Longterm-cultivation 7 + 2 days
STC: Shortterm-cultivation 2 days

further 19 sera were detected showing that their donors were viraemic when their sera were additionally tested by LTC.

Discussion and conclusions

Cattle of any age from herds with or without BVD anamnesis were randomly selected and tested for BVD virus. Calves under 6 months of age showed relatively high antibody titers presumably of colostral origin thus confirming former results on the decay of maternal antibodies [3]. In addition it had to be considered that the blood samples tested were derived from cattle with unknown status of BVD infection with transient or persistent viraemia.

The sensitivity of the STC procedure was obviously lower as compared with the results of the conventional LTC. This might be partially explained by the facts mentioned above. It needs also remembering that in the course of the pathogenetic events, BVD virus appears earlier and disappears later in the leucocytes than in blood serum. The virus might be demonstrated in leucocytes even in the presence of developing neutralizing antibody titres. Inspite of the obviously lower sensitivity, the application of STC appears to be reasonably safe if proper precautions are taken. For this sera from all animals of a particular herd should be subjected to STC testing. The same sera must be tested for BVD neutralizing antibodies, too, and the results compared with those obtained by STC.

In all questionable cases, i.e. in seropositive calves (less than 6 months of age) or seronegative, STC-negative animals of more than 6 months of age within an otherwise seropositive herd, the LTC should be applied in order to attempt virus isolation from buffy coat leucocytes. The same is true in cattle herds where BVD vaccines were applied previously.

STC proved to be safe in routine diagnosis of persistent BVD virus infections of cattle more than 6 months of age. Therefore this test can be recommended for the rapid and inexpensive detection of cattle persistently infected with BVD virus in cattle herds with or without BVD anamnesis.

References

1. Hyera JMK, Liess B, Frey HR (1987a) A direct neutralizing peroxidase-linked antibody assay for detection and titration of antibodies to bovine viral diarrhoea virus. J Vet Med B 34: 227–239
2. Hyera JMK, Dahle J, Liess B, Moennig V, Frey HR (1987b) Production of potent antisera raised in pigs by anamnestic response and use for direct immunofluorescent and immunoperoxidase techniques. In: Harkness JW (ed) Pestivirus infections of ruminants, EUR 10238, Luxembourg, pp 87–101
3. Liess B, Landelius L, Lackmann-Pavenstaedt B (1982) Bovine Virusdiarrhoe (BVD) – Ergebnisse einer serologischen Langzeitstudie unter enzootischen Infektionsbedingungen in einem mittelgroßen Rinderzuchtbestand. Prakt Tierarzt 63: 118–121

4. Orban S, Liess B, Hafez SM, Frey HR, Blindow H, Sasse-Patzer B (1983) Studies on transplacental transmissibility of a bovine virus diarrhoea (BVD) vaccine virus. I. Inoculation of pregnant cows 15 to 90 days before parturition (190th to 265th day of gestation). J Vet Med B 30: 619–634

Author's address: H.-R. Frey, Institute of Virology, Hannover Veterinary School, Bünteweg 17, D-W-3000 Hannover, Federal Republic of Germany.

Arch Virol (1991) [Suppl 3]: 261–265

Molecular characterisation of the coding region for the p125 from homologous BVDV biotypes

M. Desport and **J. Brownlie**

AFRC Institute for Animal Health, Compton Laboratory, Compton, Newbury,
Berkshire RG16 0NN, U.K.

Accepted March 20, 1991

Summary. We amplified and sequenced the p125 coding regions of a 'homologous' pair of BVDV biotypes, Pe515 cytopathogenic and non-cytopathogenic. The sequences were aligned with the published sequences of Osloss, NADL and the HCV Alfort strains, but no insertions of host sequence were observed in that region.

Key words: BVDV, HCV, pestivirus, PCR

*

Bovine viral diarrhoea virus (BVDV) is a positive stranded RNA virus, currently classified in the Togaviridae as a member of the Pestivirus genus, together with border disease virus (BDV) of sheep and hog cholera virus (HCV) [14]. Two forms of BVD virus are usually isolated from outbreaks of mucosal disease and these are differentiated principally by their different growth characteristics in vitro. The non-cytopathogenic (ncp) BVDV biotype induces little cytopathic effect on tissue culture cells, whilst the cytopathogenic (cp) biotype causes vacuolation in the cytoplasm and ensuing cell death.

Analysis of virus encoded proteins in cells infected with cp and ncp viruses also reveals that the viruses differ in their processing of a 125 kDa protein (Fig. 1). In cp biotypes the p125 is cleaved into 80 kDa and 54 kDa subunits. This cleavage and consequently the 80 kDa protein, is not observed in cells infected with ncp virus preparations. This difference is consistently observed in all of the 'homologous' pairs that have been studied to date [12, 6].

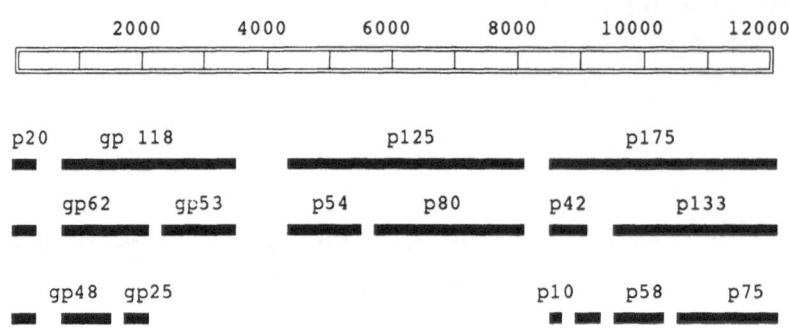

based on genome map of M.S. Collett (1990)
(personal communication)

Fig. 1

The complete nucleotide sequences of two cp strains of BVDV [13, 4] and one ncp HCV isolate [9] have recently been published and confirmed the size of the BVDV and HCV genomes to be approximately 12.5 kb. When these sequences are aligned (Fig. 1) NADL is found to have an insertion of 90 amino acids while Osloss has an insertion of 76 amino acids, both of these occurring in the predicted coding region for the p54. The NADL insertion has 99% homology with a host cellular mRNA coding for a protein of unknown function [8] and the Osloss insertion corresponds to an ubiquitin-like protein [3, 10]. No comparable insertions were observed in the ncp HCV sequence [9]. Since these insertions only occur in cp isolates it has been suggested that this uptake of cellular sequences may be associated with the mutation of ncp to cp virus [10]. We decided to sequence the p125 coding region of the Pe515 'homologous' viruses to identify any differences between the two isolates.

BVDV RNA was prepared using a total RNA extraction method for Pe515ncp and a viral RNA technique for Pe515cp. The total RNA protocol was a modification of the method of Chomczynski et al. [2] where RNA is isolated by a single extraction with an acid guanidinium thiocyanate-phenol-chloroform mixture. The viral RNA method was Chang's modification [1] of Maniatis' method [7] and involves extraction of RNA from the cytoplasm of virus-infected cells using SDS/proteinase K followed by phenol/chloroform extraction. First strand cDNA synthesis was performed on both RNA preparations using MuLV reverse transcriptase (Gibco BRL) in a total volume of 100 μl. 10 μl aliquots of the reverse transcription cDNA product were amplified using specific primers (Fig. 2) with Taq DNA polymerase (Cetus corp). The cDNA was denatured at 94°C for 1 minute and the primers annealed at 50°C for 1.5 minutes. Primer extension was at 72°C for

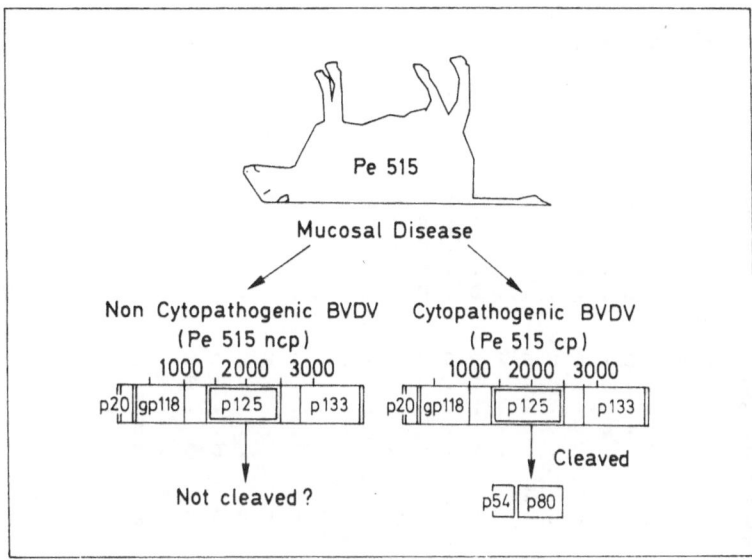

Fig. 2

2.5 minutes and the amplification cycle was performed 30 times with a final incubation at 72°C for 7 minutes.

The PCR products were analyzed by electrophoresis on a 0.8% agarose gel and were visualised with ethidium bromide under UV light. Fragments were excised from the gel and purified using the Geneclean kit. They were then either cloned into pUC9 for plasmid sequencing or sequenced directly using the amplification primers and Sequenase (Applied Biosystems). Using several combinations of primers, cDNAs corresponding to the entire p125 region of both biotypes were amplified and sequenced. As the sequence data was obtained it was aligned to the corresponding regions of the Osloss and NADL BVDV isolates and the Alfort HCV strain (Fig. 3).

The p54 region was successfully amplified in both viruses using primers that were complementary to the Pe515 cDNA. The p80 region, however, was produced in two overlapping segments using primers made from regions of good homology between the NADL and Osloss sequences. Apart from reducing the annealing temperature of the primers to 40°C in the latter case, fragments were amplified and sequenced under the same conditions. There was some difficulty in amplifying fragments of greater than 1 kb from the Pe515ncp cDNA made from the total RNA preparation. Smaller size sections were successfully amplified, indicating that this was not due to primer:template mismatch but perhaps was a problem with the quality of the RNA and, therefore, the cDNA. We amplified and sequenced the region coding for the p125 in the Pe515 'homologous' pair of viruses. The degree of sequence homology observed at both the DNA and amino acid level is particularly striking: the Pe515 viruses were 99.6% homologous with respect to each other and 97% homologous with NADL, excluding the known

```
             1460               1470               1480
NADL cp  T V A S W Y G E E E V Y G M P K I M T I I K A S T L S K S R
Ossl cp  . . . . . . . . . . . . . . V . . . . R . C S . N . N K
Pe   cp  . . . R . . . . . . . . . . . . . . . . . . . . . . N . N .
Pe   ncp A . . R . . . . . . . . . . . . . . . . . . . . . . N . N .
Alf  ncp V . V R . F . D . . I . . . . . L I G L V . . A . . . R N K

             1490               1500               1510
NADL cp  H C I I C T V C E G R E W K G G T C P K C G R H G K P I T C
Ossl cp  . . . . . . . . A K K . . . . N . . . . . . . . . . . . .
Pe   cp  . . . . . . . . . . K . . . . . . . . . . . . . . . . . .
Pe   ncp . . . . . . . . . . K . . . . . . . . . . . . . . . . . .
Alf  ncp . . M L . . . . . D . D . R . E . . . . . . . F . P . V V .

             1520               1530               1537
NADL cp  G M S L A D F E E R H Y K R I F I R E G N F E G [ 90 Amino
Ossl cp  . . T . . . . . . . . . . . . . . . . . . T . . .
Pe   cp  . . . . . . . . . . . . . . . . . . . . . D .
Pe   ncp . . . . . . . . . . . . . . . . . . . . . D .
Alf  ncp . . T . . . . . K . . . . . . . . . . D Q S G G

                    1626      1630               1640
NADL cp  Acid Insertion ]  P F R Q E Y N G F V Q Y T A R G Q L F L R
Ossl cp                    . . . . H S . . . . . . . . . . . . . . .
Pe   cp                    . . . . . . . . . I . . . . . . . . . . .
Pe   ncp                   . . . . . . . . . I . . . . . . . . . . .
Alf  ncp                   . L . E . H A . Y L . . K . . . . . . . .

             1650               1660               1670
NADL cp  N L P V L A T K V K M L M V G N L G E E I G N L E H L G W I
Ossl cp  . . . I . . . . . . . . . . . . . V . . . D . . . . . . .
Pe   cp  . . . I . . . . . . . . . . . . . . . . . D . . . . . . .
Pe   ncp . . . I . . . . . . . . . . . . . . . . . D . . . . . . .
Alf  ncp . . . . . . . . . . . . . L . . . . T . . . D . . . . . V

                                  1680               1690
NADL cp  L R                       G P A V C K K I T E H E K C
Ossl cp  . K [ 76 Amino Acid Insertion ]  . . . . . . . . . . . . .
Pe   cp  . .                       . . . . . . . . . . . . . R .
Pe   ncp . .                       . . . . . . . . . . . . . R .
Alf  ncp . .                       . . . . . . . V . . . . R .
```

Fig. 3

host insertion. No insertion was observed when comparing the Pe 515 cp and ncp sequences. The area immediately preceding the p 54 coding region has recently been amplified in both biotypes. Fragments obtained by PCR and analyzed on agarose gels appear equal in size and preliminary sequence data indicates the same high degree of homology as that observed for the p 125.

This report confirms that the techniques of reverse transcription, amplification and direct sequencing of PCR products can allow rapid analysis of specific areas of the BVDV genome. Any host sequence insertions in Pe 515 cp, similar to the NADL or Osloss insertions, should have been detected using this methodology since they only span relatively small areas. De Moerlooze et al. [5] have amplified and sequenced 10 pestiviral strains across this region of the p 54 and found that NADL and Osloss cp strains were the only viruses with insertions. The significance of these insertions in cp isolates is still unconfirmed. However, one of the disadvantages of PCR when compared to cDNA cloning, is that duplications of sequences may not be detected, particularly if they span large areas. Meyers et al. [11] reported finding a duplication of the entire p80 sequence in one of their cytopathogenic viruses and this would have been indistinguishable from the original sequence by the methods employed here. With more homologous pairs of viruses being examined, the raison d'être for the cytopathogenic biotypes will soon be resolved.

References

1. Chang T-J (1986) Molecular cloning of bovine viral diarrhea viral genome. PhD Thesis University of California, Davis, USA, pp 6–9
2. Chomczynski P, Sacchi N (1987) Single-step method of RNA isolation by acid guanidinium thiocyanate-phenol-chloroform extraction. Anal Biochem 162: 156–159
3. Clarke LE, Pocock DH, Brown DK, De Moerlooze L, Lecomte C, Renard A (1990) A variant of ubiquitin encoded in the genome of bovine virus diarrhoea virus isolate. (To be published)
4. Collett MS, Larson R, Gold C, Strick D, Anderson DK, Purchio AF (1988) Molecular cloning and nucleotide sequence of the pestivirus bovine viral diarrhea virus. Virology 165: 191–199
5. De Moerlooze L, Desport M, Renard A, Lecomte C, Brownlie J, Martial JA (1990) The coding region for the 54 kDa protein of several pestiviruses lacks host insertions but reveals a "Zinc finger-like" domain. Virology 177: 812–815
6. Donis RO, Dubovi EJ (1987) Differences in virus-induced polypeptides in cells infected by cytopathic and noncytopathic biotypes of bovine virus diarrhea-mucosal disease virus. Virology 158: 168–173
7. Maniatis T, Fritsch EF, Sambrooks S (1982) Molecular cloning: a laboratory manual. Cold Spring Harbour Laboratories, Cold Spring Harbour, NY, pp 191–193
8. Meyers G, Rumenapf T, Thiel H-J (1989) Insertion of ubiquitin coding sequence identified in the RNA genome of a togavirus. Proceedings of 11th International Symposium of the World Association of Veterinary Microbiologists, Immunologists, and specialists in infectious diseases, p 239
9. Meyers G, Rumenapf T, Thiel H-J (1989) Molecular cloning and nucleotide sequence of the genome of hog cholera virus. Virology 171: 555–567
10. Meyers G, Rumenapf T, Thiel H-J (1989) Ubiquitin in a togavirus. Nature (London) 341: 491
11. Meyers G, Tautz N, Rumenapf T, Dubovi EJ, Thiel H-J (1991) Viral cytopathogenicity correlated with integration of ubiquitin-coding sequences. Virology 180: 602–616
12. Pocock DH, Howard CJ, Clarke MC, Brownlie J (1987) Variation in the intracellular polypeptide profiles from different isolates of bovine virus diarrhoea virus. Arch Virol 94: 43–53
13. Renard A, Dino D, Martial J (1987) Vaccines and diagnostics derived from bovine diarrhoea virus. European Patent Application number 86870095.6. Publication number 0208672
14. Westaway EG, Brinton MA, Gaidamovich SYA, Horzinek MC, Igarashi A, Kaariainen L, Lvov DK, Porterfields JS, Russel PK, Trent DW (1985) Togaviridae. Intervirology 24: 125–139

Author's address: Moira Desport, AFRC Institute for Animal Health, Compton Laboratory, Compton, Newbury, Berkshire RG16 0NN, U.K.

Arch Virol (1991) [Suppl 3]: 267–271

Progeny of sheep persistently infected with border disease virus

Z. Woldehiwet[1] and **P. F. Nettleton**[2]

[1] University of Liverpool, Department of Veterinary Pathology, Veterinary Field Station,
Leahurst, Neston, Wirral, and
[2] Moredun Research Institute, Edinburgh, U.K.

Accepted March 20, 1991

Summary. Most lambs affected with border disease die early in life but those which survive gradually loose their body tremors and their fleece abnormalities become less clear. Seven female lambs persistently infected with border disease virus were reared to maturity and bred from when they were 2 to 3 years old. Two failed to conceive but five gave birth to 6 live lambs with clinical signs of border disease characterized by hairy and pigmented fleece with or without body tremors. The epidemiological significance of persistently infected sheep is discussed.

Key words: Border disease virus, persistent infection, progeny.

Introduction

Border disease virus (BDV) affects newborn lambs causing a disease characterized by tremors and abnormally hairy birth coat [7, 13]. It results from infection in utero with a pestivirus antigenically related to that which causes bovine viral diarrhoea/mucosal disease [11, 7]. The most obvious clinical manifestations of the disease occur at lambing. Lambs are weak with altered body conformation, changes in fleece and tremors [2]. Many affected lambs die early in life but those which survive may gradually lose their nervous signs and the fleece abnormalities become less obvious [1]. It is generally accepted that in the field few lambs affected with border disease survive to sexual maturity [8, 9] but field observations in the United Kingdom and Australia [8, 14] and some experimental data [3] suggest that clinically affected lambs raised to maturity can be sources of BD infection for their progeny. In the present study 7 persistently infected female lambs from a field outbreak of border disease were raised to maturity and mated to examine the possible vertical transmission of border disease virus.

Materials and methods

Persistently infected sheep

In February 1987, 10 Suffolk crossbred lambs with clinical signs of BD were obtained from a farm in Shropshire. Another group of 7 affected lambs was obtained from the same farm in March 1988. The animals were kept separate from other sheep and regularly sampled for virus isolation as described earlier [15]. All but one of the lambs had virus and no antibody and were, therefore, regarded as persistently infected. Seven of the female lambs were maintained in isolation until they were 2–3 years old. The persistently infected ewes and one control ewe were served by a normal Suffolk-cross ram, after oestrus-synchronization, in November 1989 and maintained in the isolation unit.

Virological status of ewes and lambs

Four to six weeks after delivery, clotted and heparinized blood samples were collected from all the ewes and their lambs. Heat inactivated serum samples were tested for the presence of antibodies in a microneutralisation test against the Moredun reference strain of cytopathic BDV as previously described [6]. Results are expressed as the reciprocal of the serum dilution corresponding to the 50 per cent end point of neutralisation.

The heparinized blood samples were tested for the presence of specific BDV antigen by the ELISA method of Fenton and colleagues [5]. Samples with OD values greater than 0.1 were considered positive.

Results

Reproductive performance

The reproductive performance of the persistently infected ewes is given in Table 1. Five of the persistently infected ewes produced live 6 lambs four of which were males. All the lambs had dark or black pigmentation of fleece but only 4 had detectable body tremors at birth. These tremors were very transient, disappearing within the first week of life in all but one lamb. The control ewe maintained with the persistently infected flock delivered a normal male lamb. No special nursing care was provided and four of the 6 lambs survived without bottle feeding. Two of the lambs died within 48 hrs due to trauma.

Virological and serological status of ewes and lambs

In samples collected 4–6 weeks after the birth of the lambs, BDV antigen was demonstrated in leucocyte preparations from all seven ewes and all of the persistently infected ewes were seronegative. Border disease virus was demonstrated in the blood samples of the four surviving lambs and non-cytopathic BDV was isolated from spleens of the two lambs which died (Table 2). All of the lambs were seronegative. The control ewe and its lamb were both seropositive and virusnegative.

Table 1. Reproductive performance of persistently infected ewes

Ewes	Lambs				
No	No	Sex	Tremors	Pigmented	Survived
5					
6	101	M	+	+	48 hrs
	697	M	+	+	yes
10	696	F	+	+	yes
12					
9	689	F	+	+	yes
302	695	M	−	+	48 hrs
155	698	M	−	+	yes
7[a]	690	M	−	−	yes

[a] Control
F female
M male

Table 2. Virological status of ewes and lambs

Ewes	Lambs				
No	Antibody	Virus	No	Antibody	Virus
5	−	+			
302	−	+	685	ND	+
6	−	+	101	ND	+
			697	−	+
10	−	+	696	−	+
12	−	+			
9	−	+	689	−	+
155	−	+	693	−	+
7[a]	+	−	690	+	−

[a] Control
ND not determined

Discussion

Both BVDV and BDV can cross the placenta and establish a persistent infection in the animals infected in utero. Persistently infected lambs are born either to susceptible mothers which are infected before the onset of foetal immune responses or to persistently infected mothers. The main determinant of foetal infection in acutely infected ewes is the gestational age of the foetus at the time of infection. Infection of a susceptible ewe in early gestation usually results in death of the foetus but those foetuses which survive become persistently infected and foetuses infected in late gestation develop immunity to the virus [10]. These persistently infected animals excrete virus continuously and spread the virus. It is generally thought that lambs with clinical border disease do not reach sexual maturity and, therefore, vertical transmission from persistently infected ewes is not regarded as a common source of infection. However, some strains of BDV may cause mild or inapparent manifestations of the disease. Recently a strain of virus which causes no fleece changes or body tremors has been described [4]. In the present study most of the lambs born to persistently infected ewes had fleece abnormalities characterized by abnormal pigmentation and hairiness without body tremors. Body tremors are easy to detect but fleece abnormalities are not readily apparent in certain breeds of sheep [2]. Furthermore, some strains of BDV may cause mild clinical signs characterized by abortions and the birth of small weak lambs [4]. In these circumstances the lambs without the classical signs of body tremor and fleece changes could easily be missed and be kept for breeding. Unless culled, ewes which had produced affected lambs, these carriers could continue to be a source of infection of successive breeding seasons [14, 3]. These have implications on the strategies of control of the disease. It has generally been assumed that the removal of affected lambs is sufficient to control border disease in a flock [12] but detecting persistently infected ewes requires expensive laboratory tests [4]. The present results suggest that ewes which give birth to lambs with signs of BD must be regarded as potentially persistently infected themselves, particularly when the disease occurs repeatedly in a flock. In these cases testing all the barren ewes and ewes with lambs affected with border disease will help to pinpoint persistently infected ewes.

References

1. Barlow RM (1982) Clinical border disease. In: Border disease of sheep: a virus-induced teratogenic disorder. Adv Vet Med 36: 9–12
2. Barlow RM, Vantsis JT, Gardiner AC, Linklater KA (1979) The definition of border disease: problems for the diagnostician. Vet Rec 104: 334–336
3. Barlow RM, Vantsis JT, Gardiner AC, Rennie JC, Herring JA, Scott FMM (1980) Mechanisms of natural transmission of border disease. J Comp Pathol 93: 451–461
4. Bonniwell MA, Nettleton PF, Gardiner AC, Barlow RM, Gilmour JS (1987) Border disease without nervous signs or fleece changes. Vet Rec 120: 246–249
5. Fenton A, Entrican G, Herring JA, Nettleton PF (1990) An ELISA for detecting pestivirus antigen in the blood of sheep persistently infected with border disease virus. J Virol Methods 27: 253–260
6. Gardiner AC, Nettleton PF, Barlow RM (1983) Virology and immunology of a spontaneous and experimental mucosal disease-like syndrome in sheep recovered from clinical border disease. J Comp Pathol 93: 463–469
7. Harkness JW, King AA, Terecki S, Sand JJ (1977) Border disease in sheep: Isolation of the virus in tissue culture and experimental reproduction of the disease. Vet Rec 100: 71–72

8. Hughes LE, Kershaw GF, Shaw IG (1959) "B" or Border disease: an undescribed disease of sheep. Vet Rec 71: 313–317

9. Manktelow BW, Porter WL, Lewis KHC (1969) Hairy shaker disease of lambs. N Z Vet J 17: 245–248

10. Nettleton PF (1990) Pestivirus infections in ruminants other than cattle. Rev Sci Tech Off Int Epiz 9: 131–150

11. Plant JW, Littlejohns IR, Gardiner AC, Vantsis JT, Huck RA (1973) Immunological relationship between Border disease, mucosal disease and swine fever. Vet Rec 92: 455

12. Plant JW, Gard GP, Acland HM (1977) Transmission of a mucosal disease virus infection between sheep. Aust Vet J 53: 574–577

13. Plant JW, Acland HM, Gard GP, Walker KH (1983) Variations of border disease in sheep according to the source of the inoculum. Vet Rec 113: 58–60

14. Westbury HA, Napthine DV, Straube E (1979) Border disease: persistent infection with the virus. Vet Rec 104: 406–409

15. Woldehiwet Z, Sharma R (1990) Alterations in lymphocyte subpopulations in peripheral blood of sheep persistently infected with border disease virus. Vet Microbiol 22: 153–160

Author's address: Dr. Z. Woldehiwet, University of Liverpool, Department of Veterinary Pathology, Veterinary Field Station, Leahurst, Neston, Wirral, L64 7TE, U.K.

New by Springer-Verlag

R. I. B. Francki, C. M. Fauquet, D. L. Knudson, F. Brown (eds.)

Classification and Nomenclature of Viruses

**Fifth Report of the International Committee
on Taxonomy of Viruses**

(Archives of Virology/Supplementum 2)

1991. IV, 450 pages.
Soft cover DM 110,-, öS 770,-
Reduced price for subscribers to
"Archives of Virology":
Soft cover DM 99,-, öS 693,-
ISBN 3-211-82286-0

The Fifth Report of the International Committee on Taxonomy of Viruses (ICTV), summarizes the proceedings and decisions reached by the ICTV at its meetings held at the International Congresses of Virology in Sendai (1984), Edmonton (1987), and Berlin (1990). This report includes 2,430 viruses belonging to 73 families or groups, as well as virus satellites and viroids descriptions. Each family description is providing all the information concerning particle properties, serological properties, and biological properties.

The Fifth Report of the International Committee on Taxonomy of Viruses (ICTV) has been organized in the same way as the previous ones, yet it encompasses many more families and groups of viruses than previous reports. It includes diagrams grouped by their type of host, tables listing the viruses by alphabetical order, as well as by their host type and their type of nucleic acid and keys of family and group identification. The officers and members of the ICTV study groups from 1984 to 1990 are listed, as the current ICTV statutes and rules of nomenclature. Information on the format for submission of new taxonomic proposals to the ICTV is also provided. Indexes of viruses, taxonomic and authors' names are provided at the end of the report.

Springer-Verlag Wien New York

Sachsenplatz 4–6, P.O. Box 89, A-1201 Wien · Heidelberger Platz 3, D-1000 Berlin 33
175 Fifth Avenue, New York, NY 10010, USA · 37-3, Hongo 3-chome, Bunkyo-ku, Tokyo 113, Japan